The
Renewable Energy Handbook

The
Renewable Energy Handbook

The Updated Comprehensive Guide to Renewable Energy and Independent Living

William H. Kemp

AZTEXT
PRESS

Aztext Press
P.O Box 214, Tamworth, ON K0K 3G0 Canada
michelle@aztext.com   •   www.aztext.com

Library and Archives Canada Cataloguing in Publication

Kemp, William H., 1960-
        The renewable energy handbook : the updated comprehensive guide to renewable energy and independent living / William H. Kemp.

Includes bibliographical references and index.
ISBN 978-0-9810132-1-3

        1. Renewable energy sources--Handbooks, manuals, etc. 2. Energy conservation--Handbooks, manuals, etc. I. Title.

TJ808.3.K47 2009   621.042   C2009-902812-3

Disclaimer:
The installation and operation of renewable energy systems involves a degree of risk. Ensure that all proper installation, regulatory and common sense safety rules are followed. If you are unsure of what you are doing, STOP! Seek skilled and competent help by discussing these activities with your dealer or electrician.
Electrical systems are subject to the rules of the National Electrical Code ™ in the United States and the Canadian Electrical Code ™ in Canada. Wood and pellet stove units are also subject to national and local building code requirements.
In addition, local utility, insurance, zoning, and many other issues must be dealt with prior to beginning any installation. In the words of Emerson, "People don't plan to fail, they just fail to plan." Nowhere is this more apparent than when installing renewable energy systems.
The author and publishers assume no liability for personal injury, property damage, consequential damage or loss, including errors and omissions, from using the information in this book, however caused.
The views expressed in this book are those of the author personally and do not necessarily reflect the views of contributors or others who have provided information or material used herein.

Printed and bound in Canada.
Distributed in Canada by Aztext Press
Distributed in the U.S. by Consortium Book Sales & Distribution (www.cbsd.com)

This book is dedicated to our planet Earth.
We humans, blessed with Her bounty,
relentlessly abuse what we should nurture.
There are many who say they wish to save Her,
but few who are willing to pay their share of the bill.

Acknowledgements

A book of the magnitude of *The Renewable Energy Handbook* requires technical knowledge and access to information from many individuals and companies, not to mention the patience of family while writing and editing the text.

It is customary to identify those contributors who have helped with the technical data required to put this book together and ensured that the information is the most up-to-date.

I am indebted to, and would like to thank, Steve Anderson Biodiesel; Sean Towmey and the delightful staff at Arbour Environmental Shoppe; ARISE Technologies; Michael Reed of Array Technology Inc. for information on tracking solar mounts; Mike Bergey of Bergey Wind Power for technical publications and photographs on wind power systems; Don Bishop of Benjamin Heating Products for information regarding dual energy heating boilers; Bogart Engineering, (energy meters); Liujin, Les, and Jose from Carearth Inc. for supplying a great solar thermal system for my house and providing information and photographs; Dankoff Solar Products; Vanessa Percival at Embers Wood Stoves; Energy Systems and Design (micro hydro); Bob Fisher, Enerworks (solar thermal); Nicole and Aidan Foss for being excellent critics and mentors of world energy policy; Paul Gipe for his assistance and mentoring; Harmen Pellet Stoves Company; Peter Talbot of Home Power Systems (microhydro); Ross and Kathryn Elliot of Homestead Building Solutions for help with the home air leakage tests; George Peroni at HydroCap Corporation for information regarding hydrogen gas reclaiming devices for batteries; Tania Glithero, Iogen Corporation; Madawaska Mill Works (hot tubs and saunas); Steve Howell of MARC-IV Consulting; Patrice Feldman at Morning Star Corporation for her quick response regarding PV controller design and installation; Rajindar Rangi and Tony Tung of Natural Resources Canada, Energy Technology Branch for their guidance and for providing access to reams of technical bulletins and contacts; Mike McGahern, Ottawa Solar Power; "The Boys" from Renewable Energy of Plum Hollow; Deborah Doncaster and Melinda Zytaruk, Ontario Sustainable Energy Association (OSEA); Outback Power Systems; Gerald Van Decker of Renewability Energy Inc. for assistance with the waste water heat recovery system; Sharp Solar Corporation; South West Wind Power for the great site photographs and micro-wind technical data; Snorkel Hot Tub Company; Rolls-Surrette Battery Company; Dr. David Suzuki; Jaeson Tanner for cover and web design; Govindh Jayaraman of Topia Energy Inc.; Pam Carlson of Xantrex Technology Inc. for tremendous support with technical documentation and photographs; and lastly, Stefani Kuykendall of Zomeworks Corp. for information on fixed PV mounting hardware. (If I have missed anyone, please accept my apologies and understand that your assistance was truly appreciated!)

A huge thank you must also be given to Joan McKibbin, my editor. Joan always works tirelessly under very demanding deadlines.

I am also indebted to those who have had to work directly with me in the preparation of the text. A big thanks goes out to Cam Mather for his excellent line drawings and graphics; to Lorraine Kemp for her great help behind the camera and proving the old adage that a picture is worth a thousand of my words. (All images without photo credits are by Lorraine.); to Michelle Mather, whose skill with a computer makes her the best possible research assistant; and to all of those folks who let me invade their privacy to photograph and discuss their personal experiences using renewable energy.

Table of Contents

Preface

The first edition of *The Renewable Energy Handbook* was published in 2003 and dealt specifically with living life disconnected from the electrical grid: off-grid living. It was conceived when my friend, fellow "off-gridder" and now book publisher Cam Mather suggested that we should collaborate on such a work. His enthusiasm for the project was perhaps driven by desperation, or at least a lack of access to credible information on the subject. He and his wife and two daughters had recently moved from the city to a farm several miles and tens of thousands of dollars from the nearest source of electricity. Seven years later, Cam is much more comfortable with the subject, although he still has a bit of trouble understanding the difference between "power" and "energy." I have promised to work a bit harder discussing this in the current edition.

Interest in off-grid living and in *The Renewable Energy Handbook* was considerable, so in 2005 I decided to publish *Smart Power: An Urban Guide to Renewable Energy and Efficiency* as a complementary title. As the title suggests, the intended audience was the vast majority of the North American population who live connected to the electrical grid and want to learn more about the technology and economics of the subject.

What a difference a few years can make.

Since the publication of *Smart Power*, the United States has seen the worst hurricane season on record, with Katrina alone causing damage in the amount of an estimated $81.2 billion, making it the most destructive and costliest natural disaster ever.

The opening of Al Gore's *An Inconvenient Truth* in May 2006 examined the connection between the world's insatiable thirst for carbon-based energy and climate change. Perhaps Katrina was at least partly created by man's own hand?

In July 2007 oil prices spiked to a record $147 per barrel, causing massive social and economic challenges around the world. While prices have since plunged from these stratospherically high levels, there is concern about future prices hitting even higher levels. The phenomenon of "Peak Oil" is known to most people, and whether one believes in it or not it creates a nagging fear of having to pay dearly for energy.

At the same time, governments around the world are seeking ways of using energy more efficiently and developing technologies and regulations that will push society towards a reduced-carbon energy diet. Europeans have been keen on these concepts for many years, with Germany leading the way in both energy efficiency and renewable energy deployment. Regulations that allow homeowners, farm operators, businesses, and local co-operatives to generate and *profitably sell* carbon-free energy are being exploited with great zeal. Germany alone employs over 249,000 people in its renewable energy sector, according to 2007 figures released by the government. Job growth in this sector is expected to rise over the next two years, with total employment exceeding 400,000 people.

Governments in North America have been slower to move away from so-called "cheap energy sources" such as coal and oil, but rising production costs, worries about climate change, and the geopolitical realities related to importing oil and natural gas are helping with the transition. "Green Jobs" is the new political rallying cry in North America, and federal, state, provincial, and municipal governments are falling over themselves trying to encourage clean energy generation, manufacturing, and engineering jobs in their jurisdictions.

Because of this rapid change of mindset to "green" the residential energy sector, I felt it was time to revamp *The Renewable Energy Handbook* in an expanded, updated volume. This new edition includes::

- updated chapters, expanding on the technology and its implementation
- a comparison chart detailing wind turbine specifications
- discussion of advanced off-grid developments, including "zero-carbon" living
- both on- and off-grid technology review
- larger format with improved readability and picture resolution

We live in a time where environmental ethics and economic value can be paired in such a way that neither need be compromised. It is my hope that with a little help from this book you can live just a bit lighter on the planet.

For the sake of the Earth
WHK

Chapter 1
INTRODUCTION

Energy is the life breath of modern society. Without energy, almost every event we take for granted comes to a screeching halt. If you happened to live on the eastern side of North America during the blackout of 2003, you will acutely understand how life without electricity changes everything. For a few dozen hours, modern life stopped—*completely*: people stuck in elevators; no building or traffic lights; traffic gridlocks; no air conditioning, cell phones, computers, television, or email. The entire world appeared to go black.

Likewise, if you were of driving age in 1973 you will no doubt recall the long lineups and short supply of gasoline at local filling stations. Even if gasoline was available, OPEC raised the price of oil from $4.90 a barrel to $8.25 a barrel in that year.

Energy consumption in the United States and Canada is insatiable, and it appears no one has learned from the mistakes of the past. Look around: Sport Utility Vehicles (SUVs) and minivans abound, with the result that fuel economy is nearing an all-time low. At the same time, the square footage of the average North American home has more than doubled, requiring additional energy for heating and air conditioning.

Domestic oil supplies in the United States have dropped from 100% in the early 1900s to approximately 25% of total demand as of 2009, requiring vast quantities of imported oil to make up the shortfall. Current world consumption is estimated to be 85.3 million barrels per day in 2009,[1] while U.S. consumption has fallen slightly to 19.25 million barrels per day as a result of the credit-driven economic downturn.[2] Bear in mind that the importation of large amounts of fossil fuel causes the *exportation* of equally large numbers of dollars that ultimately weakens the domestic economy further.

The United States has only 4.5% of the world's population (306 million out of 6.8 billion) and yet it consumes a whopping 30% of *all* of the world's resources. This includes not only energy but also water, steel, aluminum, timber, and just about everything else you can imagine. China, Russia, India and many other emerging-economy nations with combined populations many times that of North America will cause violent price and supply volatility of energy, food, and raw materials.

Family budgets are straining from the cost of energy to power our homes and fuel our cars. Many people are surprised to learn that they must work an average of two months per year just to pay the family energy bills. With the cost of all types of energy steadily increasing, the outlook is bleak.

Consider the Environment

What about the environment? Climate change is one of the key challenges facing worldwide development. Global warming, caused mainly by the burning of fossil fuels, deforestation, and inefficient use of energy are pushing ecosystems to the brink of catastrophic failure.

Atmospheric concentrations of carbon dioxide (CO_2) have increased dramatically since the industrialization of society over the last 150 years and are at the highest levels measured during the past 500,000 years. Global energy use amongst industrialized countries is continuing to increase unabated. Underdeveloped countries, where more than one–third of the world's population does not have access to electricity, are starting to ask for their fair share of the energy pie. "Warming of the climate system is unequivocal, as is now evident from observations of increases in global average air and ocean temperatures, widespread melting of snow and ice and rising global average sea levels" according to the United Nations report on climate change issued in 2007.[3]

This temperature increase sounds innocent enough until you realize that a change of the magnitude described in the report has never previously occurred and will amplify weather effects throughout the earth's ecosystems. Increased levels of drought, flooding, and storms in areas already sensitized to these environmental stresses will result. The press and respected scientific papers are full of articles describing the effects of climate change: melting glaciers, reduced arctic pack ice, increasing ocean water levels, and the destruction of coral reefs. These ecologically sensitive phenomena are harbingers of far more devastating events to come. Gwynne Dyer, author of the book *Climate Wars*, predicts that along with the ecological impacts of climate change,

millions of people will migrate across continents as weak governments collapse, unable to feed their starving populations. In their desperate effort to survive, the marauding millions will create havoc for richer economies.

The Status Quo Is No Longer Valid

Our current way of life includes the belief that cheap energy is our God-given right. Never mind that the cost of gasoline or imported heating oil does not include the vast subsidies lavished on the oil industry. The price of a gallon of gasoline neglects the ongoing American military presence in the Middle East, depletion subsidies, cheap access to government land, and monies invested in the development of drilling and exploration technologies. None of these "hidden" costs even touches on the environmental and health damage caused by the burning of fossil fuels. For example, air pollution, largely from the burning of fossil fuels, kills an estimated 1,900 Ontarians prematurely each year and results in hundreds of thousands of incidents of illness, absenteeism, and asthma attacks—costing the economy billions of dollars.[4]

Fossil fuels don't just power our cars. Home heating and the majority of electrical power plants rely on fossil fuels as well. Coal accounts for approximately 20% of global CHG emissions, and is so inexpensive its use is growing rapidly, as indicated by the estimated 130 new coal-fired power plants now being considered.[5]

The United States relies increasingly on oil and gas imports from Canada, which is the largest supplier of these commodities. Yet Canadian domestic reserves of natural gas are declining rapidly, with the country extracting gas faster than it can locate new reserves. The number of years of proven reserves declined from 35 years in 1985 to 9 years in January 2006.[6]

Given the demand for increased domestic consumption and exports to the United States, Canada has moved to importing liquefied natural gas (LNG) to make up the shortfall. This concept is folly, considering that LNG is a fungible

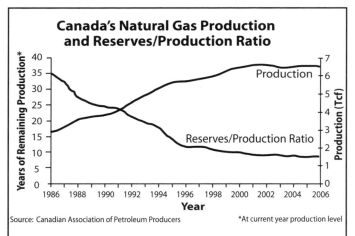

Figure 1-1. Canadian domestic reserves of natural gas are declining rapidly, with the country extracting gas faster than it can locate new reserves. The number of years of proven reserves has declined from 35 years in 1985 to 9 years in January 2008. Given that Canada is a major exporter of natural gas to the United States, which uses the fuel to produce electricity, it is clear that serious supply problems are looming. (Courtesy Energy Information Administration)

commodity that can be connected to the pipeline of the highest bidder, whether in Canada or Japan.

With much of the developing world's population connecting to the electrical grid every day, it is obvious that world energy demand will mushroom as more (and larger) appliances are brought online, further exacerbating the problem. In the developed world, middle-class families are demanding more and more appliances that were considered luxury items only one generation ago. Central air conditioning, multiple refrigerators, computers, chest freezers, hot tubs, and swimming pools, all luxury or unimaginable items in our parents' day, consume enormous amounts of energy and greatly contribute to climate change.

The World Wildlife Fund (WWF) has introduced a program to get major power utilities to switch from high-carbon fuels—especially coal—to cleaner options. The key technologies and policies of the WWF Powerswitch! program include:

- energy efficiency and demand-side management strategies;
- large-scale wind energy projects—mainly offshore;
- large-scale biomass co-firing at existing coal power stations;
- increased support for renewable energies;
- high rates of aluminum recycling;
- rapid growth of combined heat and power systems (CHP);
- movement from coal to natural gas as a transition fuel.

On the home scale, several of these options are completely viable, reliable, and just plain fun, as you will learn in the following chapters. (Receiving a check from the electrical utility puts a smile on thousands of faces each month thanks to renewable electricity generating systems.)

A New Strategy

North Americans need a new strategy, one that focuses on reducing energy use through efficiency and conservation rather than on increasing supply. Experience from across North America has proven that it is significantly cheaper to invest in energy efficiency than to build or even maintain polluting sources of electricity supply. This theory holds true whether we are describing the nuclear power plant down the road or your own home-based renewable energy system. Or, as Amory Lovins, the energy guru from the Rocky Mountain Institute, is fond of saying, "It is far less expensive and environmentally more responsible to generate *negawatts* than megawatts."

Using less energy isn't about making drastic lifestyle changes or sacrifices. Conservation and efficiency measures can be as simple as improving insulation standards for new buildings, replacing incandescent light bulbs with compact fluorescent models, or replacing an old refrigerator with a more efficient one. In fact, energy efficiency often provides an improvement in lifestyle. A poorly insulated or drafty house may be impossible to keep comfortable no matter how much energy (and money) you use trying to keep warm.

Using less energy isn't about making drastic lifestyle changes or sacrifices.

California learned firsthand that saving energy means saving money and the environment. You may recall the rolling blackouts and severe power shortages that afflicted the state a few years ago. It was predicted that dozens of generating stations would be required on an urgent basis to solve the state's energy problems. Faced with the realization that construction cycles for significant generating capacity would take several years, forward-thinking officials looked to energy efficiency instead. The state's energy–efficiency standards for appliances and buildings have helped Californians save more than $15.8 billion in electricity and natural gas costs. One–third of Californians cut their electricity use by 20% to qualify for a 20% rebate on their bill. The government introduced a renewable-energy buydown and accompanying net-metering program that resulted in thousands of clean photovoltaic power systems being installed on residential rooftops.[7]

In addition to saving electricity and reducing fossil fuel burning, California's conservation and efficiency efforts reduced greenhouse gas emissions by close to 8 million tonnes and nitrogen oxide emissions by 2,700 tonnes during 2001 and 2002.[8]

Consider the following points:

- Switching to compact fluorescent lamps will reduce lighting energy consumption and parasitic heat output by 80%.
- High-efficiency appliances such as refrigerators, washing machines, and dishwashers can reduce energy consumption by a factor of 5 times.
- Using on-demand water heaters will reduce hot water energy costs by up to 50%.
- Low-flow showerheads, aerator faucets, and similar fittings will reduce water consumption and resulting heating costs by approximately one-half while lowering well-pump energy consumption and septic or leaching bed stress.
- A well-sealed and insulated house can reduce home heating and cooling costs by 50%.
- Adding a solar thermal water-heating system can further reduce hot water heating costs by at least 50%.

Many of the items on this list are neither expensive nor difficult to implement. Best of all, these measures not only dramatically reduce smog and greenhouse gas emissions but also greatly

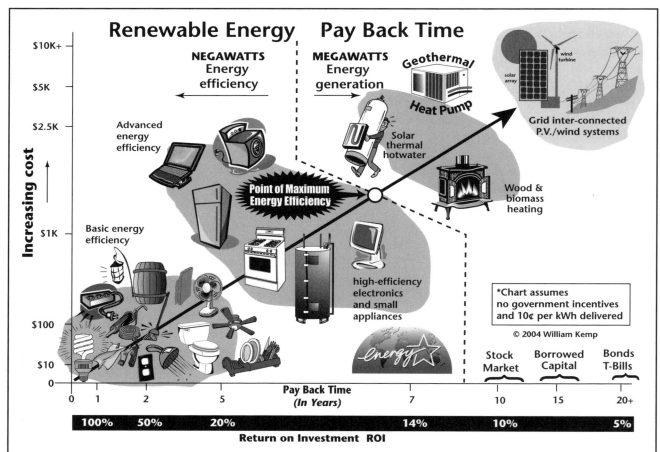

Figure 1-2. This graph relates the relative cost of different energy efficiency and generation technologies to payback time or simple return on investment without any contribution from government incentives applied.

decrease the size and capital cost of your renewable energy, off-grid power station components. Expressed in a slightly more dramatic way, the United States Department of Energy states that for every dollar you invest in energy efficiency, you can reduce the cost of generating power by three to five dollars. For those who wish to stay connected to the electrical grid, adopting energy efficiency technologies has a further benefit: phenomenal financial return on investment (ROI). Many people watched the past decade of investment returns evaporate as the market meltdown of 2008/2009 unfolded. Many of these investment returns have now proven to be illusory; not so the gains achieved through energy efficiency. I developed the graph shown in Figure 1-2 as a quick way of demonstrating the powerful impact energy efficiency has on ROI. Although changing a light bulb or adding insulation to a home is neither sexy nor headline grabbing, the impact on your wallet can impress even the dourest of market-burned bankers.

Energy Efficiency is Not Conservation
There is a major difference between energy efficiency and conservation, an issue I want to make perfectly clear before going further on our journey. Many people take both terms to mean "turning down the thermostat and lights and freezing in the dark," or simply "doing without." Nothing could be further from the truth.

I would prefer it if people would consider my slightly more upbeat approach to energy efficiency as "doing more with less." Perhaps the concept is akin to discussing whether the glass is half full or half empty, but if semantics changes people's attitudes, I'm all for it.

Consider this. If I change your old-fashioned water heater with a small, wall-mounted on-demand model, you will save about 40% of your hot water heating costs, forever. This reduction may amount to several hundred dollars per year, reduce the impact of rising energy costs, and, as an added "bonus," provide endless hot water for teenage showers. The impact on your lifestyle is

zero. You will receive an endless financial dividend, year after year, and your consumption of carbon-based fuel will be reduced by nearly one half.

Here is another example. Your wife explains that the "sharp-looking" avocado side-by-side refrigerator you brought to the marriage from your first apartment is a bit tired. You capitulate and head down to the appliance store to look at newer models and select one that suits her needs. Not only is the new fridge

There is a major difference between energy efficiency and conservation.

likely to be 25% larger and have more features, its energy consumption will have fallen by a factor of 5, resulting in energy savings of up to a couple of hundred dollars per year. Even today's large, fancy models will provide a return on investment of 20% or more.

In both of these examples, the homeowner is not being asked to reduce quality of life or "conserve" anything. This is the concept of energy efficiency—doing more with less—and it will be a central theme of this book.

If we take the concept of efficiency to the limit, an unlikely event for most people, we reach a point on the graph in Figure 1-2 called the "Point of Maximum Energy Efficiency." It is here that you will have squeezed every bit of waste out of your chosen lifestyle.

At this point, the only possible way to reduce energy consumption further is to reduce the amount of television you watch or turn down the thermostat and pull on a sweater. In short, once you reach the point of maximum energy efficiency, the only way you can reduce energy consumption any further is by doing without—in short, conserving.

While there is no question that North Americans tend to be an entitled lot, demanding excessive amounts of things that consume energy and resources, it is difficult to develop the concept of conservation when the idea of energy efficiency is completely foreign. Better to begin with the easy, financially rewarding path.

On-Grid vs. Off-Grid Clean Energy Generation

Once the point of maximum energy efficiency has been reached, the next step is to consider clean energy generation for the home.

At first glance one would assume that generating renewable electricity when connected to the electrical grid would be similar to making it off-grid. While it is true that there are similarities, it is equally true that there are major differences that need to be considered.

To assist readers who are interested in only one area or the other, I have provided a page-marking system that indicates whether information is relevant for on-grid or off-grid technologies.

The picture of an electrical meter with a cross through it, similar to a no-smoking sign, indicates an area of the book that deals specifically with off-grid systems.

An image of an electrical meter with a check mark indicates a technology that is suited for on-grid users. Where there is no symbol, the text is applicable to both technologies.

1.1
Renewable Energy Economics

A word about pricing: Many of the products described in this handbook are either manufactured in the United States or priced in U.S. dollars. To avoid confusion, all prices are shown in U.S. dollars.

As you contemplate generating some or all of your personal energy requirements, you may be eager to run out and purchase a wind turbine or install solar thermal hot water heating. Not so fast! No matter how much of a "keener" you are for eco-stuff, there is a right and a wrong way to do everything. In fact, anyone who has had a bad experience probably didn't take the time to fully understand the technology and the economics of renewable energy systems.

For off-grid systems, it is essential to understand that energy efficiency is as important as energy generation, if not more important. Remember that for every dollar you invest in efficiency measures you save three to five dollars in generating equipment. It stands to reason that you **must** perform all of the updates, installations, and conversions of an energy-efficient lifestyle **before** you invest one dollar in electrical generating equipment. Most homeowners who have abandoned off-grid living blame battery failures, excessive generator running time, or insufficient electricity—all indications of a poorly designed and/or operated system.

The delivered price of grid-supplied electricity averages 10 to 12 cents per kilowatt-hour in North America, while power produced by an off-grid system costs many times this amount. Additionally, there are practical limits on the amount of energy that can be realistically produced. Therefore, the capital cost of energy-efficient upgrades and installations has a very short payback time. Inefficient appliances require electrical generating equipment to power them that will cost many

Figure 1-3a. This house has been photographed in visible light and infrared conditions. Using camera equipment tuned to the thermal or infrared spectrum allows building auditors to analyze house construction and determine heating and cooling load efficiency. In this image, the vertical bar graph indicates a grayscale that shows the upper and lower temperatures and corresponds to the matching tones on the house. The higher the temperature (lighter shade of gray) the more heat is leaking from the house, indicating inadequate insulation or air sealing. (Courtesy Ross Elliott, Homesol Building Solutions)

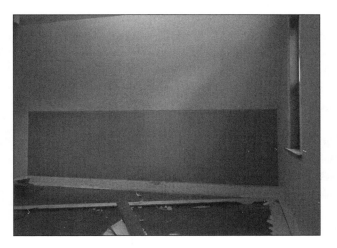

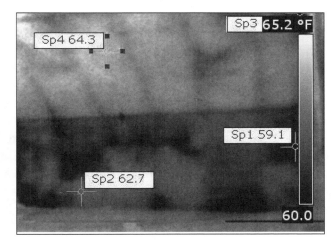

Figure 1-3b, c, d and e. This montage of images shows a century-old country home that is both uncomfortable and hard to heat due to the lack of insulation in the wall system. (Courtesy Ross Elliott, Homesol Building Solutions)

times more than the appliance itself. Replacing a 15-year-old refrigerator with a new model will **always** make financial sense because of the 1:3-5 efficiency: generation cost ratio.

Many people have an unfortunate habit of looking at "first cost" rather than "life cycle" or "true cost" when considering appliances and consumer goods. The everyday incandescent light bulb is an excellent example. Thomas Edison's mid-1800s invention was a technological marvel in its day. Replacing smoky, dim kerosene lamps with clean and bright electric lights can never be undervalued. However, just as we no longer use a horse and buggy or wireless radio sets from that era, the inefficient Edison lamp should now be relegated to the museum. The replacement is the modern compact fluorescent lamp that uses 4 1/2 times less energy and lasts 10 times longer.

At first glance, the initial cost of the compact fluorescent lamp appears uneconomical compared with the cost of the old-fashioned incandescent lamp. Typical compact fluorescent lamps have a first cost of approximately three dollars where the incandescent lamp is approximately 40 cents. But look closer. The compact fluorescent lamp lasts approximately 10 times longer, making first cost lower ($.40 x 10 incandescent lamps = $4.00, giving the same life as 1 compact fluorescent = $3.00).

Figure 1-4. This home incorporates special flexible photovoltaic panels mounted in place of roofing shingles. Modern renewable-energy-powered homes use advanced technologies that provide the same amenities and level of comfort as their utility-connected cousins. (Courtesy Sharp Electronic Corporation)

Based on first cost alone, our compact fluorescent lamp is already one dollar *less expensive*.

Now let's look at energy costs. Assuming you pay $0.12 for each kilowatt-hour of electricity delivered to your home,[9] a 23W compact fluorescent lamp (which provides light equivalent to a 100W incandescent lamp) will save over $92 in energy costs over the life of the bulb.[10] Multiply this by the average 25 light bulbs per house and you have just put a cool $2,300 *after-tax* dollars in your pocket.

Now consider these numbers from an off-grid point of view. Assume that an off-grid system installation is $25,000 based on electrical energy consumption for an energy-efficient home of 4 kWh of energy per day. The typical home using inefficient lights and appliances easily uses 7 to 10 times this amount of energy. If this inefficient home were to be converted to an off-grid design, the system would be much larger, driving capital costs in excess of $200,000!

This is real money. It's your choice: write a monstrous check to your renewable energy dealer or spend a few thousand dollars investing in energy-efficient appliances and products before you take the off-grid plunge. Why not keep the dollars in your pocket instead of making the dealer rich? (Chapter 2 and the accompanying worksheet in Appendix 7 will help you choose the economic alternative.)

There is no refuting economics in this case. Businesses, cruise ships, hotels, and our European neighbors have used compact fluorescent, color-corrected lighting for years without any degradation of lifestyle. Indeed, compact fluorescent lighting is flicker free, dimmable, of equivalent brightness and quality to incandescent lights, available in a wide selection of sizes, usable outdoors, and available with a seven-year money-back guarantee.

Single 23W compact fluorescent lamps will also reduce greenhouse gas emissions by 1140 pounds (517 kg) and acid rain-producing compounds by 8 pounds (3.6 kg) over the life of the bulb. For an entire house, a reduction of

28,500 pounds (12,927 kg) of greenhouse gas emissions and 200 pounds (91 kg) of acid-rain-producing compounds will result. As an added bonus, there will be nine fewer dead light bulbs added to your local landfill or recycling depot.

In this example, true cost is many times lower than first cost, providing a rapid return on investment far greater than any stock market analyst could ever (legally) achieve.[11]

This is the case for *eco-nomics*.

What about Off–Grid System Capital Costs?

Off-grid system owners have a unique way of looking at payback criteria. In most (but certainly not all) cases, a homeowner who wishes to build a home or cottage on a desirable lot which is a distance away from the utility lines must pay for the line extension to the home. One does not have to run the wire very far before a serious amount of cash is burned up. It may only require a line extension of 0.5 miles (0.8 km) to exceed the cost of an off-grid system. In that case, the payback time is the instant the first light is turned on!

Obviously, the further away from the utility lines, the faster the payback, to the point where an off-grid system may be the only economic way to build in that location. Islands and mountaintop locations with beautiful views and few intruding neighbors come to mind.

"Off-gridders" (that group of people who are often fond of granola and Birkenstocks) can use the cost of utility-line extensions to their advantage. A building lot that has no access to electrical power might not be worth as much in the seller's eyes, even though the potential off-grid buyer has no intention of making the electrical utility connection.

Oddly enough, I have met dozens of off-gridders whose homes are only a stone's throw away from the electrical utility lines, a situation that will never provide a positive return on investment. When queried about their situation,

some tell me they are off grid because they "want to be" or "don't want to pay a utility bill again." However, the most common answers have to do with being independent and not having to purchase dirty coal or nuclear energy at any price. The ROI model doesn't work in all cases.

1.2
What Is Energy?

All of the earth's energy comes from the sun. In the case of renewable energy sources and how we harness that solar energy, the link is clear: sunlight shining through a window creates warmth; sunlight striking a photovoltaic (or PV) panel is converted directly into electricity; the sun's energy causes the winds to blow, which moves the blades of a wind turbine, causing a generator shaft to spin and produce electricity; the sun evaporates water which forms clouds from which the water, in the form of raindrops, falls back to earth and becomes a stream that runs downhill into a micro-hydroelectric generator.

While these energy sources are renewable, they are also variable and intermittent. The sun generally goes down at night and may not shine for several cloudy days. If we just wanted to use our energy when it is available, the various systems used to collect and distribute it would be a whole lot simpler. But we humans are just not that easily contented. I know for a fact that most people want their lights to turn on at night, even though the sun stopped shining on the PV panels hours ago. In order to ensure that heat and electricity are available when we need them, a series of cables and fuses and a seemingly bewildering array of components are required to capture nature's energy and deliver it to us on demand.

The process of capturing and using renewable energy may seem far too complicated and expensive for the average person. However, while the components themselves are complicated, the theory and techniques required to understand, install, and live with renewable energy are not.

Why Are We Discussing Math?

Although it is not absolutely necessary to be an expert in heating, wind, and electrical energy, an understanding of the basics will greatly assist you in operating your renewable energy system. This knowledge will also help you make better decisions when it comes to purchasing the most energy-efficient appliances.

Trying to save money by reducing energy costs can only be accomplished by understanding your "energy miles per gallon" quotient. It is fairly simple to determine fuel economy or efficiency for your car, but it is much more difficult to determine energy efficiency for a house full of appliances and heating equipment.

The mathematical theories described in *The Renewable Energy Handbook* are simplified. Equations are used throughout the text, but they are limited to those areas that are most important. A simple calculator (powered by the sun of course) will make the task that much easier. For those who find the mathematics a bit light there are plenty of references to data sources that will make even the most ardent "techie" happy.

The Story of Electrons

If you can remember back to your high school science class, you will recall that an atom consists of a number of electrons swirling around a nucleus. When an atom has either an excess or a lack of electrons in comparison to its "normal" state, it is negatively or positively charged, respectively.

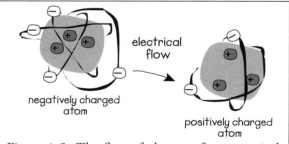

Figure 1-5. The flow of electrons from negatively charged atoms to positively charged atoms within a conductive material is known as the flow of electricity.

In the same way that the north and south poles of two magnets are attracted, two oppositely charged atoms are also attracted. When a negatively charged atom collides with a positively charged atom, the excess electrons in the negatively charged atom flow into the positively charged atom. This phenomenon, when it occurs in far larger quantities of atoms, is called the flow of electricity.

The force that causes electricity to flow is commonly known as *voltage* (or V for short). The actual flow of the electrons is referred to as the *current*. So where does this force come from? What makes the electrons flow in the first place? The trigger that brings about the flow of electrons can come from several energy sources. Typical sources are chemical batteries, photovoltaic cells, wind turbines, electric generators, and the up-and-coming fuel cell. Each source uses a different means to trigger the flow of electrons. Waterfalls, coal, oil, or nuclear energy are known as primary energy sources and are commonly used to generate commercial electricity. Fossil or nuclear fuels are used to boil water, which creates the steam that drives a turbine and generator; falling water drives a turbine and generator directly. The spinning generator shaft induces magnetic fields into the generator windings, forcing electrons to flow.

Let's use the flow of water as a visual aid to understanding the flow of electricity that is otherwise invisible.

Let's presume that a greater amount of water moving past you per second equates to a higher flow and that a river with a high flow of water has a large current. With electricity, a large number of electrons flowing from one atom to the next is similar to a large flow of water. Therefore, the greater the number of electrons flowing past a given point per second, the greater the electrical current.

The flow of water is typically measured in gallons per minute or liters per second; electrical current is measured in amperes (or A for short). If the measured current of electrons in a conductive material is said to be 2A, we know that a certain

Figure 1-6. The flow of electricity is very similar to the flow of water. A waterfall has a higher flow and greater pressure than a creek; similarly, a higher number of electrons moving from atom to atom increases electrical current.

Before electricity can be put to use, we must create an electrical circuit. We do this by connecting a source of electrical voltage to a conductor and causing electrons to flow through a load and back to the source of voltage.

The drawing in Figure 1-7 shows how a simple flashlight works. Electrons stored in the battery (we will cover that one later) are forced by the voltage (pressure) to flow into the conductor from the battery's negative contact (-) through the light bulb and back to the battery's positive contact (+). Current flowing in this manner is called *direct current* (or DC for short). The light will stay lit until the battery dies (runs out of electrons) or until we turn off the switch.

number of electrons has passed a given point per second. If the current were increased to 4A, there would be twice the number of electrons flowing through the conductor.

Gravity is a factor in water pressure. A meandering creek has very little water pressure because the water does not fall from a great height. A high waterfall has a much greater distance to fall and therefore has increased water pressure. For example, if a water-filled balloon were to fall on you from a height of 2 feet you would find this quite refreshing on a hot day. A second balloon falling from a height of 100 feet would exert a much higher pressure and probably knock you out! This increase in water pressure is similar to the electrical pressure or voltage that forces electrons to flow through a material.

Conductors and Insulators

Electricity flows through conductors in the same manner as water flows through pipes and fittings. When electrons are flowing freely through a material we know that the material is offering little resistance to that flow. These materials are known as conductors. Typical conductors are copper and aluminum, which are the substances used to make electrical wires of varying diameters. Just as a fire hose carries more water than a garden hose, a large electrical wire carries more electrons than a thin one. A larger wire can accept a greater flow of current (i.e. handle higher amperage).

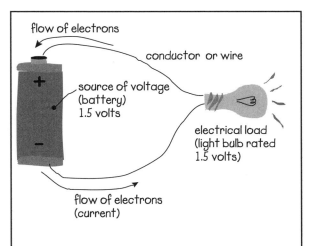

Figure 1-7. An electrical circuit consists of a source of voltage connected to an electrical load through conductors.

Oh yes, a switch would be a good idea. This handy little device allows us to turn off our flashlight to prevent using all the electrons in the battery. From our description of a circuit, we can assume that if the electrical conductor path is broken, the light will go out. How do we break the path? The flow of electrons may be interrupted using a nonconductive substance wired in *series* with the conductor. Any substance that does not conduct electricity is known as an insulator. Typical insulators include rubber, plastics, air, ceramics, and glass.

Batteries, Cells, and Voltage

You may have noticed that your brand of flashlight has two or even three cells. Placing cells in a stack or in *series* causes the voltage to increase. For example, the flashlight shown in Figure 1-8 contains a cell rated at 1.5V. Placing two cells in series, as shown in Figure 1-9, increases the voltage to 3V (1.5V + 1.5V). The light bulb in Figure 1-9 is glowing very brightly and will quickly burn out because it is rated for 1.5V. In this example, a higher voltage (pressure) is causing more current to flow through the circuit than the bulb is able to withstand. Likewise, if the cell voltage were lower than the rating of the bulb, insufficient current would flow and the bulb would be dim. This is what happens when your flashlight batteries are nearly dead and the light is becoming dim: the batteries are running out of electrons.

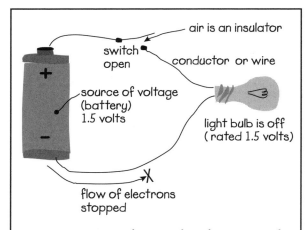

Figure 1-8. Any substance that does not conduct the flow of electricity is called an insulator. When the switch is "opened," an air gap between the conductive elements stops the flow of electrons; air is therefore an insulator.

Next time you are waiting in the grocery store lineup, put down that copy of *The National Enquirer* and take a look at the battery display. The selection will include AA, C, and D sizes of cell, which all have a rating of 1.5V. Can you guess the difference between them?

A larger cell holds more electrons than a smaller one. With more electrons, a larger battery

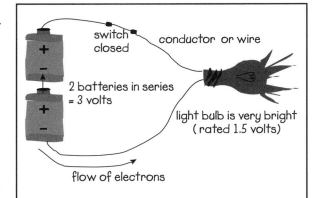

Figure 1-9. A higher voltage forces more electrons to flow (higher current) than the bulb is rated to tolerate. Always ensure that the source and load voltages are rated equally.

cell can power an electrical load for longer than a smaller one can. The circuit shown in Figure 1-11 shows a set of jumper wires connecting the cell terminals in *parallel*. This parallel arrangement creates a battery bank of 4 AA-size cells with the same number of electrons and the same capacity as the C-size cell. Any grouping of cells, whether connected in series, in parallel, or both, is called a battery bank or battery.

Obviously, a house requires far more electricity than a simple flashlight does. Off-grid homes generate electricity from renewable sources (more on that later) and usually store it in battery banks such as the one shown in Figure 1-12.

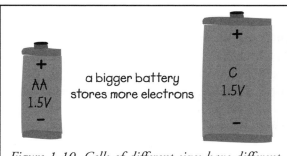

Figure 1-10. Cells of different sizes have different amounts of electron storage. Just as a 2-liter jug contains twice as much water as a 1-liter bottle, a larger battery stores more electrons and works for longer periods than a smaller one.

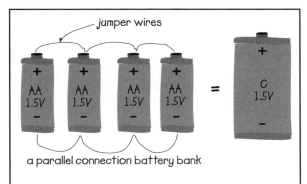

Figure 1-11. Wiring cells in parallel increases the capacity of the battery bank.

Each battery cell in Figure 1-12 is numbered 1 through 12 and has a nominal voltage of 2V. If you look carefully, you can see that each cell has a (+) and a (-) terminal. Each terminal is wired in *series* to the next battery in the manner illustrated in Figure 1-9, increasing the battery bank voltage. Therefore a series string of 12 cells rated at 2V each creates a battery bank rated at 24VDC.

You will also note that there are two such banks of batteries. The bank on the left is identical to the bank on the right. By wiring the two banks in *parallel* we create a total battery capacity that

Figure 1-12. Off-grid homes or homes equipped with emergency backup systems require much larger amounts of energy than a typical flashlight or car battery can store. These deep-cycle batteries are typical of many off-grid power systems.

is twice as large as a single bank. This is exactly the same as wiring the 4 AA cells in parallel to make the equivalent of the large C cell described in Figure 1-11. Obviously, the more electricity we use the greater the battery size and cost.

Off-Grid Houses Don't Usually Run on Batteries

You are probably aware that your house works on 120/240V and not 12V, 24V, or 48V from a battery. Early off-grid houses, small cottages, boats, and many recreational vehicles can and do use 12VDC systems. But don't consider using low-voltage DC for anything but the smallest of systems. With the limited selection of appliances and the difficulty of wiring a low-voltage, full-time home, low-voltage DC systems are not a viable option.

The modern off-grid home is supplied with electricity in the form of 120/240V *alternating current* (VAC). In a DC circuit, as defined earlier, current flows from the negative terminal of the battery through the load and back to the positive terminal of the battery. Since the current is always flowing one way, in a direct route, this is called direct current.

In an AC circuit, the current flow starts at a first terminal and flows through the load to the second terminal, like the flow in a battery circuit. However, a fraction of a second later the current stops flowing and then reverses direction, flowing from the second terminal through the load and back to the first terminal, as shown in Figure 1-13.

Alternating Current in the Home

Generating electricity in the modern, grid-connected world is accomplished by using various mechanical turbines to turn an electrical generator. In the early years of the electrical system, there was considerable debate as to whether the generator output should be transmitted as AC or DC. For a time, both AC and DC were generated and transmitted throughout a city. However, as time passed, it became clear for safety, transmission, and other practical reasons that AC was more

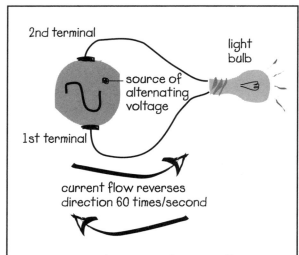

Figure 1-13. The process of continually reversing the direction of current flow is known as alternating current (AC).

desirable. As the old saying goes, "The rest is history." All modern houses and electrical appliances are standardized in North America to operate on either 120VAC or 240VAC. For this reason, it is advisable for off-grid homes to convert the low-voltage electrical energy stored in the battery bank to 120/240 VAC. The device used for this electrical conversion is the inverter, which will be discussed in more detail in Chapter 11.

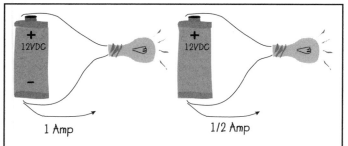

Figure 1-14. The lamp on the right uses 50% of the electrical energy of the one on the left and both lamps have equal brightness. The lamp on the right is said to be twice as efficient.

Power, Energy, and Conservation

Conserving energy and doing more with less not only is good for the planet but also helps keep the size and cost of your renewable-energy power station within reasonable limits. We discussed earlier how a bigger battery could run a light bulb for longer than a smaller battery could. This is fairly obvious. What might not be so obvious is that if we were to replace the "ordinary" light bulb with a more efficient one we might not need the larger battery in the first place.

Let's review how any typical electrical circuit operates. A source of electrons under pressure (i.e. voltage) flows through a conductor to an electrical load and back to the source. For household circuits, the voltage (pressure) is usually fixed at 120VAC or 240VAC. For battery-supplied circuits, the voltage is usually fixed at either 12VDC, 24VDC, or 48VDC, depending on the size of the load and the amount of electron flow (i.e. current) that is required to make the load operate. Let's say that an ordinary light bulb requires 12V of pressure and 1A of current flow to make it light. Now suppose that we can find a light bulb that uses 12V and only 0.5A of current flow. Assuming that both lights are the same brightness, we infer that the second light is twice as efficient as the first. Stated another way, we would need only half the battery-bank size (at lower cost) to run the second lamp for the same period of time or the same size battery bank to run the second lamp for twice as long.

The relationship between the pressure (voltage) required to push the electrons to flow in a circuit and the number of electrons actually flowing to make the load operate (current or amps) is the power (commonly expressed as *watts*, or W) consumed by the load. Using the above example, let's compare the power of the two circuits:

Ordinary Lamp:
   12V x 1A = 12W

More Efficient Lamp:
   12V x 0.5A = 6W

The second lamp uses half the number of electrons to operate. The power of a circuit is simply the voltage multiplied by the current in amps, giving us the instantaneous flow of electrons in the circuit measured in watts.

Remember that the more efficient bulb operates for twice as long on the same battery as the other bulb. How does time factor into this equation? If a battery has a known number of electrons stored in it and we use them up at a given rate, the battery will become empty over time. The use of electrons (power) over a period of time is known as *energy*.

Energy is *power* multiplied by the *time* the load is turned on:

Ordinary Lamp:
12V x 1A x 1 hour = 12 watt-hours; or
12W x 1 hour = 12 watt-hours

More Efficient Lamp:
12V x 0.5A x 1 hour = 6 watt-hours; or
6W x 1hour = 6 watt-hours

Your electrical utility charges you for energy, not power. You run around turning off unused lights to cut down on the amount of time the lights are left on. If we think about it long enough, we can also use these calculations to figure out how much energy is stored in a battery bank. For example, if a 12V battery bank can run a 10A load for 30 hours, how much energy is stored in the battery?

Battery Bank Energy (in watt-hours):
12V x 10A x 30 hours
= 3,600 watt-hours
or 3.6 kilowatt-hours of energy

An interesting thing about batteries is that their voltage tends to be a bit "elastic," dipping and rising as a function of their state of charge. For example, our nominal 12V battery bank only registers 12V when the batteries are nearly dead. When they are under full charge, the voltage may reach nearly 16V. Because of this swing in the voltage, many batteries are not sized in *watt-hours* of energy, but in *amp-hours*.

The math is similar; just drop the voltage from the energy calculation:

Battery Bank Energy (in amp-hours):
10A x 30 hours = 300 amp-hours of energy

To convert amp-hours of energy to watt-hours, simply multiply amp-hours by the *nominal* battery voltage. Likewise, to convert watt-hours of battery bank capacity to amp-hours, divide watt-hours by the battery voltage.

That pretty well covers all the math required to understand electrical energy. With a bit of practice (and you'll get plenty of that in the chapters to follow) you will know this stuff well enough to brag at your next office party.

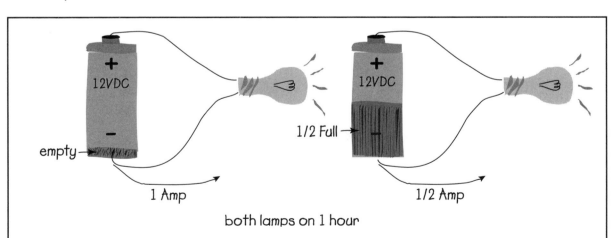

Figure 1-15. The more efficient lamp on the right consumes 50% less power than the lamp on the left. This allows the more efficient lamp to operate for a longer period on the same battery as its low-efficiency cousin. The consumption of electrical energy (power) over a period of time is known as energy.

Heating Energy

If you happen to be sitting beside a nice warm wood stove as you read this book, you can feel the heat from the fire radiating towards you. This warmth and the effects of reading the preceding section on electrons may have caused you to doze off to sleep. If not, you might consider that heat is not the same thing as electricity, for if it were, it could not reach you since air is an insulator.

In fact, the idea that heat is a form of energy baffled many earlier physicists and took a long time to be understood. It took an even longer period for the theory of heat to exert its effect on industry. Consider that builders of houses around the turn of the century gave no thought to conserving heat by means of insulating or any of a number of techniques which seem quite obvious to us now.

So what exactly is heat? Early researchers thought that heat was a substance, something you could put in a bottle—a fluid they called *caloric*. This theory persisted until the middle of the 19th century and was not nearly as silly as you might think. Consider: Heating a liquid causes it to expand, as if something is added to it; when wood is burned, a small pile of ashes is left behind, as if something has escaped or evaporated; a jar of boiling hot water placed next to a jar of cool water causes the cool water to warm up. Perhaps *caloric* flowed from the warmer jar to the colder?

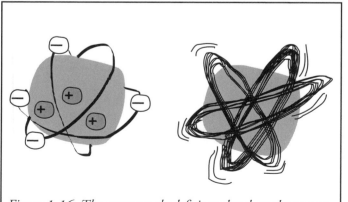

Figure 1-16. The atom on the left is cooler than the one on the right because it is moving more slowly.

Although we now understand that the notion of *caloric* is incorrect, the concept of heat flowing as if it were a fluid is not that far from the truth, as the above examples illustrate. As scientists continued to scratch their heads, current laws of heat and thermodynamics slowly replaced the theory of *caloric*.

It is now understood that heat is a form of energy caused by the motion of molecules, or groups of atoms. All matter is made up of molecules, which (like young children) are always in motion. As a substance is warmed, its molecules move faster; as it is cooled, its molecules move less. The temperature of a substance is directly related to the motion of these tiny molecules.

One of the primary practical considerations relating to heat is that it does not like to stay still (also like children). Consider any type of heat and you will notice that it always wants to move from the hotter object to a colder one. Placing a hot pan in cold water will cool the pan while warming the water. In the same manner, in winter your expensive heated air wants to get outside as quickly as possible to help melt the snow. Although melting the snow might be a useful task, the cost in dollars and energy is well beyond reach and our heat energy should remain trapped indoors.

When we wish to stop the flow of electricity in a circuit, we use a non-conductive device called an insulator. Heat energy can also be slowed down on its relentless path into or out of our homes by the use of an insulator for heat (home insulation), which is discussed in Chapter 2.

Because heat is energy, it is possible to quantify it in the same way that electricity is quantified—in kilowatt-hour units. As a matter of fact, if we were to heat, cool, and operate our homes completely on electricity, the energy usage charge would indeed be in kWh, as all of the energy used would be sourced from the electrical utility. However, as you know there are many sources of heating/cooling energy to choose from, including natural gas, electricity, propane, oil, solar thermal, wood and wood by-products, and even coal. Each of these sources of heat energy is delivered to you in a bewildering array of units, making comparison shopping very difficult. To complicate matters further, the efficiency of the heating or cooling equipment using these various sources varies greatly.

To level the playing field, we will use the English "British thermal unit," or BTU, as our standard for comparing heating and cooling energy. For readers more accustomed to the metric system, the calorie or joule measure will provide the same basis for comparison. Just make sure you don't use both systems at the same time, or you may be wondering why you have to chip ice out of your toilets next January.

One BTU is the quantity of heat required to raise, by one Fahrenheit degree, the temperature of one pound of water. Using these units allows you to quickly compare two heating sources that have different base units. For example, suppose you are trying to assess the cost of heating a house with oil as compared with propane. Which is more economical?

Assume that the current cost of heating your home is $500.00 per year and that you require 250 gallons of oil at $2.00 per gallon. Propane costs $1.75 per gallon. You assume that propane costs 25 cents less per gallon and that it will be cheaper to use. Sorry, it doesn't work like that. The first step is to find out how much heat energy in BTUs is stored in each energy source. Refer to the cross-reference chart in Appendix 1 and look up information on both energy sources. You will see that oil contains 142,000 BTUs per gallon and propane contains 91,500 BTUs per gallon. Therefore:

250 gallons of oil x 142,000 BTU/gallon = 35,500,000 BTU per heating season required to heat your house.

And, if we were to assume that we need 250 gallons of propane:

250 gallons of propane x 91,500 BTU/gallon = 22,875,000 BTU

But your house requires 35.5 million BTU, so we have a shortfall of:

35,500,000 required − 22,875,000 from

propane = 12,625,000 BTU shortfall

Now let's take a look at the costs. The current heating charge using oil is $500.00, and our assumption that 250 gallons of propane would be sufficient produced our first estimate of:

250 gallons of propane x $1.75 per gallon = $437.50 or a savings of $62.50/year

However, in order to make up the shortfall of 12+ million BTU, we have to purchase more propane:

12,625,000 BTU shortfall ÷ 91,500 BTU/ gallon propane = 138 more gallons

138 more gallons x $1.75 / gallon of propane = $241.50 additional cost

The total cost of heating your house with propane is now *$437.50 + $241.50 = $679.00.*

As you can see, your cost for using propane over heating oil went from an estimated savings of $62.50 to an increase of $179.00 per year! Of course these costs are not real, but the example shows that the amount of heat energy "stored" in a fuel is known and can be used for comparison when all fuels have the same base units of BTU or joules. This also applies to renewable sources such as firewood, wood pellets, and even solar heating systems.

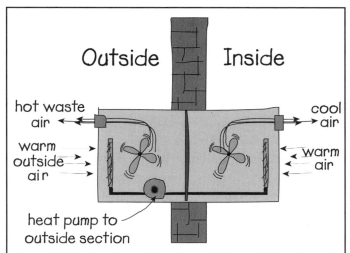

Figure 1-17. An air conditioning unit "pumps" heat from the room to be cooled and sends it outdoors through a complex compressor system. Air conditioning is a very energy-intensive operation, and alternative cooling means should be used whenever possible.

You must also consider the efficiency of the heating appliance. Suppose you find two nearly identical heating devices at the same price with output ratings of 120,000 BTUs per hour. (The fuel type is not important for this calculation.) They have efficiency ratings of 80% and 65% respectively. It stands to reason that the 80%-efficiency-rated unit is the better buy. Over the lifetime of a product, this difference in efficiency can add up to tens of thousands of dollars.

> *You must also consider the efficiency of the heating appliance.*

Not interested in doing the math? Contact a competent home energy dealer and discuss these issues. Most dealers can show how fuel prices affect the operating cost of the heating appliances they sell. You will have to go to the fortuneteller down the street to determine future energy costs. The one thing you won't need a fortuneteller for is to understand that *all* energy costs are rising.

Cooling Energy

An important consideration for homeowners is how to "make" cool air in the dog days of summer. No matter how big your air conditioning unit is, you do not make cold air. A better way to understand a mechanical cooling system is to think of it as a *heat pump*. Modern air conditioning units are complex devices that move heat from one location to another. A window air conditioner pumps heat from your warm indoor room and moves it outside. This may seem a little difficult to understand until you realize that the outside part of the A/C unit is quite hot because of the indoor heat moved there by the internal mechanism. As the outside component (the condenser) is hotter than the outdoor air, the heat wants to move from the hot condenser to the relatively cool outdoor air. This phenomenon of air-conditioning heat transfer explains why the unit has to work harder in hotter weather. As the outside air temperature approximates that of the condenser, heat transfer slows, making the unit run for longer periods. This is also why you want to place the condenser facing north or out of direct sunlight.

Regardless of the type of A/C system installed, the energy used to cool a given area is measured either by the BTUs of heat moved or by the electrical energy consumed by the unit. Commercial A/C units are usually rated both ways, with some units still carrying a "cooling tonnage" rating. This archaic rating compares the cooling capacity of the A/C unit to a ton of ice. Just in case you were wondering, this is approximately equal to 12,000 BTU of cooling capacity per hour. This cooling process is not efficient in even the best-designed air conditioners.

With all the discussion of "moving" heat from here to there, perhaps the idea of *caloric* wasn't so far off after all.

1.3

Energy, Pollution, and Climate Change

Everyone has heard about them, but do you actually know what smog, climate change, and greenhouse gas emissions really are? Ask anyone who lived in Los Angeles in the early 1980s and they will tell you about dirty skies and endless haze. For someone with asthma or other respiratory sensitivities it meant staying indoors for countless hours or relying on medical inhalers for each breath. Steve Ovett wished the air had been cleaner in Los Angeles. Although he had effortlessly beaten his rivals in 800- and 1,500-meter races for years, the air in Los Angeles was his undoing. During a race in the 1984 Olympics he collapsed as a result of smog-induced asthma and spent two nights in hospital.

Smog and soot pollution are created when emissions from burning fossil fuels combine with atmospheric oxygen and ultraviolet light from the sun. When smog and soot particles combine with raindrops, acid rain is formed. Estimates are that over half a million deaths worldwide are directly attributable each year to excessive levels of airborne pollution.

Clean rain water is neither acidic nor base, meaning that this life-giving element is non-corrosive and restorative to everything it touches. Acid rain, on the other hand, is precisely that—water which has become acidic. When it comes into contact with metal, rock, or plants the result is a corrosive action and the decay and destruction of a wide array of living and lifeless objects.

Smog has been greatly reduced in the developed world in recent years as a result of decreased sulfur concentrations in gasoline and diesel fuels. Improvements in automotive and coal-fired power-station emissions have also gone a long way in reducing smog over the last 20 years; however, the problem is far from over. Increasing demand for coal-fired electrical energy and ever-increasing numbers of automobiles on the world's highways continue to exacerbate the problem.

Carbon dioxide and methane are two greenhouse gases that are increasing in concentration in the atmosphere and are directly linked to global warming. Carbon dioxide concentrations are higher than at any time in the last 420,000 years and have directly contributed to a rise in the Earth's surface temperature of 0.6 degrees over the last century. It should also be noted that the 1990s were the warmest decade on record, with 1998 recorded as the single warmest year of the last 1,000 years.

Scientists predict that if the earth's average temperature were to rise by approximately 2°C, numerous devastating effects could occur:
- increased magnitude and frequency of weather events;
- rising ocean levels, causing massive flooding and devastation of low-lying areas;
- increased incidents of drought, affecting food production;
- rapid spread of non-native diseases.

So where do these gases come from? Carbon dioxide is a by-product of the burning of any carbon-based fuel: gasoline, wood, oil, coal, or natural gas. With few exceptions, the world's current energy economy is fueled by carbon.

We have to go back half a million years and more to learn how this carbon fuel came to be. All carbon fuel sources began as living things in prehistoric times. Peat growing in bogs absorbed carbon dioxide from the air as part of the photosynthetic process of plant life. The rolling, heaving crust of the earth entombed the dead plant material. Sealed and deprived of oxygen, the plant matter could not rot. Over the millennia, shifting soils and ground

Scientists predict that if the earth's average temperature were to rise by approximately 2°C, numerous devastating effects could occur.

heating compressed the organic material into soft coal that we retrieve today from shallow open-pit mines. Allowed to simmer and churn longer, under higher heat and pressure, oil and hard coal located in deep underground mines are created.

Provided these prehistoric fuel sources remain trapped underground there is no net increase in atmospheric carbon dioxide. However, the process of burning any of these fuels reduces the carbon stored in the coal or oil and in turn drives off CO_2. Conversely, the burning of renewable carbon-based fuels such as wood, biomass, and oilseed-based fuels does not contribute to net greenhouse gas emissions. These plant materials "recycle" carbon dioxide from the atmosphere into carbon during the growing cycle. Burning these materials releases the same CO_2 that was originally absorbed. Although the burning of prehistoric fossil fuels such as coal and oil simply gives back carbon dioxide the atmosphere lost millennia ago, these "new" emissions increase concentration levels, causing the effects of global warming.

It is not necessary for society to make a wholesale switch to a non-carbon-based fuel such as hydrogen in order to reduce the effects of global warming. By selecting a more fuel-efficient car, switching to compact fluorescent lamps, or improving the energy efficiency of our home we can drastically reduce greenhouse gas emissions and enjoy significant savings along the way.

From an *eco-nomic* point of view, doing more with less is the path to sustainable development.

Endnotes

1   Christopher Johnson, "Oil falls towards $35 after IEA demand report," Reuters, January 16, 2009, http://uk.reuters.com/article/topNews/idUKTRE50E6L9200901.

2   Energy Information Administration, Annual Energy Outlook 2009 Early Release, Report #:DOE/EIA-0383(2009), http://www.eia.doe.gov/oiaf/aeo/.

3   Intergovernmental Panel on Climate Change, Climate Change 2007 Synthesis Report, http://www.ipcc.ch/ipccrreports/ar4-syr.htm

4   Ontario Medical Association, The Illness Costs of Air Pollution in Ontario (ICAP), June 2000, http://www.oma.org/Health/smog/icap.asp

5   "Carbon Dioxide (CO_2) Inventory Report for Calendar Years 2006 & 2007, World Resource Institute, December 2008"

6   Energy Information Administration, "Natural Gas," Canada, May 2008, http://www.eia.doe.gov/cabs/Canada/NaturalGas.html

7   David Suzuki Foundation, Bright Future, September 2003:22, http://www.davidsuzuki.org/files/Climate/Ontario/brightfuture.pdf.

8   Ibid.

9   Electrical utilities are quick to point out that the cost of generation is "x" cents per kilowatt-hour. What they forget to tell you is that the electricity has to be delivered to your door using transmission and distribution systems that must be paid for. Of course we can't forget the bit of tax added for the government. On average the cost of electricity delivered to your home is approximately double the cost of generation in most jurisdictions. To calculate your actual cost of electrical energy, divide your total electrical bill charges by the number of kilowatt-hours of electricity used during that billing period.

10 Bulbs of varying wattages use more or less energy according to their power consumption:
    Incandescent Bulb = $0.12/kWh of electricity x 10,000 hours x 100 W/bulb = $120.00
    CF Bulb = $0.12/kWh of electricity x 10,000 hours x 23 W/bulb = $27.60
    Savings per bulb = $120.00 per Incandescent Bulb - $27.60 per CF bulb = $92.40

11 The website www.greenandsave.com has compiled ROI data on dozens of day-to-day products used in and around the home. This site also provides similar data for an assortment of home renovations.

Chapter 2
ENERGY EFFICIENCY

When the price of gasoline jumps for the tenth time in a month, most people tend to cut back on their driving—or at least to talk about it. Some people might even consider trading their vehicle in for something with better gas mileage than a Hummer.

Family budgets are straining from the rising cost of living: taxes, mortgages, car payments, and energy bills for homes and vehicles. As a society we rarely consider the source of these costs. People hop on the bandwagon of the middle-class dream fueled by the two incomes that make it happen. Large homes in the suburbs require enormous amounts of heat, light, and air conditioning, and many people have two cars to get to distant jobs. All of this eats away at our precious discretionary income—and our free time.

Doesn't energy conservation mean giving things up, not living the middle-class dream North Americans have come to expect, or living a spartan life of near poverty? Well, yes and no. Yes, energy conservation does mean giving things up, but they are mostly wasteful, inefficient things. And no, it does not mean that you must compromise the quality of your lifestyle.

How can we reduce our heating and electrical loads or transportation costs to embrace energy conservation without giving anything up? The answer is efficiency. As the current corporate downsizing mantra goes: "Do more with less."

I recently heard a politician asking us all to do our part to conserve energy by restricting ourselves to five-minute showers. This is a misdirected effort. People may do their part by conserving energy or resources in times of severe famine, war, or national disaster, but it is never sustainable. As soon as things are back to normal people immediately return to their old habits. Conservation is not sustainable if there is a perception that it will interfere with quality of life.

The better approach is to take a ten-minute shower but use a low-flow showerhead. Installing this five-dollar device will not reduce your quality of life and in fact might improve it, as many models incorporate a massage feature. By reducing energy and water consumption by 50% or more, such a device will put real money in your pocket.

This is the eco-nomic approach: doing more with less; adopting an energy-efficient, sustainable lifestyle.

It may sound like nickel-and-dime stuff, but in fact the opposite is true. The constant "leakage" of energy dollars here and there can become quite significant. The simple act of switching common incandescent lamps to high-efficiency compact fluorescent models will put over $2,300 after tax in your pocket over the operating life of the lamps. If you are in a 35% tax bracket this translates into a savings of over $3,100.

Figure 2-1 An energy-conserving lifestyle does not mean a spartan lifestyle. Choosing the most energy-efficient appliances and products that "do more with less" is how to make it happen. The hybrid 2004 Toyota Prius doubles gas mileage efficiency and reduces pollutants by 90%. (Courtesy Toyota Corporation)

This is just one example of dozens of tips that, added together, translate into serious energy and financial savings. What you do with the savings is up to you. Have you been trying to scrape together enough cash to top up your retirement savings plan? If so, this is one of the easiest ways to do it.

For those of you considering an off-grid lifestyle, energy efficiency is doubly important.  An interesting statistic being floated about is from the U.S. Department of Energy, which states that for every dollar you spend on energy efficiency you will save three to five dollars on the cost of generating equipment. This rule applies to both the big nuclear plant built by the mega-utility and your own personal renewable energy power station at home.

I am well acquainted with explaining this logic; on Sunday afternoons, tourists often stop by our house, after seeing our wind turbine, to have a chat about the "big fan on the pole over there." Enjoying a cup of solar-power-brewed coffee (which somehow always tastes better), they mull over why they don't consider using renewable energy as well.

At some point in our conversation, I am asked the usual question: "Our hydro bill is a couple of hundred bucks a month. How much to install a rig like yours to power our house?"

The answer is always the same: "Too much." Applying the Department of Energy rule discussed above, it is much less expensive to optimize the energy consumption of the house than to build a bigger power station, whether or not the energy comes from renewable sources.

And no, you don't have to resort to coal-oil lamps and a black and white television with a rabbit ear antenna. You will be introduced to homes in Chapter 4 that will surprise you with the number and type of appliances that can be used in an off-grid home. No one has to suffer in making the transition.

Let's start our quest for home-energy efficiency with an understanding of the house itself. First, we'll look at some of the most obvious design features that can be incorporated into the building. Next, we'll take a look around inside. Lighting, heating, electronics, and major appliances all consume energy and must be considered in an energy-efficient lifestyle.

2.1
New Home-Design Considerations

By far the largest energy use in North American homes is heating and cooling. As we learned in Chapter 1.3, heat has a nasty habit of wanting to escape outside in winter. In summer, the opposite is true. Heat from the hot summer sun just can't wait to come inside and enjoy the air conditioning with you.

It is not necessary to design your own home in order to achieve satisfactory energy-efficiency results. Many builders now recognize that energy-efficient homes can almost compete with granite countertops and hardwood floors in attracting the buying public. Look for builders that are certified for R-2000 or other high-quality construction methods.

Heat Loss and Insulation
Heat loss and gain in your home is linked to the level and quality of insulation in the ceilings, walls, and basement. Not so obvious are the losses associated with windows, doors, and the sealing in various joints and holes in the structural cavity. The key to designing an energy-efficient house is to ensure that it is well constructed and airtight and contains adequate levels of insulation.

The national and local building codes in your area set the minimum standards that your building must meet, but it is fairly easy to incorporate upgrades into your design that will pay for themselves many times over. With the world's current political climate and uncertainty over fossil fuel supplies in the coming years, these upgrades will provide a safety net against increasing energy prices.

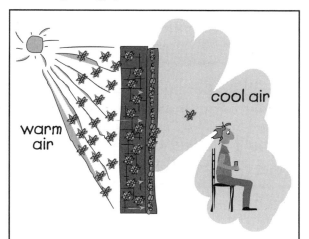

Figure 2-2. Heat loves to move from the hot summer outdoors to enjoy the air conditioning inside with you. Some materials such as brick and stone allow heat to travel fairly easily. Proper insulation slows the flow of heat.

Insulation in the walls and ceilings of our homes slows the transfer of heat into or out of a house (in much the same manner that electrical energy cannot flow through a non-conductive material). The higher the quality and thickness of the insulation, the harder time the heat has getting through. Many people believe that because hot air rises, most of the heat loss in the house will be up through the ceiling. Not so. Heat moves wherever it can, shifting from warm areas to colder ones, whether that be upwards, downwards, or sideways. Keep this in mind as we visit all areas of the home during our insulation spree.

Table 2-1 lists typical recommended insulation values for a very well insulated home. The "R" value (English system) and "RSI" value (metric system) indicate quality levels of insulation. The higher the value, the better the insulation level. If you live in colder climates where the number of heating days is high,

it may be in your best interests to increase these thermal resistance values.

As the price of heating fuel continues to rise over the coming years, any effort you make now to create a more energy-efficient home will be paid back several times over. I recall my parents' first home, built at a time when fuel costs weren't considered at all. This home was constructed using 2 x 4" lumber (63 x 125 mm) and minimal levels of insulation, with no wind barrier and no vapor barrier. But with heating-fuel costs in the $0.25/gallon range, no one cared.

How times have changed. With oil prices now ten times this price, no one can afford to let their heating dollars escape because of poor home planning or construction. Although no one has a crystal ball, I suspect that energy prices will rise considerably faster in the coming years. Our parents had little or no concern about supply, environmental issues, security, or hostile foreign governments. Today, we do not have this luxury.

Moisture Barriers

A vapor barrier consisting of 6-mil-thick (0.006") polyethylene plastic can be attached to the wall structure on the warm side of the insulation. The vapor barrier completely surrounds the inside of the house and must be well sealed at the joints and overlapped edges. During construction, care must be taken to ensure that this barrier extends without a break from one floor to the next.

The function of the vapor barrier is to seal the house in a plastic bag, controlling air intake and leakage. It also stops warm, moist air from penetrating the insulation and contacting cooler air, condensing, and causing mold and rot problems in the wall. Additionally, a vapor barrier

Table 2-1: Minimum Recommended Insulation Values				
Insulation Quality (over unheated space)	Walls	Basement Wall	Roof	Floor
R Value	23	13	40	30
RSI Value	4.1	2.2	7.1	5

ensures that the air inside the insulation remains still. These conditions are absolutely essential in making the insulation system function properly.

Uncontrolled moisture can cause wood rot, peeled paint, damaged plaster, and ruined carpets. Moisture can also directly influence the formation of molds, which are allergens for many people. Moisture control is not to be taken lightly. Left uncontrolled, it can cause damage to building components.

Some insulation materials such as urethane foam spray or sheet styrofoam are fabricated with millions of trapped air or nitrogen bubbles in the plastic material. Although more expensive than traditional fiberglass, inch-for-inch they provide higher insulation ratings and also form their own vapor barrier.

Wind Barriers

A properly taped and well-jointed Tyvek®-style wind and water barrier on the outside of the house, just under the siding material, helps to keep winter winds from whistling right through the insulation. Controlling wind pressure infiltration into the building structure is an area most people overlook. Even if the breeze isn't blowing through your hair while the windows are closed, wind leakage is important.

Insulation works by keeping dry air very still. Even the slightest movement will greatly reduce the insulation value of the system. Think about how much force you exert wrestling with an umbrella in even a light wind. Now imagine the same effect wind has on the entire surface area of a house. Quite frankly, it's amazing more homes don't blow down! The force of the wind on the house structure can penetrate the insulation. This upsets the still-air requirement and lowers insulation values.

Provided all of these insulation techniques are followed carefully, you will have a very efficient home-insulation system.

Provided all of the above insulation techniques are followed carefully, you will have a very efficient home-insulation system.

Examples of these techniques are shown in the cross-section or side view of the bungalow in Figure 2-3. This design strengthens the local building code by increasing insulation in areas that you may not have considered.

Basement Floor

This floor comprises a 3" (75 mm)-thick sheet of extruded foam board insulation installed directly on top of packed crushed stone. Over this can be applied an 8-mil layer of polyethylene moisture barrier, with all joints taped or overlapping seams caulked with acoustical sealant. Such a membrane prevents ground-based moisture from entering through the floor area and thus reduces mold potential. It will also stop ground gases, including radon, from entering. In areas where foundation footings are not subject to frost heaving, a footing insulation board can be added between the basement wall and the concrete slab floor.

Basement Walls

The basement walls can be "finished" using standard framing techniques, with the exception that the 2 x 4" stud wall is set away from the concrete basement wall by 4" (100 mm). This increases the wall cavity space for extra insulation without requiring additional framing material. Instead of traditional batt insulation, blown-in fiberglass or cellulose is suggested. These materials ensure complete coverage and packing density, especially in the hard-to-insulate areas around plumbing lines, electrical boxes, and wiring. Figure 2-4 shows the use of a tarpaper-style moisture barrier glued to the concrete wall from just below grade. It is then folded up and glued to the bottom of the interior vapor barrier.

Rim Joints

Rim joints are not an obscure arthritic condition, but rather an obscure area of your home where the floor joists meet with the exterior support walls. (They are also known as rim joists or

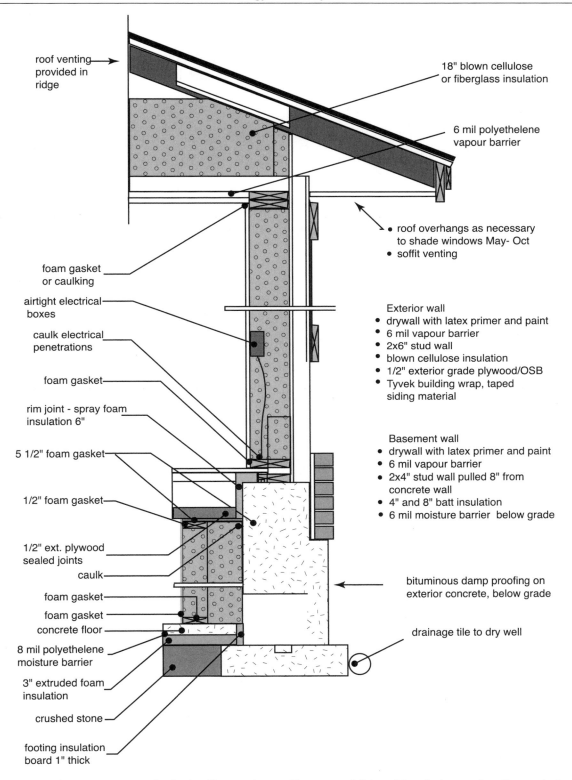

roof venting provided in ridge

18" blown cellulose or fiberglass insulation

6 mil polyethelene vapour barrier

• roof overhangs as necessary to shade windows May- Oct
• soffit venting

foam gasket or caulking

airtight electrical boxes

caulk electrical penetrations

foam gasket

rim joint - spray foam insulation 6"

5 1/2" foam gasket

1/2" foam gasket

1/2" ext. plywood sealed joints

caulk

foam gasket

foam gasket

concrete floor

8 mil polyethelene moisture barrier

3" extruded foam insulation

crushed stone

footing insulation board 1" thick

Exterior wall
• drywall with latex primer and paint
• 6 mil vapour barrier
• 2x6" stud wall
• blown cellulose insulation
• 1/2" exterior grade plywood/OSB
• Tyvek building wrap, taped siding material

Basement wall
• drywall with latex primer and paint
• 6 mil vapour barrier
• 2x4" stud wall pulled 8" from concrete wall
• 4" and 8" batt insulation
• 6 mil moisture barrier below grade

bituminous damp proofing on exterior concrete, below grade

drainage tile to dry well

Figure 2-3. This cross-section of a high-efficiency house illustrates additional insulation and sealing techniques that will greatly reduce your heating and cooling energy usage.

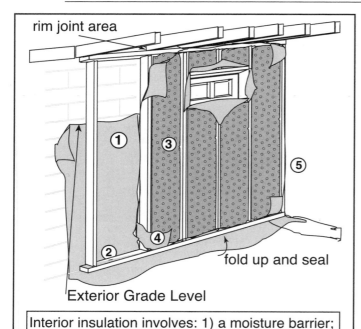

rim joint area

① ③ ⑤

④

② fold up and seal

Exterior Grade Level

Interior insulation involves: 1) a moisture barrier; 2) new frame wall; 3) insulation; 4) air and vapor barrier; 5) finishing

Figure 2-4. The technique illustrated provides excellent insulation value and ensures that your basement will also remain dry and mold free.

A much better method of insulating the rim joists is to use a spray foam material. Although more expensive than traditional methods, it actually works. As an added benefit, the foam also provides its own vapor barrier, as shown in Figure 2-5b.

Exterior Walls

Typical framed construction now employs 2 x 6 " (50 x 150 mm) stud walls. This wall-cavity thickness is suitable in all construction areas provided quality workmanship is assured. The major problem with exterior wall construction is air leakage. (Have you ever felt the breeze blowing from the electrical boxes of older homes?) The use of blown-in fiberglass, cellulose, or rock wool in the wall cavity will ensure complete coverage of the entire surface area. It also guarantees that insulation works its way around electrical boxes and other obstructions in the cavity. Make sure the contractor fills the wall with a sufficient density of insulation. The material should be tightly packed and not sagging at the top of the frame wall.

The next areas to consider are electrical, phone, and cable boxes, which penetrate the exterior structure. Ensure that airtight boxes are used, and caulk between them with a 6-mil layer of polyethylene vapor barrier. The vapor barrier must be overlapped and caulked tightly at all seams and with the top and bottom plates making up the wall section.

Air can easily blow between the top and bottom plates of the wall and the floor components. For

header joints.) They are notoriously difficult to insulate and vapor barrier correctly, given the huge amount of surface area accorded this space. Many insulation contractors make only a passing effort in this area, stuffing a piece of insulation in place and then stapling a swath of plastic on top, as shown in Figure 2-5a.

The photograph on the left (Figure 2-5a) details a too typical "stuff and run" rim joist insulation job. The photograph on the right (Figure 2-5b) shows the best way to complete rim joist insulation using sprayed urethane foam, which provides an integral vapor barrier.

this reason, these areas should have foam gaskets or caulking compound to ensure air tightness.

Windows

It would be quite easy to write an entire book on the design and treatment of windows. For our purposes, let's consider windows just from an energy standpoint. A single pane of glass is a very poor insulator. During the last thirty years, there have been many advances in window design including double- and triple-pane versions, low-emissivity coatings, argon and krypton gas fillings, and a bewildering array of styles and construction.

At the risk of oversimplifying this issue, follow these basic rules when choosing your windows:

• Purchase the best-quality window you can afford. Ensure that it is from a reputable supplier and has a good warranty. Ask for references.

Figure 2-6 Electrical boxes and other wall penetrations must be carefully sealed with the vapor barrier. You can purchase electrical boxes that are integrally sealed or fabricate your own using a piece of 6-mil polyethylene sheet.

Figure 2-7 The vapor barrier must create a "plastic bag" effect for the entire home. Ensure that this membrane is wrapped from one floor to the next and caulked at each joint.

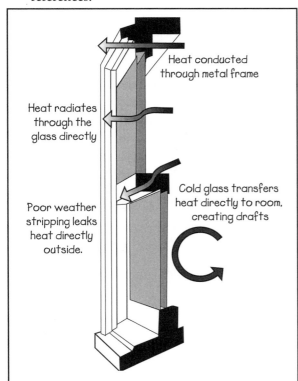

Heat conducted through metal frame

Heat radiates through the glass directly

Cold glass transfers heat directly to room, creating drafts

Poor weather stripping leaks heat directly outside.

Figure 2-8 Windows lose heat through air leakage as well as direct conduction through the glass and frame. Always purchase the best-quality window you can afford from a reputable company.

• If you don't require solar gain, purchase at least double-pane windows with low-emissivity glass (also known as "low-e"). This type of glass reduces heat gain and loss. Note, though, that low-e glass and coatings should not be used with south-facing windows where winter solar gain is desired.

• Ensure that the cavity between window panes uses argon gas, or better still krypton gas. Triple-glazed windows using low-e glass and filled with krypton gas have an insulation value 3.5 times better than that of traditional double-pane windows.

• All framing materials used in the construction of the window should have wood or vinyl cladding to prevent heat transfer through a metallic structure.

• When installing windows be sure to adequately insulate and caulk, preferably with spray foam, between the window and the house frame.

• Make certain that windows are shaded with roof overhangs or awnings or by the leaves of deciduous trees from early May to mid-

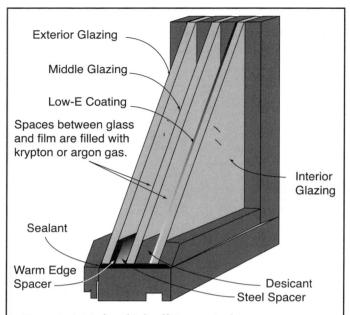

Figure 2-9 Modern high-efficiency windows use two or more panes of glass which are separated by thermal insulation and filled with argon or krypton gas which increases the insulation value. When solar gain is not a factor (such as on the north side of the home), purchase windows with low-emissivity coatings.

Figure 2-10 Solar blinds such as these sunshades block 97% of the daylight heat energy and glare while allowing you to see outside (www.shade-o-matic.com). Quilted blockout curtains or bamboo shading blinds will also work.

October to prevent undesired solar heating. In turn, ensure that windows are free to capture the winter sun for the remainder of the year.

Attic and Ceiling Treatment

Provided the attic is not finished and has a suitable vapor barrier, the best way to insulate this area is to blow in 18" of cellulose or fiberglass. To ensure that the area above the insulation remains as cool and dry as possible during the year, install adequate roof and soffit vents to provide proper air circulation: warm air rises and exits through the roof peak vents while cooler outside air is drawn in through the soffit vents. This upgrade will eliminate moisture damage and ice damming in winter.

If you live in an area where summer air-conditioning load is a higher concern than winter heating, you may wish to consider radiant insulation. This material is similar to aluminum foil stapled to the underside of your roof rafters. Working like a mirror, the film reflects heat back through the roof surface before it gets a chance to hit the attic insulation.

The ceiling vapor barrier should continue down the wall surface a few inches and overlap the wall vapor barrier. Where these overlap, ensure

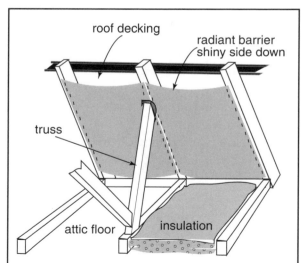

Figure 2-11 Summer air-conditioning loads can be reduced by the use of radiant insulation, which reflects heat back through the roof.

that adequate caulking is applied. The best vapor barrier job is wasted if the finishing trades rip, cut, or damage the sealed home. We are striving to have you live inside a plastic bag, so take extra care to seal and repair any cuts. Good workmanship will guarantee an energy-efficient home.

Lastly, gasket and seal the attic hatch door in the closed position. This reduces air leakage around its perimeter.

If your home has more complex design features, be sure to discuss these general guidelines with your architect during the planning stage.

Ventilation Systems

Living in an energy-efficient house is like living in a plastic bag. With all of the air leaks sealed and the vapor barrier extending from the basement floor to the ceiling, we really have created a sealed, airtight home. Contrast this with the typical older house with leaky kitchen and bathroom fans and other cracks and leaks in the building structure. While these provide ventilation, they are completely uncontrolled and cause tremendous losses in heating and cooling energy. In a typical thirty-year-old house, it is estimated that these leaks collectively equate to a one-square-foot hole in the wall!

To ventilate your sealed home, you will have to resort to alternative measures. The methods you use will depend to a large extent on where you live.

Most high-efficiency homes employ a device known as an air-to-air exchanger or heat recovery ventilator (HRV)—a marvel of simplicity which is brilliant in its execution. Stale, moist air from

the kitchen, bathrooms, and other areas is drawn into the HRV unit and passes over a membrane before being exhausted from the house. At the same time, colder incoming fresh air is pulled

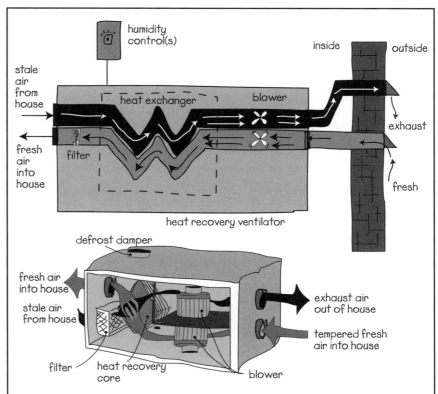

Figure 2-12 The heat recovery ventilator (HRV) controls air flow into and out of a house. Warm exhaust air is passed over a membrane, transferring this heat energy to the colder, incoming fresh air.

into the unit and passes over the opposite side of the membrane. This causes waste heat to be transferred to the colder, dry incoming air, which is then warmed and distributed in the central furnace air duct.

Controls and timers located throughout the house can be programmed to monitor humidity and smoke. Even carbon dioxide levels can be monitored, indicating that the house is occupied and adjusting ventilation accordingly. Based on rules programmed into the control system, the HRV automatically adjusts air-exchange flow to current conditions, while at the same time saving plenty of heating dollars.

Homes in areas that have a low heating load in winter do not fully benefit from the heat-recovery aspect of the HRV. However, these homes would benefit from the ability of the HRV to filter dust and pollen as well as provide proper ventilation.

If there is a disadvantage with HRV units, it is that their fan and electrical circuits can  cause difficulty in off-grid homes. Although the blower and control circuits may appear to be fairly small loads, in actual fact they consume a relatively large amount of electrical energy. (A typical unit operating 12 hours per day can consume 2,000 watt-hours of energy, which may equate to 50% or more of a total winter day's production.)

Unfortunately, there are only a few options that can substitute for a standard HRV:

- Remove the 120V fans in the HRV unit and replace them with lower-flow, high-efficiency 15W computer-style fans. Two 15W fans operating for 10 hours per day lower HRV energy consumption to 300 watt-hours, which is manageable. This modification does require someone who is handy with wire cutters and duct tape.
- Install an air loop intake in conjunction with a bathroom-style vent fan.
- Install an air loop intake in conjunction with a non-direct-venting gas, wood, or pellet stove fireplace.

When a kitchen or bathroom vent fan is activated, or a wood stove draws in room combustion air, stale air is drawn outside. This creates a partial vacuum in the house that causes fresh, dry, colder air to be drawn into the house. As the intake is placed in a little-used area, the fresh air has time to mix with warmer room air before contacting the living area. Although this system is not as efficient as an HRV, intake airflow is controlled. This is a viable alternative to relying on building leakage to make up fresh air, causing uncomfortable drafts. Adjusting the intake damper will control building humidity through the influx of drier winter air.

It is important to discuss this or any airflow plan with building officials to ensure that they are onside with the chosen design. It is also important to make them aware that an off-grid electrical system should not be used to power an HRV unit.

New Construction Summary

By following the tips provided above, you can achieve a satisfactory balance between capital cost

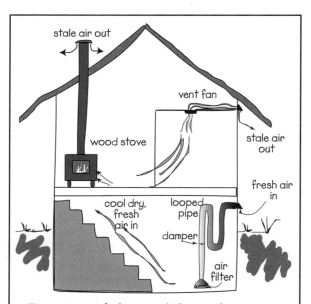

Figure 2-13. A fresh-air intake loop used in conjunction with a ventilation fan or a woodstove creates a partial vacuum to exhaust stale air and draw in fresh air.

and energy conservation. In contrast to a home built to standard building codes, a home designed using the principles provided in this (and the next) section can reduce your heating and cooling energy requirements and costs by 50% or more.

The above discussion of new home construction methods is by no means exhaustive, and the reader interested in learning more on the subject is directed to review the Resource Guide located in Appendix 3.

2.2
Updating an Older Home

If your home is more than ten years old, it is likely that many of the energy-saving features described in the previous section are not incorporated into its design. Don't despair; there is no need to tear down the old place or move. Instead, we can review the entire home systematically and determine where to put your renovation dollars to make energy sense.

Step 1 – Assessment
Every home is unique and, whether it's one, ten, or a hundred years old, what works for one house may not work for the next. As with any project, an assessment is the place to start. We need to check the general condition of the house and test what areas need to be beefed up. Over the last few decades, home designs have improved and architects, engineers, and building contractors have all learned from past practices. With this in mind, there is a general checklist we can use to determine what energy shape our house is in right now.

- Consider having a professional inspection of your home. A professional assessment will not only identify problem areas but also provide suggested corrective action and even precalculate any applicable government rebates for implementing the suggested work. By way of example, the Canadian ecoENERGY

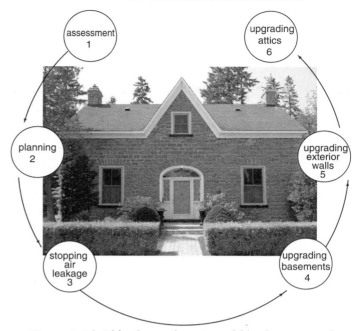

Figure 2-14 Older homes have incredible character and appeal. What they lack is energy efficiency. Proper assessments, planning, and retrofit work always make energy sense.

Retrofit program provides tax-free grants to homeowners who invest in home energy improvements.
- Air leakage is the most common problem with all older homes. Poorly fitting window

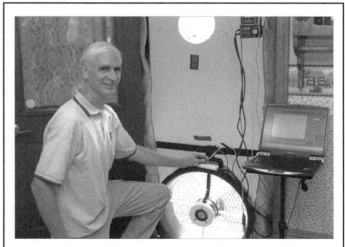

Figure 2-15 A professional assessment of your home will include a pressure door test. A computer-controlled fan will pressurize a home, calculate air leakage, and allow the assessor to identify problem areas in the home and suggest corrective action.

frames and leaks around doors and chimneys abound. Simple corrective action such as applying sealants and weather stripping stops breezes from blowing into the house. Correcting air leakage also helps to regulate humidity and reduce condensation problems in existing insulation.

- Many older homes treat the basement like an outcast relative: it's there, but leave it alone. Often dark, damp holes which are not properly insulated, basements can eat an enormous number of heating dollars. Moisture problems tend to be left to a dehumidifier, which is about as useful as putting a bucket under a leaky roof. Let's try to correct these issues with damp-proofing, moisture barriers, drain gutters, and proper insulation systems.
- Cavities between walls in older homes are woefully under-insulated or completely uninsulated. Stone walls look great on the outside, but are murder on the heating bill if not properly dressed inside. There are many ways of tackling these problems, from both the inside as well as the outside of the home.
- Everyone knows that heated air rises. Let's try to stop it before it escapes through underinsulated attics, leaky joints, and chimney and plumbing areas.

Air Leakage

Air leakage control is the most important step that can be taken in upgrading any home. Controlling air leakage provides many benefits:

- Heating/cooling costs are reduced as the infiltration of outside air is decreased.
- Insulation efficiency is increased, since still air allows insulation to work properly.
- Humidity and condensation levels can be controlled.
- Home comfort is improved as drafts and cold spots are eliminated.

It is possible to hire professional contractors to assess the quality of the air barrier system in your house. Alternatively, a simple yet effective method is to conduct the test yourself using several incense sticks left over from your psychedelic days in college.

Hold the incense near the suspected leakage area on a windy day and observe what happens to the smoke. Smoke drawn towards or blown away from a suspect area indicates an air leakage path. Mark the location down on your list. Continue testing in this manner, paying attention to:

- electrical outlets,

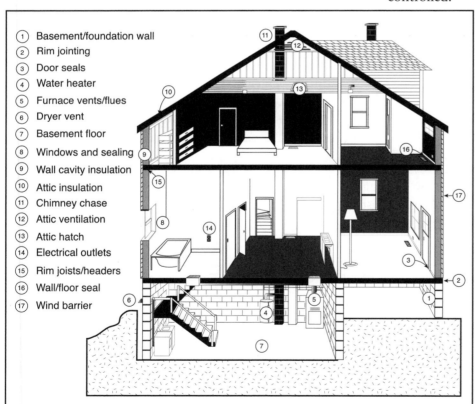

1. Basement/foundation wall
2. Rim jointing
3. Door seals
4. Water heater
5. Furnace vents/flues
6. Dryer vent
7. Basement floor
8. Windows and sealing
9. Wall cavity insulation
10. Attic insulation
11. Chimney chase
12. Attic ventilation
13. Attic hatch
14. Electrical outlets
15. Rim joists/headers
16. Wall/floor seal
17. Wind barrier

Figure 2-16 A thousand little air leaks and a poorly insulated area add up to big energy costs. A thorough assessment of these areas will let you know where you are wasting your money.

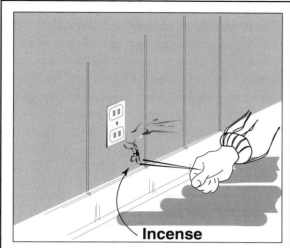

Figure 2-17 A burning incense stick held near suspected areas of leakage on a windy day will cause the smoke to move towards or away from the leak.

including switches and light fixtures

- plumbing penetrations that include the attic vent stack, plumbing lines to taps and dishwashers, etc.
- floor-to-ceiling joints and other building areas that are storied
- baseboards, crown molding, and doorway molding

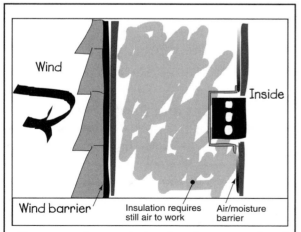

Figure 2-18 Proper wind and air barriers combined with adequate insulation keep the heat in. If any one element is missing or faulty, you may be throwing money away.

- fireplace damper area and chimney exit through the attic or wall
- attic hatch
- windows and doors (Ensure that glass fits tightly and casing area is sealed.)
- ventilation for appliances (kitchen and bathroom fans, dryer vents, gas stoves, water heaters, etc.)
- pipes, vents, wiring, and plumbing lines in the basement and attic (Move the insulation if necessary to access these areas.)

In an older home, it is quite likely that the smoke from the incense will blow when you're just standing in the middle of the room. Don't despair. Just mark all of the areas down on your sheet and perhaps record a "severity rating" from 1 to 10. This way you can tackle the tough problem areas first.

The Basement

Most homeowners don't even consider unfinished basements as a source of heat loss. Part of this mentality comes from the mistaken idea that heat only rises and that earth is a good insulator. Both are wrong. Heat travels in any direction it chooses, but it always travels from a warm area to a colder one. Additionally, older basements have large surface areas of uninsulated walls and flooring that act as a heat sink, drawing warm air to these cooler surfaces.

As we discussed in the Air Leakage section above, there is also a lot of heat loss through crevices in the walls, around and through windows, and at the top of the foundation wall where it meets the first floor. An uninsulated basement can account for up to 35% of the total heat loss in a home.

It is not possible to simply add insulation and air barriers to a damp, leaky basement without first correcting any underlying problems. Any areas that accumulate water in the spring or after a heavy rain must be repaired, as wet insulation has no energy-efficiency value and will ultimately contribute to mold and air quality issues. Check the basement for dampness, water leaks, and puddles in wet periods, and look for major or moving cracks in the foundation wall. Also ensure that a sump pit and pump have been installed in areas where persistent water accumulation occurs.

Some basement wetness problems are corrected by sloping the landscaping away from the foundation wall or adding rain gutters to the house eaves. If you are not sure that these measures will correct the basement water problems, it will be necessary to excavate around the perimeter of the house to provide supplementary drainage and to apply damp-proofing and external, waterproof insulation treatments to the foundation wall.

Wood-Framed Walls

Wood-framed walls are the easiest to insulate, as they are very similar to new construction. The major concerns relate to wall-cavity thickness and access. When assessing these wall structures, attempt to determine if the wall is empty or if there is already some form of insulation in place.

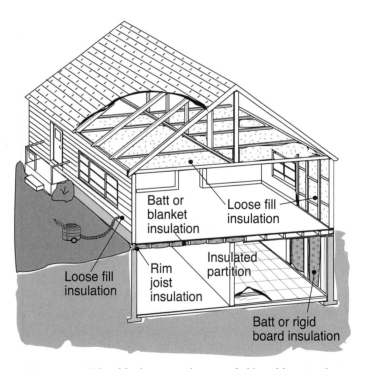

Loose fill insulation

Batt or blanket insulation

Loose fill insulation

Rim joist insulation

Insulated partition

Batt or rigid board insulation

Figure 2-19 The older home can be upgraded by adding insulation to the structure as shown above. Air and vapor barriers may also be added.

The simplest means of checking these walls for insulation is to remove the cover plate of an exterior-wall electrical plug or switch and, using a flashlight and a thin probe, check for insulation behind the electrical box. This test will have to be conducted at a few points around the house to ensure that your sample investigation is accurate.

CAUTION! Before completing this test, turn off the circuit breaker or remove the fuse for that circuit and test the outlet or switch to verify that it is off.

Brick Walls

Brick walls of homes are almost always constructed of veneer with a frame wall on the interior side. Usually there is a small air gap between the brick and the frame wall to allow air to circulate and prevent moisture on the brick from rotting the framing members.

This air gap must not be filled with insulation. However, the frame wall might allow for insulation to be added. Use the same tests as those used for a standard frame wall to determine the depth and area that may be insulated.

Stone or Other Solid Walls

Stone, concrete block, cut stone, and other solid-wall treatments are similar in nature to brick walls. Solid wall treatments are not suitable for insulating and sealing and will require extensive reframing on the inside to add an insulation cavity and vapor barrier. Alternatively, special insulating siding and coverings may be added to the exterior.

Attics and Roof Areas

Most homeowners love to dump their home insulation upgrade dollars in the attic. Perhaps this results from the mistaken idea that all heat rises and the losses must therefore be highest in the attic? Could it also be because attics are one of the easiest places to insulate first? Just dump a few bags of insulation in the old attic and, *voilà*! Your heating bill goes down by 50%?

Sorry, not so fast. The attic does lose heat, but it actually loses less than an uninsulated basement or exterior wall. Most homeowners, no matter how old their house, may have had a passing thought about attic insulation and may even have added a few inches of something up there. Adding more insulation on top of old is not a problem unless the air leakage tests discussed earlier confirm that problems exist. If this is the case, it may be necessary to remove or move aside existing insulation to gain access to the leaking area. The necessity of ensuring quality air sealing in the attic area cannot be understated.

Another potential problem in the attic is moisture, which may originate from different sources such as a leaky roof, ice damming, or frost. Moisture in the attic can also come from within the house through leaky ventilation fan outlets from the bathroom and kitchen areas.

Ventilation of the attic itself is important to provide summer cooling and winter dryness, but many older homes do not have adequate (or any) vents. Ventilation is provided by air intake vents in the soffit and roof vents in the gable or peak where hot air can exit. There should be a ratio of 1 unit of roof vent area for every 300 units of attic floor area. For example, a 1200-square-foot attic should have 4 square feet of vent area.

A common upgrading practice is to add electric vent fans to the attic area, usually in the gable or roof peak. This is not required, nor is it recommended. An electric exhaust fan increases airflow in the attic and may exceed the intake capacity of the soffit vents. If this should occur, additional air will be drawn from the main part of the house, which is exactly the opposite of what is desired.

Check the attic several times during the year, for example after a heavy rain or on a very cold day. Look for wet areas, mold, and rot or small "drip holes" on the insulation or attic floor surfaces.

Attics come in all manner of shapes, sizes, and designs. The trickier it is getting access to the existing insulation (or locating where new insulation should be placed), the more difficult it will be to upgrade, and professional assistance may be required.

Step 2 – Planning the Work

You may wish to tackle some of the upgrading work yourself. Most of the tools required are pretty common household items and the few specialized tools can be rented from your local rental depot or borrowed from a friend. One couple recently completed a blown-in cellulose cathedral ceiling while balancing on a scaffold and using an insulation blower loaned to them by the material supplier. The do-it-yourself approach might not be for everyone, but if you do it correctly, it can result in considerable savings.

Some upgrading work is best left to the professionals: urethane foam spray insulation, for example, requires a truckload of specialized equipment; and if excavating around the foundation of the house is a job that just doesn't make the top of your "list of things to do in life," a backhoe and an experienced operator will really work wonders.

Building Codes

The numerous national, state or provincial, and local building codes as well as their variations are enough to frustrate any do-it-yourselfer. But don't cut corners. The reason the codes are there in the first place is for your health and well-being. Get to know your local building inspector. Although there are plenty of horror stories circulating about these officials, the rumours are generally started by people who began the upgrade work before completing the planning and getting a building permit. You will find that most inspectors are very helpful, providing guidance on technical issues and referrals to qualified contractors or suppliers in your area. Work with them.

Safety

Climbing ladders or working with insulation and chemicals can be dangerous. Make sure you have the proper safety equipment, including work boots, dust masks, rubber or latex gloves, and eye protection. Attic and basement areas may not be well illuminated; ensure that you have a suitable light. Use caution on the attic "floor." Often this floor is nothing more than the drywall or lath and plaster finish material on the ceiling below. It will barely hold a cat, let alone your body weight.

Step 3 – Stopping Air Leakage

Now that we are armed with our "list of air leaks" from the assessment phase, we can start getting down to business.

Caulking

Seal up small cracks, leaks, and penetrations on the inside (warm side) of exterior walls, ceilings, and floors. Sealant applied on the inside lasts longer as the material is not exposed to the elements.

Caulking is done using an inexpensive gun with a tube of appropriate material. There are literally hundreds of caulking materials available. Discuss with a building supply store the type best suited to your project. After a caulking job, many people are dissatisfied with the brand they used and/or the job they did. Avoid the tendency to purchase poor-quality materials which are difficult to apply and do not last.

Remember to purchase high-temperature silicone or polysulfide compounds for areas around wood-stove chimneys or hot water heater flue vents.

- Identify the area to be caulked. Determine the compound type appropriate for the job.

Figure 2-20 It takes a bit of practice to make a good caulking joint. Go slowly and don't cut the nozzle too large; otherwise you'll end up with hard-to-clean goop!

- Never caulk in cold weather. Caulking should be applied as close to room temperature as possible.
- Clean the area to be caulked. Large cracks and holes greater than 1/4" will require a filler of oakum or foam rope sold for this purpose.
- The nozzle of the compound should be cut just large enough to overlap the crack. Insert a piece of coat hanger wire or a long nail into the nozzle to break the thin metal seal.
- Pull the caulking gun along at right angles to the crack, ensuring that sufficient compound is dispensed to cover both sides of the crack. Remember that caulking shrinks, so it's better to go a bit overboard.
- The finished caulking "bead" should be smooth and clean. The surface of the bead may be smoothed with a finger dipped in water.
- Some compounds require paint thinner or other chemicals to clean up. Check the label of the compound before using it to determine the cleanup method.

Electrical Boxes

Air leakage around electrical boxes is so common that there is an off-the-shelf solution available. Special fireproof foam gaskets and pads may be added just behind the decorative plate as shown in Figure 2-21. To ensure a superior seal, use an indoor latex caulking compound on the gasket face before applying the decorative cover.

If the room is being renovated at the same time, install plastic "hats" around the entire electrical box. These hats have an opening for the supply wires and small flaps that can be sealed to the vapor barrier. Ensure that the area where the supply wire enters the hat is well sealed with acoustic sealant.

Windows

Older homes often have single panes of glass puttied into wooden frames. When the putty dries out, air leakage occurs. Remove old putty and replace it with glazing compound to ensure a high degree of flexibility and a long life. Old-fashioned putty is not recommended.

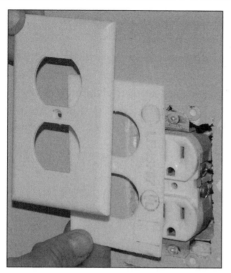

Figure 2-21 Air leakage is so common in older homes that off-the-shelf sealing pads have been designed to aid in stopping leaks.

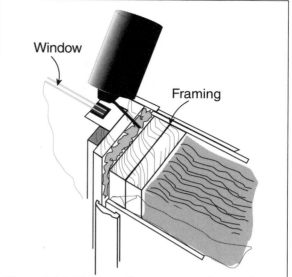

Figure 2-22 Urethane foam spray works wonders in small nooks and crannies such as those between the window and house framing.

The area between the window and the frame is another area prone to leakage. Access to this area usually requires the removal of the window casing trim. Use oakum, foam rope or, better still, urethane foam spray to air/vapor seal and insulate this area in one application.

Baseboards, Moldings, and Doorway Trim

Trim pieces such as baseboards, moldings, and door casing are used to cover the gap between one framing section and another. For example, a premanufactured door is placed in a framed section of the wall called the "rough opening." Obviously the door must be smaller than the opening in order to fit into it. If this opening is not properly sealed, considerable air leakage will result.

The best way to seal these areas is to use methods similar to those described for window framing (see Figure 2-22). For smaller gaps or areas where urethane foam may be too messy (near carpets or finished wood), oakum or foam rope may be jammed into the gap.

Fireplaces

Fireplaces warm the heart, but rob you blind. Air leakage up and down the chimney when the unit is not in use is enormous. When a fire is burning, the suction it creates draws large volumes of expensive, heated room air up the chimney, drawing in cold outside air to replace the heated air that literally went up in smoke!

What to do? Simple. Replace the fireplace with an airtight wood-burning stove or a similar controlled combustion unit. We will discuss these items in greater detail in Chapter 5.

If you really must keep the fireplace, make a removable flue plug that properly seals the chimney when it is not in use. Check to make sure the damper closes as it should. If you detected air leakage around the chimney and framing as part of your assessment in Step 1, seal any cracks with heat-resistant sealant and mineral wool or fiberglass batting.

A quick word on glass fireplace doors; don't waste your money. If you like the look of them or are worried about flying sparks, so be it, but be warned that the majority of these door units are cheaply made and will not provide any air sealing capacity.

Attic Hatch

Seal the attic hatch in the same manner as any exterior doorway trim. Use hook and eye screws to ensure that the hatch fits snugly against the weather stripping (see Figure 2-23).

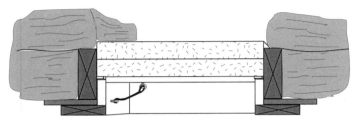

Figure 2-23 The attic hatch is often forgotten when it comes to air leakage and insulation. Use exterior doorway sealing gaskets and hook and eye screws to ensure that the hatch door fits snugly, preventing air leakage.

Air and Vapor Barrier

So far in our discussion of air leakage we have been concerned with the worst but most easily corrected areas. If you are undertaking more extensive renovations, it may be possible to improve large sections of the air and vapor barrier. Before we continue with how to perform this type of major upgrade, let's quickly review the function of each component to be sure we understand how this work can be implemented (see Figure 2-18).

On the exterior of the house, directly under the siding material, is a layer of spun-bonded olefin, which is often sold under the trade name Tyvek®. This material stops wind from penetrating into the insulation area and repels rainwater. Interestingly, it is also permeable to water vapor, which allows trapped moisture in the insulation cavity to escape outdoors, preventing rot.

On the warm side of the insulation is the vapor barrier. In newer homes, this barrier consists of 0.006" (6-mil) polyethylene sheet (think food wrap on steroids) affixed to the wall studs and carefully sealed.

The vapor barrier acts as a secondary wind barrier, stilling the air trapped in the insulation. It also prevents warm, moist household air from penetrating into the insulation. (Remember— heat moves toward colder areas.) Moist air

contacting cold insulation and wood siding on the exterior side of the insulation will quickly condense, forming water. Should this water be present for extended periods of time, it will cause structural rot and mold problems. Many a house has been seriously damaged as a result of high inside humidity coupled with an inadequate vapor barrier.

Your Options

During renovation, it may be possible to install both an air barrier and a vapor barrier. This type of upgrade can only be completed when the framing members are accessible, which typically occurs when a major rebuilding effort is in progress or an addition is being constructed. Major renovation involving more than just a few sheets of vapor barrier is discussed in Step 5 – Upgrading Walls. It doesn't matter if your renovation work is concentrated on the exterior or interior side of the wall; in either case there are ways to complete the work.

Exterior Wall Upgrade

If you are upgrading, removing, or adding to existing siding, it is possible to increase the air barrier with little difficulty. Removal of the old siding reveals the wood sheeting covering the framing members. A layer of Tyvek® air barrier can be taped to the exposed wall sheeting or over

Figure 2-24 A home wrapped with a suitable air barrier decreases air leakage and increases insulation values.

existing smooth siding materials. The air barrier is similar to the vapor barrier in that all joints should be carefully taped to ensure a continuous barrier to wind penetration.

If the interior wall is not being upgraded, follow Step 3 – Stopping Air Leakage. Although a polyethylene film is not being applied, a reasonably good vapor barrier can be made using multiple layers of latex paint over well-sealed drywall. When an external air barrier is mixed with air-leakage sealing and latex paint, you can achieve the next-best thing to new construction.

Interior Wall Upgrade

Your choices for upgrading air and vapor barriers increase if you are updating an interior wall. We will discuss this work further in Step 5 – Upgrading Walls.

Step 4 – Upgrading Basements

In the section on basements in Step 1, we discussed the need to ensure complete dryness before insulation and/or vapor barriers can be added to the inside. Persistent moisture or water leakage problems will ruin even the best-quality work.

Dampness

Minor dampness causes staining or mold growth, blistering and peeling of paint, moldy smells, and efflorescence (whitish deposits) on concrete. These problems can be corrected from the inside

Figure 2-25 Installing rain gutters and sloping the grade away from the house are two simple but effective means of keeping the basement dry.

by cleaning up any mold and then applying damp-proofing to the foundation wall. More serious problems will have to be corrected from the outside.

Major Cracks

If your foundation wall has large cracks or cracks that are getting bigger, seek professional help prior to upgrading to determine if structural repairs are necessary.

What Are My Choices?

There is no doubt that insulating from the inside of the basement is the easiest and least costly method, provided that the basement is dry. Insulating from the outside will do the best job from a technical point of view. What to do? Let's begin by weighing the pros and cons of each method.

Insulating Inside

Insulating inside usually involves the addition of a wood-framed wall and adding some form of insulation material, just like a standard wall. There are several advantages to insulating inside:

- The work can be done at any time of the year.
- A completed job will increase the value of your home.
- Indoor finished space will be increased.
- It is the lowest-cost way to update the basement insulation and vapor barrier.
- The landscaping, porches, walkways, and other obstructions outside will not be disturbed.

There are also some disadvantages:

- Persistently damp or wet basements cannot be upgraded.
- Furnaces, electrical panels, vent pipes, and plumbing obstructions make do-it-yourself framing more difficult.

Insulating Outside

Insulating outside the home involves more complex and expensive excavation but at the same time provides an opportunity to correct existing foundation defects. There are several advantages to insulating outside:

- The outside wall is generally straighter and simpler to insulate once the excavation has been completed.
- Most moisture and water leakage problems can be corrected at the same time.
- Foundation cracking and other damage can be inspected and repaired.
- There is no lost space inside the home as a result of the added wall thickness.
- The weight or mass of the foundation is on the warm side. This mass absorbs heating and cooling energy, helping to balance temperature fluctuations.

There are also some disadvantages to insulating outside the home:

- Excavation work is costly and may be difficult if there are porches, finished walkways, or decks abutting the foundation wall.
- Storing the excavated dirt will damage lawns and bring mud and sand into the house.
- Work cannot be done economically in the winter season.

Once you have determined which system to use, the next step is to examine how the work should be done.

How to Insulate Inside

Insulating a basement that is known to be dry is not much different than insulating a new house. The major differences relate to the type of wall structure that is used. A fairly new poured-concrete or block wall should present few problems. Older rubble or cut stone walls tend to be "wavy" and vary in height somewhat. This makes framing more difficult, but does not change the methods involved. The dry-basement insulation method is just a repeat of new-home construction:

- A tarpaper-style moisture barrier is glued to the concrete wall starting from just below grade and then folded up and glued to the bottom of the interior vapor barrier as detailed in Figure 2-4.
- The basement wall is built using standard framing techniques except that the 2 x 4" stud wall is "pulled" away from the foundation wall by 4" (100 mm). This increases the wall-cavity space for additional insulation without using additional framing material.
- Instead of traditional batt insulation, blown-in fiberglass or cellulose is suggested. These materials ensure complete coverage and packing density, especially in the hard-to-insulate areas around plumbing lines, electrical boxes, and wiring.
- All other finishing details are exactly the same as in a new home.

Insulation	Type	R-Value (approx.)	R.S.I Value
Batts	Fiberglass or Rock Wool	3 1/2" (R-11) 5 1/2" (R-19) 9 1/2" (R-30)	90mm 1.9 140mm 3.3 240mm 5.0
Loose Fill	Fiberglass or Rock Wool	R-2.7 per inch	1.9 per 100mm
	Cellulose	R-3.7 per inch	2.6 per 100mm
Rigid Board	Expanded Polystrene (Beadboard)	R-4 per inch	2.8 per 100mm
	Extruded Polystrene	R-5 per inch	3.5 per 100mm
	Polyurethane or Polyisocyanurate	R-7 to R-8 per inch	4.4 - 5.6 per 100mm
	Polyurethane	R-7 to R-8 per inch	4.9 - 5.6 per 100mm

Table 2-2 Common Insulation Materials and Their Heat-Resistance Values in R/RSI.

How to Insulate Outside

Step 4 - 1 - Preparation

- Prior to beginning the insulation work, remove any outside features that will get in the way. This includes decks, stairs, trees, walkways, and so forth.
- Determine where power, water, sewer, septic tank lines, gas, telephone, and other services enter the building. Contact your utility to locate unknown pipes and wires; this is usually a free service.
- Determine where the excavated dirt will go. Placing a polyethylene sheet or tarp on the grass to hold the dirt will make the cleanup easier.

Step 4 - 2 – Excavation

- USE CAUTION! The soil may be unstable and fall back into the excavation, causing injury or death.
- The excavation must extend down to the top of the footing
- Never dig below or near the base of the footing (see Figure 2-26, Item 8), as this will cause the foot to sink and the house to drop.
- USE CAUTION! Older rubble stone walls may collapse without the support of the surrounding soil. Seek expert help if you are in doubt.

Step 4 - 3 – Preparing the Foundation Wall

- Brush and fully clean the foundation wall. Scrape any loose concrete or rubblework from the wall (Item 2).
- If concrete is missing or damaged in places, it is best to apply a coating of parging (waterproof masonry cement) to the damaged or missing sections. Allow this to dry.
- Apply damp-proofing compound to the foundation wall (Item 7). Damp-proofing is a tar-like substance that is painted on the foundation wall from the top of the footing to grade level.
- Inspect the footing drain system (Item 4). If

there is no footing drain, determine whether one can be added. The footing drain is made by installing a flexible pipe that has been manufactured with thousands of holes along its length. The pipe is normally "socked" with a pantyhose-like liner to prevent sand and dirt from entering.

- This pipe is laid down adjacent to the footing without affecting the undisturbed soil in this area.
- A 12" layer of "clear stone" is spread on top of the footing drain.
- A strip of filter fabric or "gardener's cloth" should be applied on top of the clear stone to prevent sand and dirt from plugging the footing drain.
- The footing drain should be routed downhill away from the house to a drainage ditch or dry well. A dry well can made simply by excavating a hole 3-4' in diameter to a depth lower than the house footings. Place the end of the footing drainpipe in this hole and fill with crushed "clear stone" to within 12" of the top. Cover the top of the dry well and

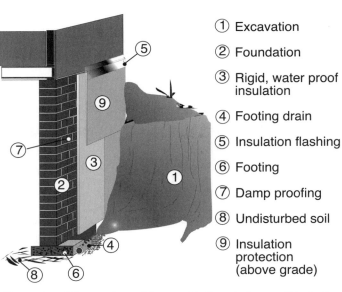

① Excavation
② Foundation
③ Rigid, water proof insulation
④ Footing drain
⑤ Insulation flashing
⑥ Footing
⑦ Damp proofing
⑧ Undisturbed soil
⑨ Insulation protection (above grade)

Figure 2-26 Excavation of the outside foundation wall is the best way to solve water leakage and foundation structural problems while upgrading insulation.

supply ditch with soil to finish.

- Check that all services penetrating the foundation wall are well sealed with a suitable caulking compound. Remember to allow room for the insulation.

Step 4 - 4 – Installing the Insulation

- There are many types of insulation in use today, but the most common and easiest to work with is polystyrene rigid board. This material is supplied in sheets that are typically 2 x 8' long and have interlocking grooves running the full length. As indicated in Table 2-1, the minimum recommended insulation level of rigid board is R13/RSI 2.0. Your building supply dealer will be able to recommend which brand of insulation is available in your area. Just make sure you explain your R/RSI rating requirements.

- Rigid board insulation is very fragile and will break when exposed to wind or undue flexing. Make sure the board is protected prior to and during the installation phase of the work. A special flashing trim is installed at the top of the foundation wall to clip the insulation board in place. Alternatively, pressure-treated plywood may be applied on-site to hold the insulation to the header joist at the top of the foundation wall (Item 5).

- Make sure that all insulation joints are well-sealed and clipped together. It is important that you discuss your specific needs for a flashing and insulation-clip system with your material supplier as there are many variations available.

- Ensure that insulation overlaps at the wall corners, preventing areas of concrete from being exposed.

- A covering is required to protect the insulation from sunlight and damage from traffic and animals where it protrudes above grade (Item 9). Coverings may be purchased for the application or fabricated from any of the following:

- pressure-treated plywood

- vinyl or aluminum siding to match the house
- metal lath and cement parging

Step 4 - 5 – Backfilling the Excavation

- After backfilling the drainage pipe as discussed in Step 4-3, it is time to refill the excavation. If the soil that was removed earlier is heavy clay or drains poorly, it is better to remove it and use clear-running "pit run" sand as the backfill material. This will greatly assist in encouraging water to drain away from the foundation wall, further ensuring a dry basement.

- When the excavation is refilled, ensure that the finished grade slopes away from the house. This promotes drainage and allows runoff to move away from the foundation wall. This is a good time to remind you to install eavestroughs with downspout pipes that lead away from the house.

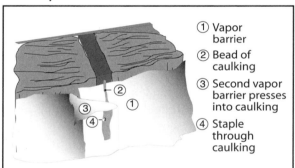

① Vapor barrier
② Bead of caulking
③ Second vapor barrier presses into caulking
④ Staple through caulking

Figure 2-27 A vapor barrier is installed in large continuous sheets. Where the sheets meet, they must overlap and be well sealed with acoustical sealant.

Problem Basements

There are a number of obstacles that can be encountered in older buildings, including:
- packed-stone or dirt floors;
- crawl spaces or basements with very low clearance heights;
- no-basement, slab-on-grade construction;
- building on piers or blocks.

While it is possible to insulate problem basements, you should seek a professional contractor to review your specific requirements.

There are too many variations in climate and construction type to generalize about how to deal with each scenario.

Step 5 - Upgrading Exterior Walls

Because of the large surface area they cover, walls account for a sizeable percentage of the heat loss in houses. Older homes, constructed of solid material such as stone, brick, or log, often have no interior insulation. Many of these designs have a small air space between the exterior covering and the small inner frame wall. This area must not be insulated, as it is used as a drainage cavity for water leakage and condensation.

Hollow concrete-block walls should not be filled with insulation. The quality of the insulation and the "thermal bridging" effect of heat and cold passing through the block does not warrant the trouble or expense.

Traditional frame walls are easily insulated as there is usually an accessible cavity. Using various construction techniques, you can determine if there are cables, duct work, or other obstructions inside the wall cavity that may interfere with the application of insulation.

Solid walls of stone, brick, or log may be insulated from either the inside or outside, depending on several factors:

- The building may have heritage appeal. Homes built of traditional stone, log, or brick may be too beautiful to cover. Insulating from the inside may make the most sense in these situations.
- The outside may need refinishing. If your exterior siding or building material is looking a little tired, there are a number of ways to insulate from the outside and refinish the exterior siding at the same time.
- Interior walls may need new lath and plaster or general updating. When the inside wall surface is cracking and the wallpaper is starting to get to you, perhaps insulating from the inside is the right choice.
- Some exterior and interior finishing may be required. If energy efficiency is the ultimate

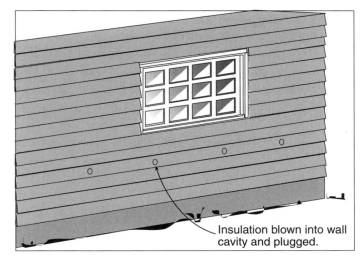

Figure 2-28 Cellulose or urethane foam insulation may be blown into the wall cavity from either the interior or the exterior side of the wall.

quest, why not consider doing both? It is possible to add insulation to both the outside and inside exterior walls, creating a "good as new" insulated home.

Construction type, material costs, and your skill level will determine the path you take on this upgrade.

Frame-Wall Cavities

Frame-wall cavities are by far the easiest to upgrade provided they are empty. A wall that is half-filled with insulation from an earlier job makes it almost impossible to do a good job and it ultimately will not be worth the expense.

During the assessment stage, we determined how much insulation was in the wall by using a flashlight and poking around through electrical outlets (with the power off!) and other access points. If there is little or no existing insulation, a contractor will introduce either cellulose fiber or polyurethane foam into the cavity. These materials are applied through small holes drilled into either the exterior siding or the interior drywall finish.

The holes drilled into exterior siding are plugged using wood dowels. Once the siding is repainted, the work is almost invisible. Brick homes that are suitable for blown insulation have selected bricks removed. The insulation is sprayed and the bricks are replaced.

If the work is being done from the inside, a hole is drilled into the drywall or lath and plaster surface at strategic locations. After the insulation is applied, the holes are filled, primed, and repainted.

If the work is being done inside and the interior drywall or lathwork is poor, consider drilling holes into the existing surface and then covering it with a full vapor and air barrier. New drywall can then be placed directly on top of the existing surface. This technique uses up very little space and is cheaper than a fully framed interior wall.

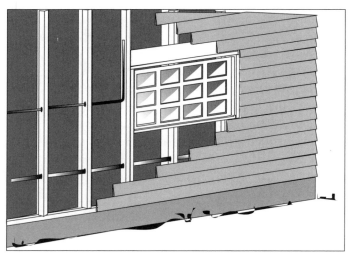

Figure 2-29 Make sure your contractor checks for obstructions or blockages in the wall, including window and door frames.

Upgrading Insulation on the Exterior

If it's time to upgrade the old siding, this is an excellent opportunity to upgrade exterior insulation as well. There are several methods available that allow you to add insulation under the new siding, significantly increasing the overall "R/RSI" value of the home. If you are applying new siding, using blown-in insulation will eliminate the need for filling and repainting access holes. Consider these points as well:

• It is possible to add a generous amount of insulation by using high-density rigid insulation sheets or creating a new wall cavity on top of the existing siding.

• If the house is poorly insulated or made of stone, solid brick, or masonry, add a vapor barrier directly to the inside of this surface, under the new insulation. Follow the rules outlined above regarding proper vapor barrier installation and sealing.

• Remember that door and window openings will have to be extended to allow for the additional insulation thickness.

• Ensure that water runoff from the roof will not drip between the old and new walls, ruining the insulation and causing structural rot damage. An eaves extension or flashing will prevent this.

• Make sure that the new insulation is well air-sealed. There is not much sense in doing a great insulation job with the winter winds howling between the old and new work.

Applying Exterior Insulation

There are dozens of ways of applying exterior insulation. We will review the most popular methods and then point you to your local building supply contractor for more detailed information about materials for your specific application. Regardless of the application method you choose, be sure to wrap the entire upgrade work in spun-bonded olefin (Tyvek® brand) air barrier and ensure that it is carefully wrapped and taped at all joints.

1. It is possible to add exterior insulation by simply purchasing insulated siding materials. This is the easiest approach, although the insulation values are somewhat limited owing to the small thickness. Check with your building supplier on the many finishes available. These preinsulated siding products install in the same manner as their uninsulated counterparts.

2. Rigid board insulation can be applied directly over existing surfaces using appropriate fasteners and adhesives. Your building supplier

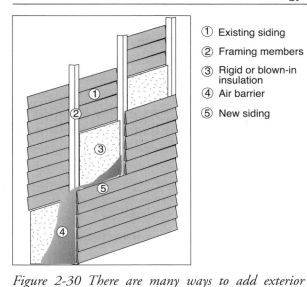

Figure 2-30 There are many ways to add exterior insulation to your home. Wood strapping used to support either rigid or batt insulation works very well.

① Existing siding
② Framing members
③ Rigid or blown-in insulation
④ Air barrier
⑤ New siding

or contractor can review which materials will work best for your application. If rigid board insulation is used, make sure that all joints are tight and well taped to reduce air leakage.

3. Rigid board or batt insulation can be added to a new wall framed on top of the existing siding structure. Batt insulation requires a thicker wall to achieve the same insulation value as rigid materials.

   a. Frame the desired wall thickness on top of the existing wall siding.

   b. Add the insulation, ensuring complete coverage between studs.

   c. Apply an air barrier of spun-bonded olefin, using well-taped edges and seams.

   d. Ensure that water runoff from the roof will not drip directly onto the top of the new wall extension.

   e. Make sure that wind cannot enter the wall cavity from the top or bottom framing plates.

Upgrading Insulation on the Interior

Determine the best application method by assessing your specific requirements. Your choices include:

- Upgrading an existing wall. This is often done when the existing wall material is damaged or lath and plaster is falling off.
- Adding rigid board insulation directly to an existing wall surface. Once the wallboard is added, a new drywall finish surface is commonly added.
- Building a new frame wall. The new wall takes more space from the interior but provides lovely window boxes. It also allows you to increase home insulation values and apply a proper vapor barrier.

Upgrading an Existing Wall

A common upgrade for older homes is removing the lath and plaster wall (a messy job at best) and replacing it with modern drywall:

- With the internal wall studs exposed, you can easily upgrade wiring, plumbing, ductwork, central vacuum pipes, and, of course, your insulation.
- Use additional horizontal strapping as shown in Figure 2-31 to increase the wall-cavity depth from a typical 2 x 4" to 2 x 6".
- Electrical boxes and window frames must be extended to allow for the increased wall-cavity depth. Building supply stores carry box extenders for such work.
- Use batt insulation layered vertically for the existing wall and horizontally for the new wall section. Alternatively, blow in cellulose

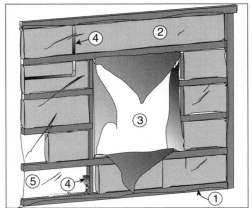

① New framing wall
② Insulation
③ Rough window opening
④ Electrical/ plumbing runs
⑤ Vapor barrier

Figure 2-31 Adding a new, non-loadbearing wall will allow you to achieve increased home-insulation values.

insulation after the vapor barrier has been installed.

- Install a 6-mil vapor barrier over the new framing studs.
- Apply the drywall or surface finish.

Adding Rigid Board Insulation to an Existing Wall

Rigid board insulation may be applied directly to an existing wall surface:

- As with upgrading an existing wall, windows and doorframes must be extended and electrical box extenders added.
- Your building supply store will be able to provide suitable fasteners to hold the rigid board insulation to the existing wall surface.
- Vapor and air barriers are not required when rigid board insulation is installed provided the seams are snug and well taped.
- Drywall or other finish materials may be applied directly to the rigid board insulation.

Building a New Frame Wall

You can build a new frame wall by using exactly the same techniques described earlier in Upgrading an Existing Wall and illustrated in Figure 2-4.

Insulating Both Sides of an Exterior Wall

A final word is required on adding insulation to both the inside and the exterior of the home. If you wish to add some of the insulation on one side of the wall and additional insulation on the other, be careful to limit the ratio to two-thirds inside and one-third outside. This mix is required to ensure that condensation does not damage the insulation system during the heating season.

Step 6 – Upgrading Attics

Air Leakage (a quick review)

Earlier, in Step 3 – Stopping Air Leakage, we discussed the need to prevent air from leaking into the attic space from the warm areas below. Eliminating air leakage stops the warm, moist air from condensing when it reaches colder winter air inside the attic. Condensing moisture may cause mold and structural rot. In minor cases, moisture buildup greatly reduces insulation effectiveness.

A proper vapor and air barrier MUST be installed on the warm side of the insulation, NOT ON THE COLD SIDE! From a practical point of view, adding a large polyethylene film under the insulation is very difficult. It requires moving large amounts of existing insulation and installing the barrier around obstructions such as duct work, plumbing pipes, and wiring.

Should the interior ceiling of the house also be under renovation, you have the opportunity of adding the air and vapor barrier directly to the attic-framing members under the drywall.

In the majority of cases, it is just not possible to add an effective air and vapor barrier into a retrofit attic. In these cases, a well-sealed and caulked attic must suffice.

Figure 2-32 Soffit vents form the air-intake section of the attic ventilation system.

Figure 2-33 Roof vents along the roof peak vent hot air out. A minimum of 1 square foot of roof venting is required for every 300 square feet of attic floor space.

Attic Ventilation (a reminder)

Attic ventilation is mandatory. A home will typically have a row of vents under the eaves called soffit vents (see Figure 2-32). These vents form the air intake for the attic ventilation system. Roof vents are installed along the roof peak (see Figure 2-33) or at the gable ends and allow hot attic air to exit. Air exiting the roof or gable vents causes suction within the attic, drawing in cooler air through the soffit vents. You will recall that a ratio of 1 unit of roof vent area to 300 units of attic floor area is required (approximately 3 square feet of roof vent space for every 900 square feet of attic floor).

Before starting any insulation work in the attic, ensure that the vents described above are present and of adequate size. If in doubt, ask your building contractor to assess their effectiveness.

Installing Attic Insulation

Once the attic has been well sealed against air leakage and the soffit and roof vents checked, you are ready to begin insulating. You have a number of choices as to which material to use. Rigid board insulation is almost never used because of the difficulty of working in confined spaces. Blown or poured fiberglass and cellulose are perhaps the easiest insulation materials to use. Most building supply stores will even provide a blower unit free of charge when you purchase the insulation from them. However, many homeowners simply carry bags of material to the attic, pour it, and rake it flat.

Adding insulation to the attic is not difficult. The major concern is to ensure that attic vents are not plugged with the stuff. Figure 2-34 illustrates how cardboard baffles can be stapled or pressed into the space between the rafters or joists. These baffles prevent insulation from rolling down into the soffit vent area.

Ensure that insulation is not applied around electrical light fixtures or bathroom vent fans that are not specifically approved for insulation coverage. Overheating light fixtures are a sure way to start a house fire.

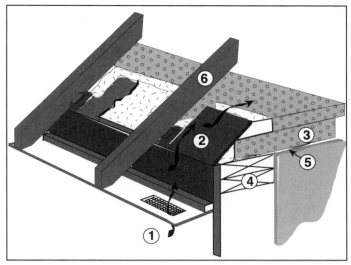

1. Soffit vent
2. Cardboard baffles
3. Insulation (R-40/RSI 7.1)
4. Top of wall
5. Vapor barrier
6. Roof truss or rafter

Figure 2-34 A well-sealed attic with plenty of ventilation and insulation is the final step in keeping your house warm in winter and cool in summer.

Figure 2-35 This eighty-year-old home boasts the world's largest icicle, indicating severe heat loss. A professional assessment of this home demonstrated to the owners how to reduce their heating bill by at least one-third while at the same time making their home more comfortable.

2.3
Appliance Selection

What's All the Fuss About?

A home without major appliances, computers, and entertainment systems is a lifestyle that few people would embrace. There is nothing fundamentally wrong with having air conditioning or a large-screen TV (other than watching it 18 hours a day). The use of energy to power these devices is not in itself a problem. The difficulty occurs when homeowners purchase inefficient models and use polluting, non-renewable resources to power them. There is no free lunch. At some point in the future (some would say right now), these non-renewable resources will start giving up

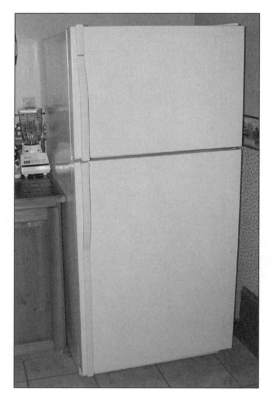

Figure 2-36 A standard Sears 18.5 cubic foot, 2-door refrigerator like this one will save thousands of dollars and kilowatt-hours of energy over its twenty-plus years of operating life.

(remember the blackout of 2003?) or kill us with their exhaust.

The good news is that governments and appliance manufacturers are beginning to understand these issues. Manufacturers are actively working to lower the energy requirements of their products. For renewable-energy-system users, these improvements have exponentially increased the number of appliances available to them. For energy conservers on the grid, these same appliances are helping put money in the bank and enabling them to live just a little bit lighter on the planet.

The average 25-year-old refrigerator (the one keeping a six-pack of beer cold in the basement) uses approximately 2,200 kilowatt-hours of energy per year to operate. Electrical rates vary greatly around North America, but the average daily rate in California (when there is power) is approximately 15 cents per kilowatt-hour or $330 per year operating cost. A new Sears 18.5 cubic foot, 2-door unit uses 435 kilowatt-hours per year or $65.25 to operate, at a savings of $265 per year. At that rate, it would take only three or four years to pay for itself (assuming that rates don't climb any higher). Put another way, if you converted the savings into beer you could probably supply suds for the whole block on the savings alone! If that isn't incentive to switch, what is?

How to Select Energy-Efficient Appliances

It's not necessary to carry around an energy meter like the one shown in Figure 2-37 to measure the energy consumption of each appliance you wish to purchase, although it wouldn't hurt.

These meters (which are available from many sources, including www.theenergyalternative. com) plug into a wall outlet, and the appliance plugs into a receptacle on the front of the meter. The meter display indicates the electrical power consumed. Remember that power (watts or W) is a measure of the flow of electricity (amps or A) multiplied by the pressure of electricity (volts or V). Most major appliances such as washing machines, refrigerators, dishwashers, and food processors plug into a standard 120V outlet (pressure). The flow of electricity in amps multiplied by the pressure (volts) results in the wattage. The calculation for a typical food processor drawing 2.4A is:

$$120V \times 2.4A = 288W$$

Compare this with the power used by an electric kettle drawing 12.5A:

$$120V \times 12.5A = 1,500W$$

What does this mean? Let's suppose that you and your partner are shopping for a new television set. You compare all of the models and find two that are about equal on your list of requirements. You whip out the power meter, or, if you are more conservative, authoritatively inspect the electrical ratings label, and find that model "A" uses 162W of power while model "B" requires 1.9A.

At this point most people would run screaming out the door having flashbacks to those grade nine "A car is moving west…"-type problems. You, on the other hand, have studied Chapter 1.3 – What is Energy? and recall that power (in watts) is the voltage multiplied by the current.

$$120V \times 1.9A = 228W$$

You close the deal by smoothly informing the salesperson that model "A" is the better TV as it will save you loads of dough over its operating life, reduce greenhouse gas emissions, or require less electrical-generation equipment if you are using off-grid energy sources. You might even save money on the purchase with the salesperson trying to get rid of you.

But power is only half of the equation. If for some strange reason you were to watch model "A" television 2 hours per day and model "B" for 1-1/4 hours per day, the energy consumption for model "B" would be lower:

Model "A" energy consumption = 162W x 2 hours
= 324 watt-hours

Model "B" energy consumption = 228W x 1.25 hours
= 285 watt-hours

Time factors into the energy cost. This is why you should turn off lights in empty rooms.

Figure 2-37 An electronic energy meter such as this one will tell you how much power an appliance requires to operate. Most models can even be programmed with your utility rates to tell you cost over a period of time.

Figure 2-38 outlines household power consumption in a format that is simpler to understand. The vertical axis of the graph shows power in watts (volts x amps), while the horizontal axis indicates relative time of day. The graph starts at midnight and shows the lowest power consumption during the day, with most appliances either turned off or in their "standby" mode. This is known as the "base load" of your house. You would assume that with all appliances "off" the base load level would be zero. Unfortunately, many products consume energy at all times. Instant-on televisions, cell-phone chargers, microwave ovens, even your doorbell—all consume energy even when they are not in use. This continuous waste of energy that provides minimal (if any) benefit is known as phantom energy and will be discussed presently.

The electric water heater consumes a great deal of power when it is active. Sadly, it is consuming power in the middle of the night while everyone is sleeping—yet another example of a phantom load. The smart consumer might consider adding a water heater thermostat (to ensure the heater operates only when required), or use insulation blankets designed for the job, or change the unit to a more efficient model.

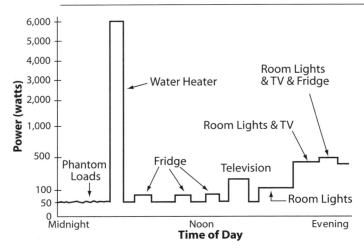

Figure 2-38. This graph shows simplified household power and energy consumption throughout the day. Multiplying the amount of power a device consumes by the number of hours of operation provides the energy consumption of the device.

You will recall that electricity is sold in units of energy (watts x time) and not power (watts only). When the water heater activates it draws power from the electrical grid for a given period of time. It stands to reason that if the unit is active for two hours it will consume twice as much energy as it would if it ran for only one hour.

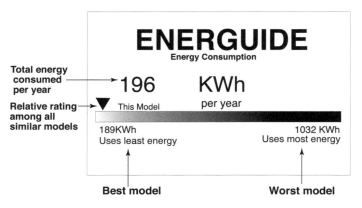

Figure 2-39. This detailed view of an "EnerGuide" label was attached to a Sears front-loading washing machine. The label allows consumers to quickly compare the energy consumption of different models of appliances. Many appliances also carry the ENERGY STAR logo, indicating to the consumer that the appliance meets minimum energy efficiency ratings.

Because electricity is sold based on power and time, switching to more efficient appliances and turning them off or unplugging them when they are not in use will help reduce total energy consumption.

There Has to Be a Better Way

Most people are not so fanatical as to carry an energy meter with them when they go shopping for appliances. Some people may take a look at the electrical ratings label, provided they can keep the watts and volts straight without having to carry a copy of *The Renewable Energy Handbook*. For the rest of us, the government has made life a little easier, at least for the larger appliances. Figure 2-39 shows a Canadian "EnerGuide" label that is affixed to a high-efficiency washing machine. The US EPA uses a similar label for its "EnergyGuide" program. Both programs require that labels be affixed to appliances, providing comparison data for appliance models of similar size with similar features and indicating the appliance's total energy consumption per year.

A closer look at an EnerGuide label reveals the following features:
- The bar graph running from left to right represents the energy consumption of all models of similar appliances.
- At the left side of the graph is the energy consumption of the most efficient appliance in kilowatt-hours per year.
- At the right side of the graph is the energy consumption of the least energy-efficient appliance in the same class.

The bar graph on the EnerGuide label indicates energy consumption of 189 kWh per year for the most efficient appliance and 1032 kWh per year for the least efficient. Think about that for a moment. Two appliances of the same size and class, one consuming almost five-and-a-half times the amount of energy to do the same job!

Efficient washing machine = 189 kWh per year
    x 10 cents per kWh = $19 per yr.

Inefficient washing machine = 1,032 kWh per year
    x 10 cents per kWh = $103 per yr.

Figure 2-40 This Sears high-efficiency clothes washer uses five-and-a-half times less energy than a similar-sized unit. Also consider the savings on soap and reduced waste water being pumped into the sewer or septic tank.

Assuming a delivered price of 10 cents per kilowatt-hour of electricity, the difference in operating expense is $84 per year. Over the life of the machine (say twenty years), that's a whopping $1,700!

The triangle-shaped pointer over the bar graph on the EnerGuide label shows the energy consumption of this particular model in relation to that of the most and least efficient models in the same class. The closer the pointer is to the left side of the bar graph, the lower the operating costs.

The sample label used in Figure 2-39 is from a Sears front-loading washing machine, straight from the catalog. The most efficient model is the Staber horizontal-axis machine. The Staber is almost twice as expensive as the Sears model but only 1% more efficient. On the other hand, the least efficient model is marginally more expensive than the Sears high-efficiency model.

Small Appliances

The EnerGuide program was designed to take care of major "white goods" appliances. What about smaller appliances and electronics that don't carry the program label?

Our choices at this point become a little more difficult. We have to revert to using an energy meter or reading the electrical-ratings label on the device. An electrical meter such as the one shown

in Figure 2-37 is a great device provided that you have access to the appliance for a sufficient amount of time to conduct a test analysis. This is important because energy consumption requires time to evaluate (energy = power x time).

Let's take a look at a conventional coffee maker to see how time factors into our assessment. A typical coffee maker such as the one shown in Figure 2-42 has an electrical rating label showing 120V and 10.5A. Whipping out your calculator you correctly arrive at a power consumption of 1,260W.

120V x 10.5A = 1,260W of power

At 7:00 o'clock on Saturday morning you stumble down the stairs and get the brew going. By 11:30 you have slugged back your third cup, draining the machine and shutting it off. Applying your caffeine-honed mathematical skills, you determine that the coffee maker was on for 4 1/2 hours, giving you an energy calculation of 5,670W.

1,260W of power x 4.5 hours
= 5,670 watt-hours of energy or 5.7 kWh

You realize that at 10 cents per kilowatt-hour for energy your morning coffee has just cost you 57 cents worth of energy, right?

5.7 kWh x $0.10 per kWh = $0.57

Figure 2-41 Not only is the solar-powered clothes dryer energy efficient; clothes also last longer, smell better, and don't require yet another artificial fragrance sheet.

Figure 2-42 A coffee maker uses a lot of energy to boil water, but considerably less to keep the pot warm. In the absence of an Ener-Guide or EnergyGuide label, use care when calculating energy consumption based on the manufacturer's label.

Wrong. The math is correct, but the assumptions are wrong. You have to be very careful when calculating the energy consumption of an appliance, as it may in fact change with time. Yes, the coffee maker label does say that it requires 10.5A or 1,260W of power. What it does not tell you is that it only needs that much power to boil the water. Once the coffee is brewed, it uses less power to keep the pot warm.

So what is the correct answer? Actually, I don't have a clue. Without access to the coffee pot and an energy meter, it's anybody's guess. When trying to calculate the energy consumption of any small appliance with no electrical rating label, follow these general guidelines:

- Appliances that draw a lot of wattage, typically over 300W, should not be used unless it is only for a short period of time. This applies to coffee makers, hair dryers, curling irons, electric kettles, clothes irons, car block heaters, and space heaters. Crock pots which slowly simmer food for hours at a time are actually more efficient than ovens. Consider using a plug-in timer to reduce energy consumption.
- Coffee makers such as the model described above can be used, but transfer the fresh coffee

to a thermos. The coffee stays warm and tastes better without using unnecessary energy.

- Consider boiling water using an electric kettle (instead of the stove) and a drip coffee basket. Transfer the fresh coffee to a thermos to keep it warm.
- Refer to *Consumer Guide to Home Energy Savings* for electrical consumption ratings of desired appliance models before you buy. (The book is available through Amazon book sales at www.amazon.com).
- Refer to Appendix 2 for a list of electrical appliances and tools with their power and electrical ratings.
- All major heating appliances such as cook stoves, ovens, electric water heaters, electric clothes dryers, furnaces, and central air conditioners draw an enormous amount of electrical energy. Always purchase the most efficient model and consider the addition of timers or other appropriate controls to regulate their operating time.
- Microwave ovens are much more efficient than regular electric models. This is an especially important consideration in the summertime when waste heat from the oven must be removed with expensive air conditioning.
- Try to purchase appliances that are not equipped with an electronic clock or "instant-on" anything. These devices are considered phantom loads and consume a large amount of power without doing anything for you. Cell phones, PDAs, MP3 devices, and other electronic items that require battery charging can be plugged into a power bar. Turn the power bar off once the device has been fully charged.
- Convert as much lighting as possible to high-efficiency compact fluorescent lamps (see below).
- Use high-efficiency front-loading washing machines. The Staber HXW-2304, for example, is so efficient that you require just one ounce of soap per wash load. You will use less electrical energy to operate the washer,

less water per cycle, and less energy to heat the smaller amount of wash water.

- If you are not sure about an appliance or tool, borrow one from a friend and plug it into an energy meter like the one shown in Figure 2-37. Even if the appliance model is not exactly the same, you will have a general idea of its power and energy consumption.

Computers and Home-Office Equipment

Many people are finding that self-employment is the way of the future, especially after the last round of corporate downsizing. An essential part of working and playing at home now revolves around computers and related home-office equipment.

With your computer, desk lamp, printer, modem, and monitor working overtime (maybe all the time if you have teenagers), how will your electrical energy consumption be affected?

Modern electronics are a marvel of efficiency and they keep getting better all the time. For example, let's compare a 7-year-old, 15-inch color computer monitor with a new flat-screen model. The older unit may use as much as 120W of power, while the new one draws about 30W. That's a 75% reduction in power usage.

You should also consider that few people use a computer for only a minute or two and then turn it off. The average user may have the computer and support system operating for many hours, which increases energy consumption (power x time). It makes sense to purchase the most efficient products because of this longer operating time.

Don't think only about the computer. You will almost certainly have a printer, monitor, desk lamp, room lights, fax machine, and possibly a photocopier to fully equip your home office. Teenagers (and aging rockers) can add to this list a 400W surround-sound system. Running an energy-efficient system for forty hours a week will still burn up a lot of juice.

Fortunately, the government has come to our rescue once more. The ENERGY STAR program

Figure 2-43 A wall full of home-theater equipment may not seem very energy efficient, but in fact it is. Modern electronics are a wonder of efficiency and work well with any renewable energy system. Just remember to switch off the power input using a special plug or power bar to eliminate phantom loads.

has been adopted across North America and is the equivalent of EnerGuide and ENERGYguide labeling. ENERGY STAR provides guidelines and requirements to manufacturers who seek compliance for their electronic stereo, TV, or data-processing products. Most compliant products use high-efficiency electronic components and specialized power-saving software. For example, a compliant monitor will enter a low-power sleep mode if the image has not changed within a given

Figure 2-44 Purchasing computers and home-office equipment with the ENERGY STAR label ensures the lowest energy usage.

time frame. Another example is a laser printer which adjusts its heater temperature when idle. A laptop computer will always be more energy efficient than the equivalent desktop model, owing to the limited energy stored in the batteries. Likewise, a bubble jet printer is more efficient than a laser printer.

As with everything else in life, there are choices to be made. If you want lower per-print costs, then laser-printer toner is cheaper than bubble-jet cartridges. Laptops are more expensive than a desktop system. Just remember to shop the energy labels and look for the ENERGY STAR logo.

Figure 2-45 We owe a nod of thanks to Mr. Edison and an entire industry that developed a replacement for the kerosene lamp. As technology marches forward, however, it is time to say goodbye to our old but wasteful friend—the incandescent lamp.

2.4
Energy-Efficient Lighting

Possibly the single most important invention to touch our lives is the incandescent light bulb. Prior to Mr. Edison's 1879 discovery, it was almost impossible to stay up past sundown. (My wife still has this problem.) Life before the modern light bulb meant filling a kerosene lamp and enduring poor lighting and air quality. In the early days of electrical-power production, little thought was given to energy efficiency or environmental concerns. As a result, light bulbs were—and are—inefficient and wasteful. The bulb in your floor lamp has remained essentially unchanged for over a hundred years. However, times have changed and so have lighting technologies.

The incandescent lamp and its many variations create light by heating a small coil or filament of wire inside a glass bulb. Anyone who has watched a welder working with metal knows that hot metal glows a dull red color. If the metal is heated further, it glows brighter and with an increasingly whiter light. At some point, the metal will vaporize with a brilliant shower of white sparks. The various stages of glowing metal are known as incandescence.

In a similar manner, electrical power applied to a light bulb causes the metal filament to glow white-hot. If you decrease the amount of power reaching the bulb (by using a dimmer switch, for example) the bulb dims and glows with a progressively redder color.

Making an incandescent lamp glow requires a large amount of energy to heat the filament. In a typical light bulb, 90% of the energy applied to the filament is wasted in the form of heat. Therefore, only 10% of the energy you are paying for is making light. What a waste!

To make matters worse, we need to think about the heat component for a moment. If it is summertime, this waste heat contributes to warming the house and increasing the air-conditioner load. An incandescent bulb hits you twice in the wallet: poor efficiency resulting in high operating cost and waste heat as well as the expense of getting rid of the waste heat using air conditioning.

Let's put this into perspective. Assume it's a warm summer evening. You have a total of fifteen 100W lights on in the house. The waste heat is:

15 bulbs x 100W x 90% heat loss
= 1,350W of heat

This is about the same amount of heat output as you get from a large electrical space heater in the winter.

You may argue that this waste heat can be used in the wintertime to help warm the house. True, except that electricity is the most expensive means of heating a home anywhere in North America and this byproduct of lighting cannot be controlled unless you want to place your house lights on a thermostat. If you are generating your own electricity off-grid, this waste is not even manageable. So what are the alternatives and how do they save energy and money?

There are numerous alternative lighting technologies, including semiconductor-LED, mercury vapor, halogen, fluorescent, halide, and sodium. These technologies have a range of applications and different efficiencies. The most common and efficient technologies for home applications are discussed below.

Before we continue it would be prudent to say a few words about semiconductor or LED lighting. Most people have seen LED lights installed in automotive applications or high-efficiency flashlights. Currently, LED lighting is only one-half as efficient

as compact fluorescent technology. Additionally, the color quality of these bulbs is very poor and would not be acceptable for home lighting. It is their long life and low energy consumption compared with standard incandescent lamps that give LED lights their passing grade.

Compact Fluorescent Lamps

Mention fluorescent lamps to my wife and all she can think of are the pasty-faced girls applying makeup in the washroom at the high school prom. In the 60s and 70s some companies used the term "cool white" on their fluorescent lamps, as if this were a good thing.

Enter the compact fluorescent lamp (CF). These marvels of efficiency may look a little odd, yet they offer many advantages over standard

Figure 2-46 The compact fluorescent (CF) lamp is the bulb of choice in modern energy-efficient homes. The choice has never been better, as indicated by the variety available at a typical hardware store. Clockwise from top left: 30W floodlights (available for indoors and outdoors), an 11W CF which looks like a regular incandescent (handy for lamp shades), a 12, 20 and 26W Trilight, a standard 13W CF bulb (with the frozen ice cream swirl design), a 15 W chandelier (narrow base), a 13W outdoor bug light (yellow in colour to discourage flying insects at night), a track and recessed flood light, a chandelier (large base), another brand of 14W CF that approximates the traditional bulb shape, and a dimmable 25W bulb (also available in the traditional light bulb shape). If your hardware store doesn't offer this selection, it's time to ask them to, or to find another hardware store.

incandescent lamps. A CF lamp is designed to last ten times longer than an incandescent lamp. Shop around when looking for these lamps. Stores such as Walmart carry the electronic CF lamps for about $3.00 each. Many stores still treat CF lamps as "special," which includes a "special" price of about four times this amount.

Light output and energy efficiency are further considerations. Contrary to popular belief, light output is measured not in watts but in lumens. The wattage rating of a bulb is the amount of electrical power required to make it operate. The standard 75W light bulb gives off approximately 1,200 lumens of light. A 20W CF lamp provides the same intensity but uses one-quarter the energy. Translate this to cost savings and the CF bulb will save $55.00 over its life span, assuming delivered energy costs of 10 cents per kilowatt-hour. This does not even take into account the savings in air-conditioning load and environmental pollution.

Lastly, CF lamps won't give you headaches or the Bride of Frankenstein look first thing in the morning. Advances in phosphor coatings and electronic ballasting eliminate ghastly color and flicker. The lighting industry applies a rating system called the Color Rendering Index (CRI) to all bulbs, although it can be a bit hard to locate the information. The closer a light source's CRI

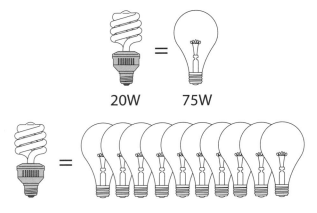

Figure 2-47 CF lamps last ten times longer than regular light bulbs and use approximately four times less energy. With no flicker and color quality similar to incandescent light bulbs, you won't get headaches or see that awful pasty color when you look in the mirror.

rating is to 100, the more natural and comfortable it will be to the eyes. The typical incandescent lamp has a CRI rating of 90 to 95 compared with a CF-lamp rating of 82. Compare this with a Frankenstein fluorescent bulb rating of 51. Still not convinced? Buy a couple of CF bulbs and try them out. It will be almost impossible to tell the difference in color or operation.

The Down Side

The shape and size of many CF lamps are not identical to those of standard bulbs. This may cause problems fitting the CF lamp into a conventional socket. Take note that shapes and sizes vary by manufacturer. Check several different brands to see if one will fit your application. Before you resort to changing the light fixture, see if your local hardware store is able to suggest an alternate base, harp, or socket extension.

Dining rooms often have overhead fixtures that are connected to a dimmer switch. CF lamps must never be placed in these sockets unless they are specifically designed for dimming. Dimmable CF lamps are available, but at a premium price.

CF lamps can be used in outdoor lighting systems even in winter. In this application, the lamps typically require one to two minutes to warm up and produce useful levels of light. This is not a problem for security or perimeter lighting on timers. Garage-door-opener lights or those applications where the light is required quickly should not use CF lamps.

T8 Fluorescent Lamps

Large-area or kitchen-cabinet indirect lighting often uses standard fluorescent lamps with magnetic ballasts. Although more efficient than incandescent lamps, these models are still only half as efficient as CF lamps or new "T8" fluorescent lamps with electronic ballasts. T8 lamps are more expensive than CF units, but they are excellent replacements for older 4' and 8' standard fluorescent tubes. It's easy to recognize T8 lamps, as they are approximately half the diameter of conventional tubes. Likewise, electronic ballasts are very light for their size, especially when compared with older, less efficient magnetic ballasts.

Fluorescent Lamp Safety

There has been a fair amount of discussion in the news regarding the safety of CF and fluorescent lamps. Many of these news items describe the need for increased studies on this or that problem in the misguided belief that the lighting technology is new. While it is true that CF lamps have only been available in North America for the last 20 years, they have been in use in Europe and Asia for considerably longer. Safety concerns have been well researched and I would like to provide this summary to reassure people that the technology is safe for all to use.

Mercury Content

All fluorescent lamps including CFLs contain trace amounts of mercury to produce light. The amount used is approximately 0.005 grams (a drop smaller than the head of a pin) per CFL. To put this in perspective, a typical watch battery has 5 times as much and a common dental filling has 0.5 grams of mercury, according to the U.S. Food and Drug Administration—nearly 100 times as much as a CFL.

CFLs in the Home

CFLs are perfectly safe in the home; there is no concern about mercury contamination. The main consideration from a safety standpoint is being cut by broken glass should a bulb break.

In the event that a bulb does break, it is recommended that you sweep—not vacuum—the glass and phosphor powder. Place the pieces in a bag and wipe the area with a damp disposable towel to clean up any remaining shards of glass.

CFL Disposal

Most major retailers, including Home Depot, offer recycling facilities for defective or broken CFLs. If a recycling facility is not available, check with your local municipality for details on how to dispose of these lamps. Most townships offer household hazardous waste days to dispose of items such as paint, CFLs, and batteries.

Track Lighting and Specialized Lighting

There are places where even a diehard energy conserver would never put a CF lamp. Lighting your Rembrandt or Picasso collection is one example that comes to mind. For these applications, low-voltage, high-intensity MR-16 lamps are ideal. Jewelry and watch stores use these miniature spotlights for clear, intense lighting of small objects, artwork, and display cabinets.

MR-16 halogen gas incandescent lamps are fabricated with specialized internal reflectors that direct light in a unidirectional flood pattern. MR-16 bulbs operate on 12V, which in turn requires a transformer (supplied with the light fixture) to drop the household 120V supply. The low wattage of MR-16 lamps is similar to that of a larger CF lamp, so they can be considered efficient in these special applications.

Figure 2-48 MR-16 flood lamps are excellent for lighting your priceless jewelry or baseball cap collection. Although they are incandescent lamps, their low wattage and high intensity make them acceptable for these applications.

Large-Area Exterior Lighting

Large outdoor areas require major lighting muscle. While it is possible to use a few million CF lamps to light your yard, an easier and even more energy-efficient method is to install high- or low-pressure sodium lamps. These have a characteristic yellow color that makes them suitable for perimeter and security lighting. They also work well in horse-riding arenas and for street lighting. Watt-for-watt, a sodium lamp produces twice as much light with the same power as a CF lamp and is up to ten times more efficient than an incandescent lamp.

Free Light

Perhaps the best lighting is free—100% energy efficient. No, don't just open the drapes. Even the best-designed house may have an area or two that suffer from low light levels during the daytime and require lighting even on the sunniest days. If you're thinking skylights, remember that typical skylights are not energy efficient, are prone to condensation problems, and often require reframing and finishing from the inside. A product available from www.sunpipe.com offers

an excellent alternative. The SunPipe is similar to a fiber-optic device on steroids. An intake piece is mounted on the roof in the same fashion as the metal chimney shown in Figure 2-49a. A reflective supply pipe is fitted down through the roof opening into the living area. A trim kit is installed and that's it.

The SunPipe directs outside light into the immediate area, even on cloudy days. A sample installation is shown in Figure 2-49b, with the SunPipe illuminating a narrow second-floor hallway. If the SunPipe were to be covered, this hallway would be almost completely dark. A 9"-diameter pipe provides illumination equivalent to that of a 400W incandescent lamp on a sunny day.

Figure 2-49a This view details the "intake" portion of the SunPipe unit.

Figure 2-49b The hallway in this picture is completely illuminated with a SunPipe. Covering the intake to show the difference without the light from the SunPipe created an almost-black photograph.

2.5

Water Supply and Energy Conservation

Turning on a tap for a glass of water is a luxury that few people give any thought to. If you live in the city, all it takes is writing a check to your local utility a few times a year and presto! Clean, clear, life-sustaining water.

Society's lack of concern for or ignorance about this natural resource is often beyond comprehension. One only has to drop by a golf course in the desert to see firsthand the waste of a precious resource. Many people counter with: "Why not use water as we wish? Isn't water a renewable resource, like the solar energy we are capturing?"

Water is a plentiful and renewable resource, within limits. The majority of the earth's water is stored as moisture in the air, salty seawater, or ice and snow. The rest is in lakes and streams or in wells drilled to reach the underground water table.

Lake- and ground-water levels have been dropping for many years; hardly a day goes by

Figure 2-50 An average person in North America consumes 100 gallons (382 liters) of water per day. The conserver can easily reduce this consumption by 50%.

without some newspaper headline screaming about the subject. Consider that the Colorado River does not always flow into the ocean, partly because of over-extraction in the upstream watershed. The reduction in water levels is largely the result of increased human population and water usage. Industrial and climatic changes are also being blamed for an ever-increasing reduction in our second most important natural resource after air.

The average person in North America consumes 100 gallons (379 liters) of water per day. If we include the water required for industrial and agricultural processing, consumption rises rapidly to over 1200 gallons (4543 liters) per day. North Americans consume water at a rate 5-1/2 times greater than that of people in Sweden. As with gasoline, our wasteful habits can be traced back to costs. According to the OECD Factbook 2005 and the UNESCO Water Assessment Report, efficiency and price are interrelated, with Germany charging $2.00 per cubic meter of water and Canada practically giving it away at 40 cents.

Besides saving money on water bills and energy bills for hot water, reducing our consumption will lighten the load (electrically and from a societal viewpoint) on water- and sewage-treatment facilities. This is a significant consideration when you factor in upgrades in capacity to aging infrastructure that would almost certainly be adequate if conservation of water and sewage were undertaken. The savings to these municipalities (and property taxpayers like you and me) would be enormous.

Water is also very heavy, requiring large amounts of energy for the operation of the municipal supply and sewage pumps as well as home-based deep-well pumping.

The Conserver Approach

It is estimated that 75% of our water consumption occurs in the bathroom and 20% involves laundry and dishes. Figure 2-51 indicates that the remaining 5% is used for cooking and drinking.

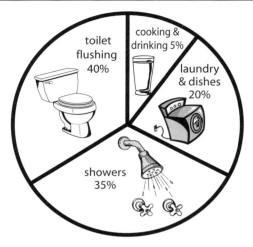

Figure 2-51. This pie chart shows how the average North American consumes water. Lawn and garden water is subject to high variability between homeowners and apartment dwellers and has been averaged into each section of the chart.

Conserving water and the energy used to heat and move it requires the same approach as with lighting and appliances: do more with less. This does not mean that the whole family has to get in the shower at the same time. It means getting the same results with more efficient appliances and methods.

Toilet Flushing

Conserving water when flushing a toilet is not only simple, it is actually the law in many areas plagued by inadequate water supplies. Simply remove the old clunker and replace it with a certified low-flush model. This conversion will reduce water usage by 66% or more. Conventional toilets can consume over 4.7 gallons (18 liters) per flush while low-flush models use only 1.6 gallons (6 liters).

Low-flush toilet models are available in a mind-numbing array of styles, colors, and shapes to suit even the most discriminating derriere. Prices keep coming down, with economy models available in the $50 range. Another good reason to consider these models is the lack of condensation dripping from the tank. If your old clunker drips every time the humidity rises, that's the excuse you need to make the swap.

No matter which brand of toilet you own, if it leaks it's a water waster. A toilet that leaks four gallons of water per hour (fifteen liters) would fill a large swimming pool if allowed to continue for a year. Not sure if the toilet leaks? Place a couple of drops of food coloring in the water reservoir. Check the bowl after fifteen or twenty minutes and see if the color shows up. If it does, the next book you need is An Introduction to Plumbing to fix it.

Figure 2-52 Low-flush toilets require only 33% of the water required by earlier models and come in a variety of styles to suit even the most discerning derriere.

Bathtubs

If you and your family prefer baths rather than showers, there is not much you can do except change an older bathtub to a new fiberglass model. Old cast iron or metal tubs that are six feet long and very deep require a lot of water and a large amount of fuel to heat it.

A fiberglass soaker bath can be ordered in a 5-foot model that is fine as long as you aren't related to Magic Johnson. A shorter bath requires less water volume and the fiberglass model resists heat loss through the sides.

If you are having a bath during the heating season, leave the water in the tub until it cools down, and then pull the plug. This little trick will transfer the heat energy from the water into the room, providing you with free warmth and humidity to boot.

Showers

Aside from turning off the water when the teenagers decide to live in the shower, the best solution is to add a low-flow showerhead. These units, such as the one shown in Figure 2-53, have built-in water-flow restrictions and special "needle orifices" to reduce flow without making you feel as if you're just standing under a leaky roof. Look for models that carry a label from a certification agency indicating a flow rate of less than 2.6 gallons (10 liters) per minute. Many models come equipped with a built-in massage feature, providing an additional incentive for making the switch.

Figure 2-53 Showerheads such as this model reduce water consumption and hot water heating costs by 50%.

The Kitchen

A very quick upgrade of the kitchen sink may be in order. Try replacing a standard-flow faucet with one that has an "aerator" nozzle, such as the one shown in Figure 2-50. These devices mix air with the water stream, giving the appearance and feel of more water flow. This apparent increase in water flow offers more coverage when washing dishes and hands.

Dishwashers

Like every other major appliance in the house, dishwashers are subject to the government's EnerGuide and ENERGYGuide programs. When purchasing a new model or updating a tired machine, review the EnerGuide or ENERGYguide label and purchase the most efficient model available. Better yet, try washing your dishes by hand. The energy savings provided by "manual" dishwashers are quite high, and you might find this a good time to catch up on family gossip.

Clothes Washers

Washing machines are in the same boat as dishwashers. Check the EnerGuide or ENERGYguide labels and purchase the most efficient machine you can afford. The washing machine shown in Figure 2-40 consumes five-and-a-half times less energy than a similar new machine. Shop carefully and save big!

Water Heaters

Besides changing your water heater completely, there are a few things that can be done to help conserve energy:

• Turn down the thermostat of the water heater to the lowest temperature you can accept. The thermostat on most gas units is visible and well marked. Electric units often have two thermostats underneath a removable cover. When adjusting them, ensure that the power to the unit is turned off. If you are unsure, contact your plumber for assistance.

• Install an insulation blanket as shown in Figure 2-54. These blankets are available from most building supply stores or plumbing contractors. The insulation batting helps to stop heat transfer from the hot water inside the unit. If you create a homemade blanket, make sure it is fireproof and does not block any air intakes or vents on a gas water heater.

Flush the water heater at least twice per year. Built-up sediment in the bottom of the tank reduces heater efficiency. Draining a few gallons from the bottom of the tank will reduce calcium and mineral buildup.

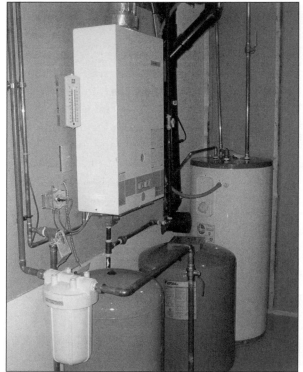

Figure 2-55 Instant water heaters such as this model from Bosch do not waste energy by storing hot water all day long. They work by rapidly heating the incoming cold water, supplying enough hot water for the entire house. Do not be misled by the old wives' tale that on-demand water heaters cannot supply sufficient hot water. This is a fallacy from days gone by which is propagated by ignorance of the technology.

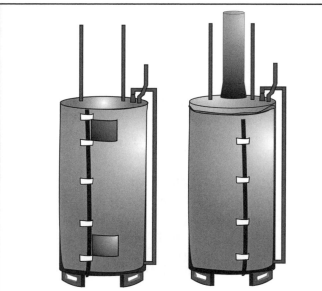

Figure 2-54 Water heater blankets save a great deal of the heat that is typically lost from hot water storage tanks. Make sure the blanket does not block any air intakes or vents on gas water heaters.

• Install pipe insulation on all hot water lines. This material is inexpensive and very easy to install (provided that the pipes are exposed).

• Add an active solar heating system that captures the sun's rays to warm the water in your tank. Solar thermal systems can easily provide 50% of your annual hot water heating requirement and provide financial returns on investment of 10% or better. Chapter 5 covers this equipment in detail.

Water heaters have an average life span of ten years. Before you discover a new indoor swimming pool courtesy of your leaking unit, install an upgraded model. A water heater that is reaching the end of its useful life is a bit like a refrigerator: the savings resulting from the energy efficiency of newer models will easily offset the purchase price and cost of installation.

New to North America are instantaneous water heaters such as the Bosch model shown in Figure 2-55. These units have long been used in Europe where energy costs prohibit wastefulness. Your plumber can easily replace your old storage unit and give you back a bit of floor space at the same time. Do not be misled by the old wives' tale that on-demand water heaters cannot supply sufficient hot water for the entire house. This is a fallacy from days gone by which is propagated by ignorance of the technology.

Solar Hot Water Heating

Solar water heating systems are a natural extension of efficient water heating systems and are discussed in Chapter 5.

Landscaping

Lawns

Possibly the greatest waste of water (and energy) is using it to maintain the perfect lawn. If there was ever a sinkhole for energy, this is it. We start by trucking in tons of topsoil and adding fertilizers, pesticides, and lime to get it growing. Then we use more fuel to cut, trim, and edge our lawns, all the while wasting energy. Added to this is our need to use purified drinking water when rain would almost certainly suffice. Why water in the middle of a hot summer day and have 50% of the water evaporate or run along the sidewalk? If we must have perfect lawns, then we should try to water them more efficiently.

Watering at dusk or early in the morning is best. Apply only as much water as your local conditions require. Check with a nursery or garden center to find out how much water is appropriate. Over-spraying the lawn and watering the sidewalk is just plain wasteful. Spend the $20 on a better sprinkler unit and maybe a few dollars more on a timer. Program the sprinkler to suit the size and area of your lawn. If you are planting a new lawn, consider low-maintenance ground covers or drought-resistant white clover grasses instead.

Are you in the market for a lawnmower? Consider a rechargeable electric model: virtually no noise, minimal maintenance compared to gas models, and negligible pollution.

Figure 2-56 Try switching to drought-resistant white clover which has the added benefit of slow growth, reducing the amount of cutting required. Clover's long root system (left) compared to typical grass (right) ensures its drought resistance.

Figure 2-57 Capturing rainwater is a time-honored tradition. Automate the process by connecting the barrels to a weeping hose for an automatic and labor-saving watering system.

Flower Gardens

The roof of your house captures a lot of rain during a downpour. Try containing some of this water in a series of rain barrels that are interconnected to supply flowerbeds. A few 50-gallon (200-liter) barrels interconnected with poly-pipe as shown in Figure 2-57 make a nostalgic-looking storage system. Connect the rain barrels with weeping hose and you have an energy-efficient and labor-saving watering can. Remember that mulch doesn't just look good and suppress weeds; it also keeps the soil from drying out, further conserving water and energy—and your time.

Although garden lighting does not help with water conservation, it helps to showcase your pretty flowers and hard work. This lighting will better harmonize with nature if it is solar powered. The price of solar-powered lights has been dropping while their quality has steadily increased. The model shown in Figure 2-58 was purchased for under $5.00 and will continue to run long after you should be in bed.

Figure 2-58 Solar-powered garden lights won't help with water conservation, but they certainly accentuate the beauty of the flowers.

Water Conservation Summary

The tasks described above will easily reduce your water consumption by more than 50%. Add to this the decreased electrical energy required to pump the water, reduced water heating costs, decreased soap consumption, and diminished load on the municipal supply and sewage system, and you simply cannot justify putting off these upgrades.

Figure 2-59. Although the urban lawn is ubiquitous, it consumes an enormous amount of energy and water as well as the time you spend trying to make it look better than your neighbor's. Xeriscaping (reducing resource requirements in landscaping) offers a beautiful and low-maintenance alternative, as shown above.

Harvesting Our Water Supply

The vast majority of off-grid homes are located in rural areas that are rarely supplied with municipal water connections. As with energy, the off-grid homeowner is generally independent of many municipal conveniences, with the possible exception of property taxes. Water harvesting is a major consideration, usually requiring many thousands of dollars in investment as well as being one of the largest electrical power consumers in the home.

The two most common methods of collecting water are pumping from a well and pumping from a surface supply such as a lake or cistern. Let's start with the well, as this is the most common method in use.

Deep Well Submersible Pumping

Referring to Figure 2-60, you will see an installation diagram of a typical deep well submersible pump. For any new or retrofit application the deep well submersible pump is the simplest, most reliable, and easiest-to-service design. The deep well submersible pump is available as a standard 120/240VAC centrifugal unit or in high-efficiency direct-current designs such as the Conergy Submersible pump shown in Figure 2-61 (www.conergy.us). Both designs install in the same manner, the major difference being the technology used in the pump design and the resulting efficiency. The Conergy submersible pump is so efficient that it can be used as a remote well pump for livestock with only one small photovoltaic panel powering the unit, as shown in Figure 2-62.

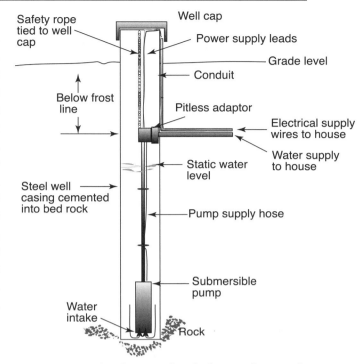

Figure 2-60. This diagram details the installation of a typical deep well submersible pump.

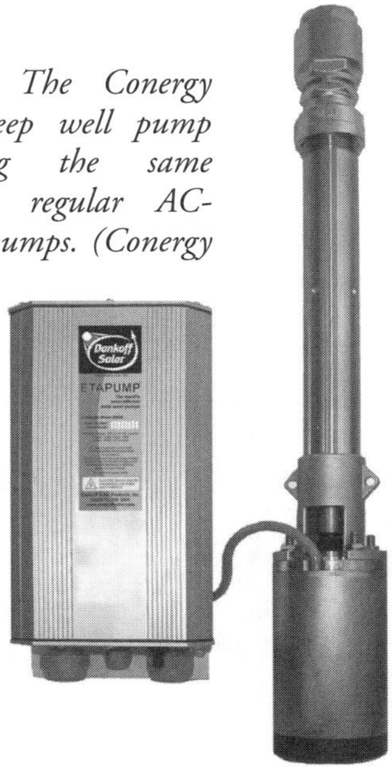

Figure 2-61. The Conergy submersible deep well pump installs using the same techniques as regular AC-powered well pumps. (Conergy USA)

Figure 2-62. The Conergy submersible pump can be used for stand-alone pumping applications such as this remote livestock watering well. Very efficient, but not nearly as picturesque as the old western windmill. (Conergy USA)

2.5.1
Off-Grid Water Supply

For off-grid applications, efficiency is important. However many homeowners (including the author) use standard AC submersible pumps. Although the efficiency is lower and the power requirements higher, they are less expensive and it is easier to find service parts in the middle of the night. Couple this with low-water-usage appliances and large storage/accumulator tanks and the resulting pump usage is reduced to 10 minutes per day.

Slow-Pumping Techniques
The deep well submersible pump is the brute force method of pumping water for a home pressure system. The "do more with less" method involves a different approach that is not nearly as popular as deep well pumping but works just as well.

A slow pump is simply an ultra-high efficiency, low-voltage pump that has low-flow characteristics compared to a standard deep well submersible pump. This lower-flow characteristic requires a means of buffering water consumption, usually by the addition of a storage tank. Figure 2-64 shows a typical slow-pumping submersible pump supplying water to a storage tank. A float located in the storage tank turns the submersible pump off when the tank is full and back on when the level drops.

A second pressure-booster pump is installed between the water storage tank and the household supply. This pump pressurizes an accumulator (see below) and turns off when the water pressure reaches 40 to 50 pounds per square inch (psi). This is the normal maximum pressure for domestic water systems. The switch is also designed to turn the pump back on when the pressure drops below the 20 psi cut-in pressure.

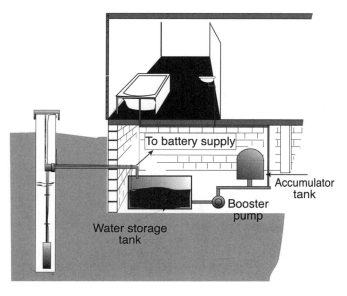

Figure 2-63. A slow-pumping system relies on high efficiency, low-flow pumping into a storage tank to increase efficiency. This design will reduce water pumping energy requirements by 50%. (Conergy USA)

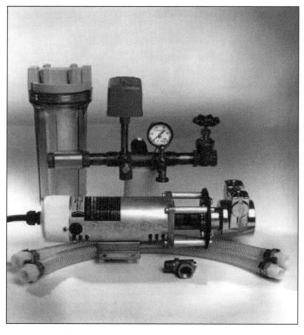

Figure 2-64. A booster pump can be used to pressurize water from a storage tank or from a lake or stream provided freezing is not an issue. (Conergy USA)

Surface Water Systems

Surface water systems do not require a submersible pump unless the installation is subject to freezing. For temperate or seasonal use the booster pump can supply the water and pressurization at the same time. Placement of the pump is very important, as pumps are not primarily designed to "suck" water; they push it. A typical installation is shown in Figure 2-65. An intake pipe is provided with a screened inlet (to keep the fish out) and supported inside a simple crib of rock or wood. The intake is then routed to a 10 micron sediment filter to stop grit and sand, but not bacteria, from entering the pump. As water supplied from an open source is not potable, water purification equipment is required.

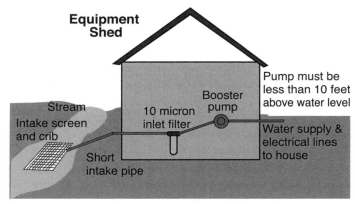

Figure 2-65. A typical surface water pressure system will operate in areas where freezing is not a concern.

The pump inlet must be positioned not higher than 10 feet above the water line to ensure proper operation. The booster pump then provides pressurized water to the accumulator. A built-in pressure switch turns the pump on and off at the desired operating pressures in the same manner as the slow-pumping method described above.

Installations such as the one shown in Figure 2-65 can vary greatly. One area to be aware of is the vertical height or "head" distance that water must be pushed uphill. Assuming that your cottage or home is at approximately the same level as the pump, there is no problem. For houses with solar panels that are located above the pump, water pressure cut-in and cut-out settings must be increased by 1 psi for every 29 inches (74 cm) in vertical lift, assuming the pressure gage is located at the pump.

This additional pressure is due to the weight of water. For example, if the house is located 13 feet above the level of the pump, the pressure switch setting will have to change by:

(13 feet above pump x 12 inches per foot) ÷ 29 inches per psi ≈ 5.4 psi increase

If the pressure switch is located in the house this change is not required, although you will have to be sure the pump can work with the increased pressure requirement. In addition, be aware of the maximum vertical height a pump can push water and still supply adequate water pressure. It's fine for the manufacturer to say that

Figure 2-66. An intake sediment filter is required for all slow and booster pump applications. This filter needs to be changed, as sediment has accumulated in the filter housing.

the pump delivers 50 psi, but if the cottage is 120 feet above the lake surface, the pressure at the top won't be enough to water your pet goldfish.
(120 feet above pump x 12 inches per foot) ÷ 29 inches per psi ≈ 49.7 psi water pressure

Pressure Accumulation

Anyone who has been to a rural home or a cottage with an improperly installed water pressure system knows the rub. The shower pressure varies from a trickle to your being blasted against the wall. Getting a drink of water sometimes means waiting two minutes to fill the glass while the next time the whole kitchen gets flooded from the spray. During this time the well pump is kicking on and off like a wild mustang being broken.

An improvement on this comical water supply system is to even out the flow and pressure

Figure 2-67. A large water accumulator will provide city-quality water pressure and flow.

by installing a proper water accumulator tank. When your plumber offers you a coke-bottle sized accumulator, decline the offer and purchase one that works. The accumulator shown in Figure 2-67 is enormous compared to the laughable, tiny units normally supplied. This model (the water accumulator, not Lorraine) holds approximately 40 gallons (152 liters) of water and will provide an even flow throughout the longest shower (teenagers notwithstanding).

In addition to smoothing out the water supply, the large tank will greatly reduce the pump cycling by storing most of your daily water usage. After applying all of the water conservation rules outlined above, Lorraine and I often manage with the pump cycling only once per day. A further advantage of the large accumulator is water temperature buffering. Well water is very cold, often entering the house at 48°F (9°C). This water has to be heated, often with propane. The accumulator acts as a large heat sink, absorbing room air heat. (Remember from Energy 101 that heat likes to go to cold things.) This free heating will warm the water to near room temperature, saving more money by lowering the amount of fuel needed by the water heater.

Water Supply Summary

There are dozens of methods available for installing a domestic water supply system, many of which require specialized design support. If simplicity and ease of service are primary concerns, then stick to standard deep well submersible AC pumps with large accumulator tanks. This is pretty standard stuff that any plumber can understand, and when used with proper water conservation techniques this method is efficient enough for all but the smallest of renewable energy off-grid systems.

Chapter 4 will demonstrate a number of off-grid systems including a typical AC-powered deep well system as well as a lakeside surface water design used with a camping trailer.

2.6

Phantom Loads

As the name implies, phantom loads are any electrical loads that are not doing immediate work for you. This includes items such as doorbells (did you know that your doorbell is always turned on, waiting for someone to push the button?), "instant on" televisions with remote controls, clock radios, and power adapters.

Figure 2-68. Phantom loads are unnoticed and often useless electrical loads that burn holes in your energy conservation plan. Examples are doorbells, "instant on" electronics, microwave ovens, or anything plugged into a "power adapter" like the one shown above.

So what is the big deal? First, these devices are consuming energy without doing anything for you. That electric toothbrush you used this morning was probably charged about 15 minutes after you put it back in the holder. During all the in-between hours when the device sits idle it is wasting energy and money. A television set that uses a remote control is actually "mostly on" all the time, just waiting to receive the "on" command from your remote.

While the total dollar cost for these luxuries is small on grid (in the order of $10 to $20 per month for an average house), this consumption off-grid is unacceptable, as it can easily equate to 15% of total energy generation or more. For example, if all of the devices shown in Figure 2-69 were installed in an off-grid home and allowed to remain "on standby," the total daily electrical energy consumption would equal approximately 6.8 kilowatt-hours. This is just about the same amount of electrical energy my energy-efficient, off-grid home consumes running all the appliances I want to operate!

Another reason for concern in off-grid installations is that the inverter unit (described in Chapter 11) goes into a sleep mode when the last light is turned off at night, conserving a fair amount more energy. Any phantom load will keep the inverter "awake," consuming more energy than it otherwise would.

Phantom Load Management

Some phantom loads can simply be eliminated. Try a door knocker instead of a doorbell and a rechargeable-battery-powered digital clock instead of a plug-in model. I have even heard that some people use manual toothbrushes.

Television sets and CD, DVD, and VHS players with instant on and remote control functions can be wired to outlets that can be switched off. You could have an electrician wire in a switchable outlet for use with stereo and other equipment, or you could use a power bar with an integral switch.

Many people cannot live without some phantom loads like fax machines or chargers for cordless phones, cell phones, PDAs, and laptops. For these items, consider using a programmable outlet timer that turns the connected devices off according to usage patterns.

Take note that even small devices plugged in will consume large amounts of energy over time; a 50W load could easily equate to 25% of the total energy produced in an off-grid household:

50 watts x 24 hours per day
= 1,200 watt-hours or 1.2kWh

Tread lightly and limit the use of phantom power loads to truly essential items!

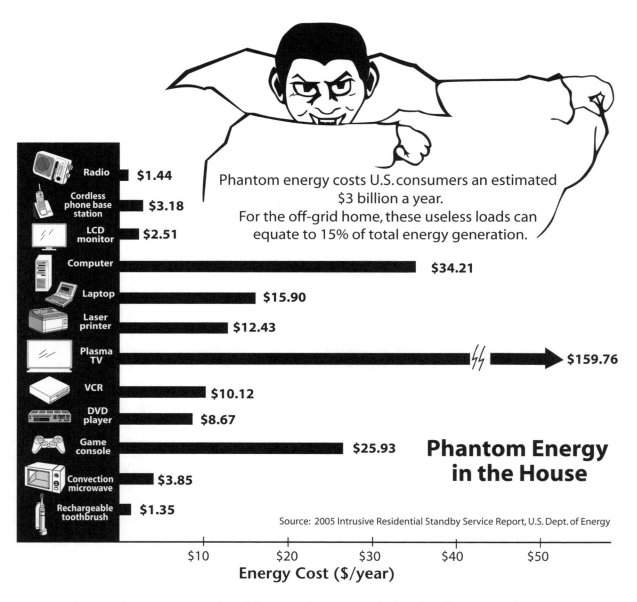

Phantom energy costs U.S. consumers an estimated $3 billion a year.
For the off-grid home, these useless loads can equate to 15% of total energy generation.

Device	Energy Cost
Radio	$1.44
Cordless phone base station	$3.18
LCD monitor	$2.51
Computer	$34.21
Laptop	$15.90
Laser printer	$12.43
Plasma TV	$159.76
VCR	$10.12
DVD player	$8.67
Game console	$25.93
Convection microwave	$3.85
Rechargeable toothbrush	$1.35

Phantom Energy in the House

Source: 2005 Intrusive Residential Standby Service Report, U.S. Dept. of Energy

Energy Cost ($/year)

Figure 2-69. This chart shows a number of electronic devices typically found in the home. Collectively, they waste nearly $280 of electricity per year, the equivalent of nearly three large power stations in the United States alone.

2.7
A Kaleidoscope of *Eco*-nomic Ideas for Renters and Homeowners Alike

Just because you don't actually own a piece of terra firma doesn't mean that you can't or shouldn't get on the energy-efficiency bandwagon. How? This photo collage will start you on the road to a lifelong relationship with eco-nomics:

☐ Replace all incandescent bulbs with compact fluorescents

☐ Let mother nature dry your clothes

☐ Get out of your car and onto a bus

☐ Use ceiling fans to cool your living space

☐ Vacuum or replace your furnace filter

☐ Keep tires inflated to the proper pressure

☐ Program your thermostat for when you are not at home

☐ Regularly clean the lint trap in your dryer

☐ Drive an economical automobile.

☐ Shop locally.

☐ Vacuum the coils on your fridge

☐ Add a timer and or photocell to control exterior lights.

☐ Get out of your car and onto your bike

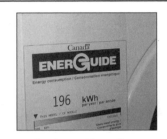

☐ Replace old appliances with effecent ones

Money Isn't All You're Saving

☐ Look for the ENERGYSTAR logo when purchasing appliances

☐ If you must use a car, fill up with ethanol

☐ Replace computers with laptops and LCD displays

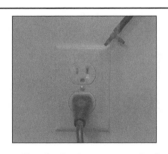

☐ Smoke test your receptacles and install outlet seals

☐ If you use a block heater, put it on a timer

☐ Use dimmer switches

☐ Use Sola-tubes to bring natural light into your living space

☐ Stop using heat producing halogen lamps

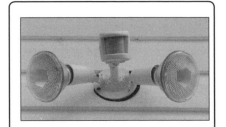

☐ Use motion sensitive exterior lighting

☐ Make sure all your toilets are low flush.

☐ If you use nightlites, use ultra efficient electroluminescent ones.

☐ Insulate hot water pipes

☐ Use a push mower to cut your clover

☐ Use a rain barrel to capture rain water for garden watering

☐ Put aerators on all faucets

☐ Leave cloth shopping bags by the door so you don't forget them

☐ Use drought resistant ground covers like clover

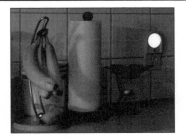

☐ Use a low flow shower head

☐ Plant a vegetable garden for your 100 "foot" diet

Chapter 3
A RENEWABLE ENERGY PRIMER

I remember the day I popped the "big question" to Lorraine. The anticipation and excitement were reaching their peak; she was clearly caught off guard. After a moment's pause, the reply: "But can I still use my hairdryer?"

That was her only concern? The rest, I was sure, would be easy….

Fifteen years have passed and yes, Lorraine can still use her hairdryer. Living off the grid is both comfortable and satisfying. Our motivation for leaving "life on the grid" was simple. Lorraine wanted to move closer to her family and still have the room and privacy to support her "addiction" to animals. The lot at the back of the family farm fit the bill (and the wallet). There was only one downside. It was about $13,000 from the nearest electrical wires.

My work as an electrical power system designer made me think. Why not try to make our own juice? Surely I could whip up something for around $13,000. In hindsight, that was naïve, but ultimately our system has grown to the point where it supports our *desired* lifestyle and is far more reliable than the power utility. (Our neighbours have suffered through a 16-day power outage during the ice storm of 1998 as well as enduring the blackout of 2003 and dozens of other "miscellaneous" outages.)

What is the trick to living off grid? A willingness to give it a try, a sense of adventure? Perhaps a bit like harvesting your own wood to heat your home or growing some of your own food—it takes a bit of work but the reward repays the effort. Author Garth Turner stated in his 2009 financial Armageddon book *After the Crash* that "people who exhibit rugged individualism and self-sufficiency are the new cool." Lorraine and I were only trying to save a bit of the environment and some money at the same time, but if this is the new cool, I'm in!

Those of you who wish to remain connected to the grid and embrace clean energy technologies may be motivated by an interest in environmental sustainability or an alternative financial investment. But the first step along either path is energy efficiency.

3.1
Summarizing the Case for Energy Efficiency

The previous chapters have been drilling efficiency into your head. Just in case you skipped all of that, I will take a moment to reiterate its importance.

Energy efficiency on or off the electrical grid does not equate to Spartan living. A big screen TV, computers, a treadmill, and a cappuccino maker are examples of devices that can be found in any energy-efficient home. In our off-grid home you can add to the list outdoor lights and stereo in the horse stable, a hot tub on the deck, and a garage with electric door openers. You might think that this house is a large electrical consumer, but in fact the opposite is true. We operate our house on between 3 and 6 kWh of electrical energy per day, depending on the season. Contrast this with other North American homes that consume 6 to 10 times this amount of electrical energy.

In addition to electricity, houses require thermal energy for space and domestic hot water heating, most often supplied by natural gas, oil, or propane. As with electricity, fossil fuel usage and related costs can be reduced through simple energy-efficiency techniques such as draft-sealing, improved insulation and the like.

An energy-efficient house has much of the same "stuff" as a regular house, but it can use many times *less* energy than an average poorly constructed or outfitted home. Off-grid homes tend to be more *electrically* efficient still, with owners keeping in mind that they must build and pay up front for the power station that generates all of their power.

This is not to say that all or even most off-grid or urban houses powered by renewable energy are automatically the most energy efficient or environmentally friendly. It is important to remember that all forms of energy must be used wisely to make economic and environmental sense. A homeowner who installs a solar thermal water heater while ignoring leaking faucets or an under-insulated water tank is not going to receive the financial return on investment or environmental dividend that was anticipated.

I would also like to discuss an interesting anomaly related to off-grid homes. Many people like to gush eloquently about how energy-efficient or environmentally sustainable off-grid living is. While there is no doubt that some off-gridders live very efficient lifestyles, most simply trade limited electrical energy for fossil fuel; they are not so much living off grid as living on propane.

With an off-grid system it is imperative to reduce electrical or "plug loads" to the point where these appliances can be supported by the size of renewable energy system you have chosen or can afford to build. Every other energy-consuming device normally moves to propane (or a combination of propane and wood heat).

If you consider this for a moment, you will realize that it is the most energy-intensive devices that are moved to propane in an effort to reduce the capital cost of the electrical generating system. The cook stove, clothes dryer, water heater, and backup space heating all end up connected to the big gas bottle out back. (Later, I will present the results of work that I have undertaken to develop an off-grid lifestyle that requires no carbon-emitting fossil fuel.)

Now consider that most people in the country will add to this list of fossil-fuel-consuming devices one or more cars (or more often pickup trucks) to commute to a distant town, gas lawn mowers or tractors, weed trimmers, snow blowers, gas barbeques, chainsaws, ATVs, and snowmobiles.

No, off-grid living is not sustainable living unless you as a homeowner understand that you are a living being interacting with a natural ecosystem. Consider that the couple who live in a downtown condo, walk to work, and have no yard to maintain will win the environmentalist award over most off-gridders hands down.

In order to avoid this dirty little secret, it is necessary to take complete stock of your life, tools, habits, and interactions with the ecosystem to determine if you are living sustainably. It is fine to grow a lovely organic garden, but your efforts are completely negated if you are using a '72 GMC pickup truck that belches oily smoke to purchase supplies in a distant town.

Most people who live off grid and try to confront this issue will find that it is almost impossible to reduce their net impact on the environment to the same level as that of the couple living downtown. This does not mean throwing away the shovel, purchasing an Armani suit, and moving to the city to sell insurance, but it does mean that you have to consider every aspect of your lifestyle and separate "needs" from "wants and ego."

The purpose of these words is not to be condescending to rural off-grid people (after all, I have been one for over 15 years), but rather to make you aware that sustainable living is not the "in thing" of the day. It is a process that requires awareness and careful attention to everyday decisions. I hope that one day all of society will understand this message.

Consider the above as a de facto requirement for sustainable living, whether you live on or off the electrical grid. Proceeding further along the energy generation path without the benefit of energy efficiency is akin to driving a dump truck to the corner store for milk. It can be done, but it makes absolutely no sense.

3.2
Generating Thermal Energy

Referring back to Figure 1-2, you will recall that once you have adapted your lifestyle to the point of maximum energy efficiency, the next stop along the economic energy path is thermal energy generation. This includes dozens of technologies for house, hot water, pool, and spa heating. The technologies that provide this energy generally provide a better return on investment and require less capital than those that are used to generate electricity and therefore should be considered first.

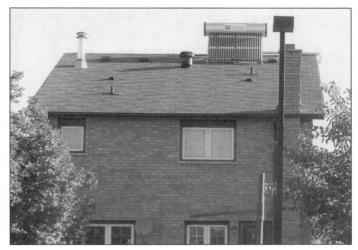

Figure 3-1. Solar thermal water heating is a technology that can be used whether you live on or off the electricity grid. The high-efficiency, low-power pumps that operate these systems fit well within the electrical energy budget of even the smallest off-grid system.

Technology	On-Grid	Off-Grid
Solar Domestic Water Heating	Yes	Yes
Solar Pool/Spa Heating	Yes	Yes
Home Heating		
Wood (may include domestic hot water heating)	Rural - Yes Urban - No	Yes
Geothermal (includes domestic hot water, air conditioning, and pool heating)	Yes	No
Air-to-Air Heat Pump (includes air conditioning)	Yes	No
Alternate Biomass	Possibly	Yes
Passive Solar	Possibly	Yes
Electricity	Yes	No
Cooking - Electric	Yes	Possibly
Clothes Drying - Electric	Yes	No

Table 3-1. The list of renewable thermal energy technologies available to all homeowners is rather extensive, but their usage can be grouped depending on whether you are planning to remain connected to the electrical grid or not.

Figure 3-2. In theory, anyone should be able to use wood or alternate biomass fuel (dried corn, wood pellets, etc.) as a renewable energy source, but in practice it is generally reserved for those living in rural areas. Although wood heating is a renewable energy source, harvesting, storing, and burning it is pretty hard to imagine in a downtown apartment.

Figure 3-3. This grid-connected home is being built with an air-to-air heat pump unit that provides both space heating and air conditioning. Due to the large electrical power demand of these units, they are not suitable for off-grid living.

As with any set of general guidelines, rules are always being bent and broken. Wood, biomass, and passive solar space and water heating are definitely not the norm (or even considered) in most North American urban settings, although this trend has started to reverse in recent years.

Likewise, most off-grid homeowners move large electrical-power-consuming appliances to fossil fuel propane in an effort to reduce the capital costs of electricity-producing equipment. Therefore, non-renewable, polluting fuels often power hot water heating, cook stoves, clothes dryers, and some space heating, a situation that most "off-gridders" find offensive and counter to their rationale for living off grid in the first place.

Fortunately, with a bit of planning and education, these inequities can be modified, as you will learn in the coming chapters.

3.3
Electrical Energy Generation Overview

There are numerous methods for producing electricity from renewable resources, including tidal power, solar steam turbine technology, and of course wind turbines. For the urban grid-connected homeowner interested in producing renewable electricity, the most common means by a wide margin uses photovoltaic (PV) technology.

Figure 3-4. Unlike fossil fuels or nuclear energy, renewable energy sources are just that: renewable. No matter how much sunlight we capture with our photovoltaic panels, there will always be more for everyone else, and we are not dumping today's pollution on tomorrow's children. This photo shows a small grid-interconnected photovoltaic array with a 700W peak output.

Rural grid-connected or off-grid homeowners have more options than their urban counterparts owing to the diversity of land use and the ability to access the primary energy sources. As with urban systems, PV technology is the number-one choice for electricity production, with small-scale wind turbines making up a distant second choice. Further down the list are micro-hydro installations and biodiesel generating units.

On-the-grid or grid-tie systems capture renewable energy directly from the source (sun, wind, stream), convert it to direct-current electricity, and feed it to a device known as an inverter. The inverter converts the direct current to alternating current, which in turn supplies the distribution grid with electricity.

Before discussing the specifics of the various technologies, I would like to review how grid-connected system owners are paid for energy generation.

Payments: Net Metering Programs

Depending on government policy, you are typically offered one of two payment programs for your energy.

With the so-called "net metering" scheme, any electricity you produce is first used to offset household electrical consumption. Any surplus energy produced is "banked" or stored on the electrical grid. Banking energy causes your electrical meter to run "backwards,"[1] providing you with a credit in your electrical energy "bank account."

When you require more energy than you are producing, for example at night, the electrical grid provides the energy supply, causing your meter to debit your account in the normal manner. Every once in a while your utility will send you a statement indicating whether you owe money or the present balance in your energy account is positive. The selling of electrical energy back and forth is called net metering. It is the law in many North American jurisdictions, requiring the utility to purchase your excess energy at the same retail price that you pay for theirs.

Although this sounds like a great idea, there are numerous problems with the concept. Firstly, most programs will not allow you to sell any excess electricity—energy beyond what your household consumes—into the distribution system. Therefore, an energy-efficient household is actually penalized by the fact that its members are efficient consumers of energy. This may cause people using the net meter program to actually become less energy efficient in an effort to use 100% of the energy that is produced by the renewable source.

Further, if a given homeowner has banked energy in his or her account at the end of a payment cycle (which varies from utility to utility), there is no credit or payment for this energy and the account is reset to zero. In effect, the utility is stealing the "excess" energy.

Lastly, if excess energy is purchased at the nominal going rate, you are not being accurately reimbursed for its true value. Renewable energy produced by a PV system is non-polluting, has zero-carbon emissions once installed, and produces power during periods of peak electrical demand. All of these features give renewable energy much higher value than dirty coal or nuclear energy, yet the net metering program in no way credits people for these intrinsic values.

Payments: Feed-in Tariffs

Approximately twenty years ago European politicians realized that net metering programs had not resulted in the mass adoption of renewable energy technologies. After much tweaking and poking, European governments developed the Feed-in Tariff (FIT) Program,[2] which solves the problems of net metering and encourages all of society to participate in clean energy generation.

The concept is very simple. The FIT Program provides an inverse sliding tariff payment scale for various renewable energy technologies, creating a level playing field for homeowners, small business, first nations, and communities. In essence, the program pays more money for small-scale electricity generation and specifically prevents "gaming" where large utilities and business monopolize power production.

Although it may seem counter-intuitive to oppose the "larger is better" scale of economy, in fact the benefits to society outweigh the disadvantages of this pricing model.

To understand the FIT program, one must first realize that the centralized model of electricity production is a relic of the early 19th century when distributed or "micro" and "small-scale" energy systems either did not exist or were considered too small to matter. This mindset resulted in the concept of selling electricity in only one direction (from a centralized plant to remote, passive consumers) using a network of transmission and distribution wires that lose approximately 10% of the total energy generated. Because of the distances between the generating plant and all potential consumers, an ultramodern coal-fired generating facility can achieve an efficiency of only 30%.[3]

The centralized power generation/distribution model is somewhat analogous to the development of large mainframe computers in the 1950s and 60s. The control and wealth generated by these two industries rested with a powerful few.

With the advent of powerful micro-computing systems and the Internet, the power and wealth associated with computing has been distributed across the entire economy; a child living in a remote northern community can participate in the same educational activities and discussions as any urban child.

So it is too with distributed electricity generation. The development of micro-scale PV systems, grid-connected inverters, and programs such as FITs allows homeowners, communities, small businesses, farming co-operatives, and First Nations to generate and consume locally produced clean electricity. This reduces transmission system losses, puts revenues earned from power-generating projects into the local economy, and creates skilled jobs.

The one downside constantly sited is that energy consumers will have to pay a higher price for energy and that "the poor" will be forced further into debt. (When I hear people speak of "the poor," I generally assume they are speaking to avoid increased costs themselves.) If folks are truly worried about the poor, governments can easily use the tax system and carbon tariffs from dirty-energy-source producers to subsidize those who cannot afford to pay their energy bills, or better yet, help them to become more energy efficient and thereby reduce their electrical costs.

As to the issue of higher energy tariffs, so what? North Americans waste so much energy on inefficient appliances, under-built houses, and gimmicky junk (four TV sets per household and outdoor lights suitable for a ball field) that I have little compassion for this argument. By following the guidelines in *The Renewable Energy Handbook*, homeowners can easily reduce their energy consumption by 50% without any impact on lifestyle. Accordingly, *if* electricity prices were to double *and* they followed the energy-efficiency advice in this book, their energy costs would remain stable and society would only have to generate half as much energy as it currently does to meet residential demand.

In reality, FITs will not cause energy prices to double for the simple reason that fossil and nuclear generated electricity fuel and capital costs

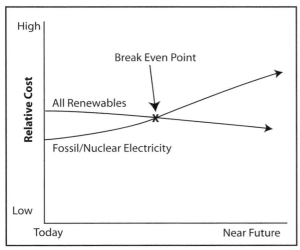

Figure 3-5. *This graph outlines the trend in ongoing cost of production (including capital and operating costs) for traditional and renewable energy technologies. Fossil and nuclear energy are very capital intensive with long design/build periods, so their financial viability is at the mercy of capital markets. Fuel costs to operate these plants continue to rise at an average level above core inflation, and carbon taxation will increase costs further still. Renewable energy capital costs are declining, with no on going fuel expense. Investment in renewable energy technologies will ensure long-term cost control and provide energy security.*

are rising fast and carbon taxation is coming in any event; it will not be long before "cheap coal" will be more expensive than a basket of renewable energy sources. Once a renewable energy-generating source is installed, costs remain fixed; the cost of wind or sunlight is currently free.

So how much does the FIT program pay for various technologies? This question cannot easily be answered because energy is a state and provincial matter, meaning that North America could easily have over 60 different policy programs. To take one example, the province of Ontario, Canada is currently reviewing[4] its current price of C 42 cents per kWh for all PV systems under 10 MW to a tiered structure where small, home-based systems mounted on roofs would receive C 80.2 cents per kWh, while commercial ground-mounted systems of less than 10 MW would receive 44.3 cents. To put this in perspective, homeowners would be paid approximately seven times more for energy they produce than for energy they consume, with no limits on the amount of energy sold back to the grid.

This type of policy provides a clear message to the markets that clean energy generation will be a way of life in the province, providing a significant ROI for homeowners, businesses, first nations people and farm operators alike.

Rebates and Incentives

On- and off-grid homeowners can participate in energy-efficiency rebates, although incentives for renewable energy generation technologies are generally limited to on-grid systems. For a starting point in your search for financial assistance, consider these two major website portals:

For readers in the United States, the website www.dsireusa.org (Database of State Incentives for Renewables & Efficiency) will provide current information regarding federal, state, municipal, and utility incentives for both energy efficiency and renewable energy generation.

Canadian readers can review Natural Resources Canada, Office of Energy Efficiency Rebates and Incentives department at http://canmetenergy-canmetenergie.nrcan-rncan.gc.ca/eng/about_us/ottawa/funding/incentive_program.html. The Canadian Solar Industries Association (CanSIA) also provides a list of federal and provincial programs at: http://www.cansia.ca/Default.aspx?pageId=139888.

Readers should also be aware that utilities, manufacturers, and banks may also provide discounted home equity loans or special discounts on capital equipment for many of the products discussed.

3.4

Electrical Generation Technology

All renewable energy electrical systems work in much the same manner by collecting energy from a renewable source (sun, wind, stream) and converting it to direct-current electricity. Basic systems are designed to use the energy as soon as it is produced, operating a water pump or driving an electric fan, for example (Figure 3-6). More advanced systems require voltage regulation, battery storage, inverters, and fairly complex wiring to allow electricity storage and conversion to more convenient forms for use. Figures 3-7, 3-8 and 3-9 detail typical renewable energy generating systems that are often used for camping, off-grid houses, and selling energy to the utility grid, respectively.

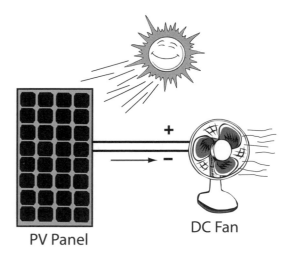

DC Direct

Figure 3-6. Renewable energy electrical systems rely on one or more technological systems to capture solar energy. Voltage regulators, batteries, and inverters store and convert this energy so that it may be used to power household appliances on demand.

The basic system shown in Figure 3-6 consists of a photovoltaic panel or PV panel (more on these units later) that converts sunlight directly into direct current electricity. When the PV panel is exposed to sunlight and the voltage (pressure) rating of this source of electricity matches that of the fan motor, the fan will spin. Obviously a design such as this is limited to very basic applications where a "use it or lose it" condition is tolerated. Operating a fan to cool an attic or a pump to supply water for an out-of-the-way garden are good applications for this system.

Figure 3-7 is a slightly more advanced system that allows solar electricity captured during the daylight hours to be used long after the sun has gone down. In this application, electricity from the PV panel passes to a charge controller and storage battery. The charge controller is designed to ensure that the battery is properly charged based on manufacturer specifications and to prevent over-discharging caused by excessive loads draining power from the battery. As with the DC direct system, the voltage rating of the PV panel, charge controller, battery, switch, and DC light must match to ensure correct operation.

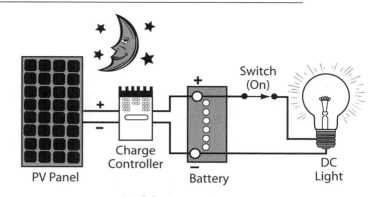

DC with Battery Storage

Figure 3-7.

Simple DC systems with storage batteries are commonly used in mobile camping trailers as well as recreational boats and small off-grid cottages to power lights, water pumps, stereos, and other assorted small appliances. Small handheld versions of this technology are also incorporated into sunlight rechargeable flashlights, radios, and calculators. Today this reliable yet simple design is used by NGOs to provide electric lighting to poor areas of the world where power grids do not extend. Many third-world families can breathe

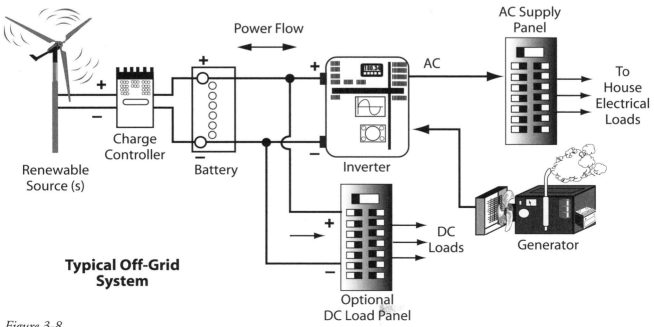

Typical Off-Grid System

Figure 3-8.

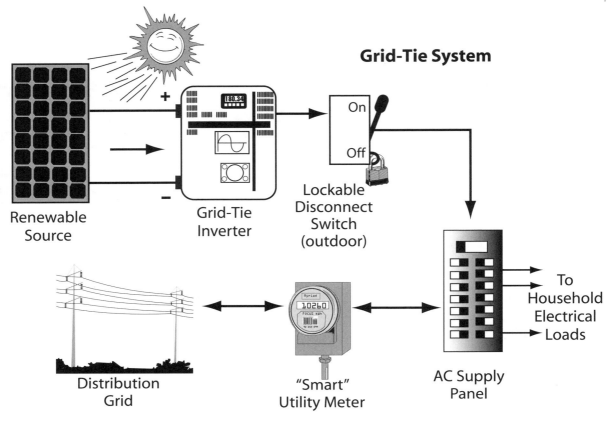

Grid-Tie System

Renewable Source

Grid-Tie Inverter

Lockable Disconnect Switch (outdoor)

To Household Electrical Loads

Distribution Grid

"Smart" Utility Meter

AC Supply Panel

Figure 3-9.

easier by using simple direct-current lighting rather than relying on smoky oil or wood fires for light.

Modern off-grid houses build on the concept of the basic DC system with battery storage by adding several additional devices that allow the house to operate in much the same manner as a "normal" grid-connected house.

Figure 3-8 comprises a source of renewable energy, a charge controller, and a battery. An optional DC load panel is sometimes provided to power small direct-current appliances in the home. Small off-grid systems with limited energy storage often use direct-current automotive stereos, television sets, and cell-phone battery chargers, all of which may be connected to this type of load panel.

The vast majority of off-grid houses now incorporate a device known as an inverter (more

on these devices later as well) that converts low-voltage direct current stored in batteries into 120/240V alternating current. Electricity from the inverter supplies a standard AC supply panel that is ubiquitous in every home, in turn feeding your unsuspecting appliances. The obvious advantage in using an inverter is that most standard electrical appliances can be used in the off-grid house. This increases the number of different appliances available and generally means increased quality and selection. However, not all electrical appliances may be plugged into the off-grid house. Factors such as battery capacity and the generating capability of the renewable source(s) place limits on the electrical loads that can be operated. If you are expecting electricity to heat your home or run a central air conditioning unit all summer, off-grid living might not be for you.

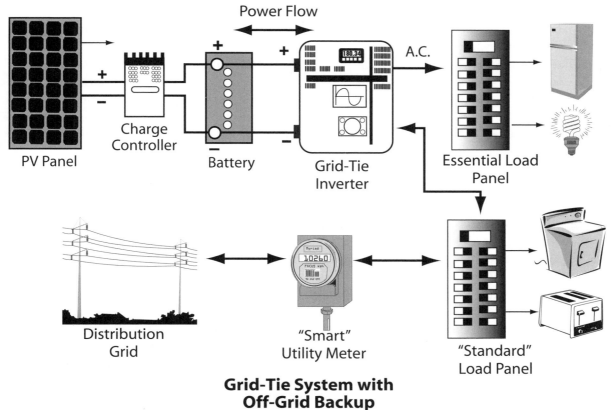

Grid-Tie System with Off-Grid Backup

Figure 3-10.

Many off-grid homes come equipped with a backup generating unit that operates on fossil fuels. The generator is used during periods of unusually high electrical demand (arc-welding the wheelbarrow back together, for example) or during periods when your renewable energy source is just not co-operating. In the Northeastern United States, cloudy fall weather will stop most PV-based systems cold, requiring you to add a touch of support from the generator to assist in charging the battery bank.

For those of you who live connected to the electrical grid, one option to consider is the grid-tie system shown in Figure 3-9. This design takes electricity from your renewable resource, converts it to utility-quality alternating current, and feeds it into your household electrical circuits. Any excess energy is "sold" to the grid based on the payment program used in your area. (See Section 3.3 above for information on net metering or FIT payment programs.)

Grid-tie systems do not require a battery bank and voltage regulation equipment or backup generator. This lowers the installed cost of the equipment compared to that of off-grid systems, and the ability to sell energy to the electrical utility provides an obvious return on investment. On the other hand, if you have no batteries or backup generation source, your home will be just as dark as your neighbor's home during the next electrical blackout. The irony of this happening might be too much to bear!

The precise layout of the grid-tie design will vary depending on the payment program provided. For example, the location of the revenue meter shown in Figure 3-9 causes the meter to "see" only the energy that is purchased from the grid, or imported, and the net energy, which is the electricity produced minus consumption, or exported. In this example, you will be paid only for the net energy exported to the grid; thus the term "net metering."

Under the FIT program, either two meters or a special "smart meter" is installed, causing all of the energy produced to be sold to the grid at the premium "green energy" rate. The homeowner then purchases electricity at the lower current retail rate, making the return on investment better still.

Perhaps the most complex system is based on a hybrid of the grid-tie and off-grid configurations, as shown in Figure 3-10. This configuration provides the benefits of being able to sell renewable electricity to the grid while maintaining the ability to operate the house in off-grid mode should utility power fail.

To achieve this capability, the renewable energy source(s) are connected to a charge regulator and battery bank in the same manner as any off-grid system. Direct current electricity is supplied to the inverter, which feeds a first "standard" electrical supply panel powering non-essential electrical loads. These loads tend to be large power-consuming devices which are not of critical importance during periods of utility blackout and may include a clothes dryer, toaster, or air conditioner, for example. Power from this panel also supplies a utility revenue meter, which records energy sales and consumption.

Essential electrical items such as lights and a refrigerator connect to a second load panel.

When the electrical grid is functioning correctly, the system operates in the same manner as a standard grid-tie unit, buying and selling energy back and forth with the utility.

During periods of utility interruption, the inverter automatically disconnects itself from the grid, leaving the non-essential load panel off-line. The balance of the system components are then powered in the same manner as a typical off-grid system, with the inverter drawing direct current electrical energy from the renewable source(s) and battery while supplying alternating current

to the essential load panel. During periods of off-grid operation, it is possible that the renewable resource(s) will not be able to supply sufficient energy to maintain battery capacity. Should the battery become depleted, the inverter will enter a "low-voltage" protection mode and disable all load operation until the battery is recharged, preferably from the renewable energy source although an optional fossil-fuelled generator can also be used.

When grid voltage returns, the inverter resynchronizes[5] itself with the utility and the process of selling energy back and forth begins again. Additionally, the inverter will check the battery state of charge and begin recharging from the AC supply as required.

Figure 3-11. With advances in materials technology, you will find wind power used in every region on earth. The Bergey XL1 small wind turbine is specifically designed for home-size off-grid and grid-interconnected applications.

Digging Deeper into the Details

In theory, all of this technology may seem simple enough, but just like everything else in life, the devil is in the details. Let's take a closer look at how the various devices described operate at the system level before we examine each part of the technology in detail.

All of the earth's energy comes from the sun. In the case of renewable electrical sources, the link is very clear: the sun's rays strike the photovoltaic panel, creating direct current electricity. The sun's energy causes the winds to blow, moving the blades of a wind turbine and causing the generator shaft to spin and produce either AC or DC electricity. Heat from the sun causes water to evaporate, creating the rains that fall in the mountains. The rain becomes a stream that runs downhill into a micro-hydro turbine, which in turn causes a generator shaft to spin, producing either AC or DC electricity.

As well as being renewable, these energy sources are also variable or intermittent. In order to ensure that electricity is available when we need it, a series of conducting wire cables, fuses, and disconnect switches delivers the energy to a battery storage bank or in the case of grid-connected systems, the energy is sold directly to the electrical utility company as described above.

Although there are many different types of storage batteries in use, the most common by far is the deep-cycle, lead-acid battery. You may be familiar with smaller batteries used in golf carts or warehouse forklift trucks. Batteries allow you to store energy when there is a surplus and hand it back out when you are a bit short.

So why are we using a battery bank? What other means do we have of storing electricity? Great questions, simple answer: there is no other feasible method of storing electrical energy. Maybe down the road, but if you want an off-grid or battery-backed system now, batteries are the only way to go. Today's industrial deep-cycle batteries are a solid investment and should last up to 20 years with a minimum amount of care. At the end of their lives, the old batteries are recycled (giving you back a portion of their value) and new ones installed.

Storing electrical energy is reasonably simple: provided the energy source supplies direct current, just connect the source and battery through a charge controller and the battery will immediately start charging. The charge controller will monitor the battery state-of-charge or "health" and adjust the charging program to suit the immediate requirements.

If the renewable energy source provides alternating current, it can be converted to direct current using a device known as a rectifier bridge connected to the battery as described.

Getting the energy back out of the battery or renewable energy source and into a useful state is a bit more complex. First, electricity is stored in a battery at a low voltage or "pressure." You will probably know that most standard household appliances use 120V from a wall plug, whereas off-grid or emergency backup batteries commonly store electricity at 12, 24, or 48V. Second, the electricity stored in a battery is in direct current (DC) form. This means that electricity flows "directly" from one terminal of the battery to the other. (Direct current loads and batteries are

Figure 3-12. An off-grid system requires a means of storing the electrical energy generated by the renewable sources for later consumption. A sizable battery bank provides the needed storage, although it is completely optional when considering a grid-tie system.

Figure 3-13. The modern off-grid world runs its electrical appliances on alternating current supplied by an inverter such as this.

identified by a red "+" and a black "-" symbol marked near the electrical terminals.)

Electricity generated from most renewable energy sources is similar to that of a battery, providing low-voltage, direct current output.

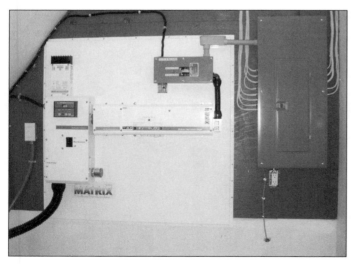

Figure 3-14. A renewable energy system should be neat, simple, and well laid out, ensuring smooth sailing with electrical inspectors and your insurance salesman. This small system is installed in a year-round home and supplies sufficient energy for two adults.

The electricity supplied by the utility to your home is "alternating current" (AC). This means the flow on the supply wires changes direction at a rate of 60 cycles per second or 60 hertz. (Many of the terms used in electrical engineering are named after the early inventors who discovered the physics, for example James Watt, Count Volta, and Heinrich Hertz.)

In order to increase the voltage (pressure) of the electricity stored in the batteries or generated by the renewable source and convert it from DC to AC, a device known as an inverter is used. Without an inverter, your choices of electrical appliances and lighting would be reduced to whatever 12V appliances you could find at the local RV store. Early off-gridders did in fact choose this path, but you should not consider it except for the smallest of summer cabins or camps. A house full of middle-class dreams means a house full of 120VAC appliances. Standard electrical power also means standard wiring, standard electricians, and happy electrical inspectors who enforce safety standards.

That is the basic system. A supply of electrical energy from wind, water, or sunlight feeds low voltage electricity into a battery bank. The batteries store the electrical energy within the chemistry of the battery "cells." When an electrical load requires energy to operate, current flows from the battery and/or the renewable energy source at low voltage to power the inverter. The inverter transforms the direct current (DC) low voltage to higher 120V alternating current (AC) to feed the house electrical panel and the waiting appliances.

The Changing Seasons

As a bad July sunburn will remind you, the amount of sunlight received in summer is much greater than in winter. Simply put, the longer the sun's rays hit a PV panel, the more electricity the panel will push into the battery. The months of November and December tend to be dark and dreary by contrast. How does this affect the system and will there be enough energy in the winter?

Seasonal variability is extreme in the northeastern section of North America. The two maps shown in Appendices 5 and 6 show the average amount of sun hours in September and December across North America. The amount of sunlight in December is approximately half what it is in September and even less than half the amount we get in June. Obviously, the PV panels' output will reduce accordingly, and the amount of stored electricity will vary with it. This creates an odd paradox for the off-grid homeowner. There is plenty of electricity in summer, when everyone is outside more and lighting loads are less, and not enough in November and December when everyone is huddled inside with TVs and lights blazing. How do we design around this problem?

North American Sun Hours per Day (Worst Month)

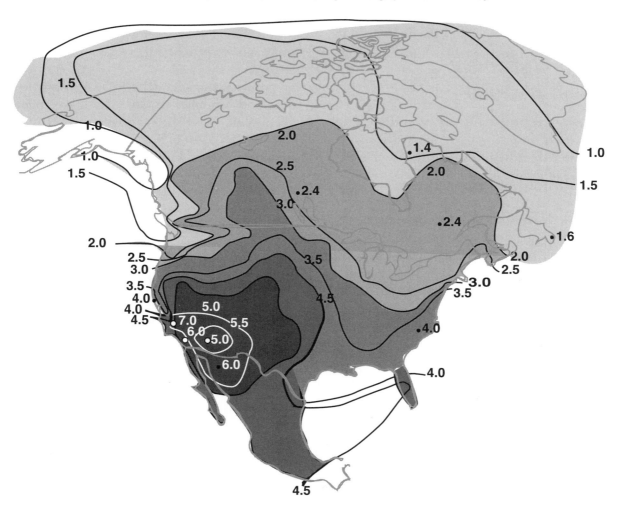

Figure 3-15. Sunlight hours per day are subject to extreme variability in most of North America. Phoenix, Arizona receives double the amount of sunlight of Rochester, New York or Toronto, Ontario. Seasonal variation can also play havoc with the level of solar energy received. Careful attention to energy efficiency and system design will ensure that off-grid homeowners have sufficient energy to power their homes.

For the grid-interconnected homeowner, seasonal variation has no effect on lifestyle, except perhaps a varying cash flow from the size of the utility cheque received. When the weather does not co-operate, electricity is supplied by the utility. Likewise, when renewable energy generation is at its peak, producing more energy than required, the excess electricity is sold to the utility.

Off-grid homes do not have the luxury of buying extra energy from the grid, so they must deal with the situation directly, using either hybrid electrical sources or generator backup.

> *Off-grid homes do not have the luxury of buying extra energy from the grid, so they must deal with the situation directly, using either hybrid electrical sources or generator backup.*

Hybrids (Winter Season)

Hybrid design simply means adding more than one power source into the energy mix. In the example in Figure 3-16, we have PV, wind, and microhydro, plus a backup generator. This design is not typical, as most off-grid systems usually start with PV as the main renewable source, a backup generator second, and possibly a wind turbine third. For those of you lucky enough to have a year-round stream sufficient to operate a microhydro system, that may be the only energy source you will require.

Grid-tie systems are typically PV-based. Wind- and water-based sources are usually not connected owing to the rural nature of these energy sources and the difficulty with building permits in urban areas.

Back to watts and numbers for a second. Remember that I mentioned my house consuming 3 to 6 kilowatt-hours (3-6 kWh) of electricity per day. We can now look at my system power production to see how well energy generation and consumption match.

The rating of our PV panels is 1700W peak power output (48V x 35ADC). In reality, they tend to output approximately 1450W under ideal conditions, less if it is hazy, and nearly zero if the day is overcast. In our home, the entire solar collector assembly is mounted on a sun tracker unit, which allows the panels to face perpendicular to the sun as it moves from early morning through late afternoon, winter and summer, ensuring that they capture the maximum solar energy available.

Referring to the worst month sun hours map (Figure 3-15) for our location, we find the average yearly sun hours to be 3.9 per day:

3.9 sun hours per day x 1,450W output
= 5,650 watt-hours per day or 5.7 kWh per day

With 5.7 kWh of production and an average consumption of 4 kWh per day, the system will have an average excess capacity of approximately 1.7 kWh per day. Either this excess energy is not used or additional equipment is installed to "dump" the energy into some other form of storage device such as a hot water heating unit, which is exactly what I do (see Figure 3-16).

That is the upside. During the winter months, the PV system is greatly reduced, which would cause the battery bank to lose 1,000 watt-hours (1 kWh) per day. If this were your bank account and you kept taking out more money than you put in, guess what would happen? Depending on how deep your pockets are you would run out of cash. The off-grid system batteries are no different and when the batteries are drained the lights go out! In fact, typical battery sizing assumes that you can run your house "normally" for three to five days without having *any* input from your renewable energy sources. For our household, running average loads means that the batteries need a *usable*[6] supply of:

5 days' supply x 4 kilowatt-hours per day
= 20 kilowatt-hours' usable capacity

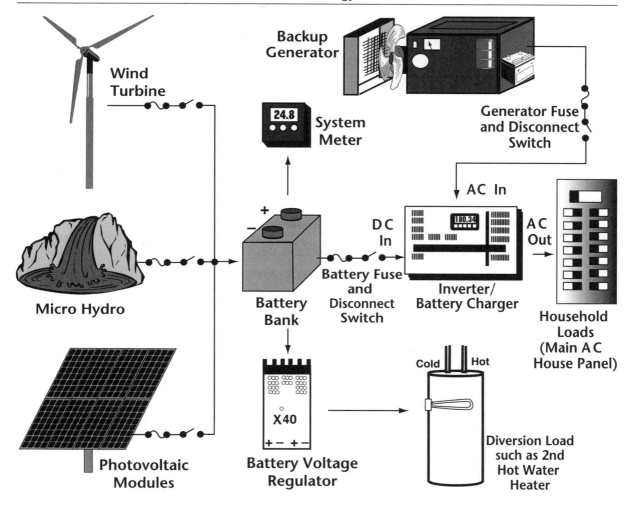

Figure 3-16. Renewable energy electrical systems often have more than one source of power. In this hybrid off-grid example, a wind turbine, hydro, and photovoltaic modules all provide energy to charge the battery bank. A backup fossil-fuel generator is also available to provide electricity during times of renewable energy drought.

So what happens at the end of four days? This is where the hybrid design comes in. You either have another renewable source pick up some of the load when the PV panels run short, thus preventing the battery bank from running low, or you rely on a backup genset (generator-set). A wind turbine or microhydro generator can operate in parallel with the PV system, thereby increasing total power generation.

If the secondary power source is a fossil-fuel generator, it will have to make up the deficit. Depending on the degree of automation in your system, either you can manually start the backup generator (gas, diesel, or propane) or an automatic generator control device will start the generator for you. The inverter then switches to battery-charging mode and fills the batteries back up. The house electrical loads automatically receive power from the generator during this charging time.

Once the batteries have reached full charge, the generator turns off automatically or you run out in your housecoat and slippers to shut it down. (I think the automatic feature is definitely worth the few extra bucks!)

If your system contains more than one renewable source, you will find that they tend to be complementary. A dull day in November or February often has brisk winds and conversely the air on a sunny summer day is hot, still, and stifling. However, do not believe that having PV and wind will eliminate a backup generator; it will not. The combination of multiple renewable sources will reduce the running time of the generator considerably, but it will not eliminate it, especially when you have many guests over at

Figure 3-17. This large 7.5 kW diesel backup generator set (or genset) will last several lifetimes and be economical to operate as well. If you are considering an off-grid lifestyle and require a generating unit, forget about anything you can purchase at a hardware store and consider only slow-speed (1800 rpm or less) industrial units. Although industrial units are more expensive than the economy models, the savings in fuel cost and aggravation when the cheap model will not start in winter is well worth the price difference.

Christmas. Our house still requires approximately 100 hours of generator time per year.

Hybrids (Summer Season)

During the summer months, the increase in sunlight hours coupled with a lower need for lighting and less time spent indoors creates a surplus of energy based on our lower consumption levels of 3.5 kilowatt-hours per day.

Production:

> *6.0 sun hours per day x 1,450W output*
> *= 8,700 watt-hours per day*
> *or 8.7 kWh per day*

Surplus:

> *8.7 kWh/day produced – 3.5 kWh/day used*
> *= 5.2 kWh/day surplus*

We may or may not need this surplus, depending on whether or not we require any air conditioning that day or do more laundry than usual. As mentioned earlier, even a relatively small room air conditioning unit uses an enormous amount of energy, approximately 1,100W per hour operated. Based on a surplus of 5.2 kWh/day, we should be able to operate the air conditioner for up to 4.5 hours per day without dipping into the energy bank or activating the genset.

On days when we do not need air conditioning, the surplus energy must go somewhere. You must consume this energy or the batteries will reach a fully charged state and eventually "overcharge." To prevent this from happening, a battery voltage regulator or charge controller either "disconnects" the renewable source from the batteries or connects it to a diversion load. A typical diversion load consists of an electric water heater supplied with low-voltage heating elements, which helps to reduce fuel requirements for domestic hot water. When the batteries are ready to accept more energy, the charge controller automatically begins a recharging cycle.

During operation, the battery voltage regulator monitors the battery voltage or *state of charge*. When the batteries become full, it starts to divert surplus electricity to the electric water heater. The water starts to heat as it absorbs the extra energy produced by the renewable energy sources. Over the course of a day or two, the water can easily reach 140°F (60°C), and it then flows into the cold side of a regular gas water heater. As the incoming water is already hot, the gas heater remains on standby, thus conserving propane gas, energy dollars, and the environment. Energy is never wasted in this system!

Another Look at Phantom Loads

As the name implies, phantom loads are any electrical loads that are not doing immediate work for you. This includes items such as doorbells (did you know that your doorbell is always turned on, waiting for someone to push the button!), "instant-on" televisions with remote controls, clock radios, and power adapters left plugged in while the cell phone remains glued to your hip.

So what is the big deal? First, these devices are consuming energy without doing anything for you. That electric toothbrush was probably charged about 15 minutes after you put it back in the holder. During all the in-between hours when the device sits idle, it is wasting energy. A television set that uses a remote control is actually "mostly on" all the time, just waiting to receive the "on" command from your remote.

While the total dollar cost for these luxuries is small on grid (in the order of $5 to $10 per month for an average house), this consumption off-grid is unacceptable. By the way, this "small" bit of waste equals a few hundred million dollars a month in North America alone and is equal to the total output of several large-scale coal-fired generating plants.

Another reason for the concern in off-grid installations is that the inverter unit cannot enter sleep mode when the last light is turned off at night. This would conserve a fair amount more energy but any phantom load will keep the inverter "awake," consuming more energy than it otherwise would.

Phantom Load Management

Some phantom loads can simply be eliminated. Try a door knocker instead of a doorbell and a battery-powered digital clock instead of the plug-in model. I have even heard that some people use manual toothbrushes.

Television sets and CD, DVD, and VHS players with instant-on and remote control functions can be wired to outlets that can be switched off. This could mean having the electrician wire a wall-mounted switch to control these items or using a power bar with an integral switch to do the same job.

Many people cannot live without some phantom loads like fax machines or chargers for cordless phones, cell phones, PDAs, and laptops. For these items, you have two choices: allow your main inverter to remain "on" at all times or use a separate wiring circuit which connects specially marked outlets reserved for ESSENTIAL "always-on" loads like the ones described above.

The power for the essential outlets can be an inexpensive 100W inverter that is wired directly to the batteries through a fuse or circuit breaker panel. Such an arrangement ensures that the main inverter can go to sleep at night and still provide power for the small devices you require. If your home is already built, it may be possible to group all of these special loads at one central location and run the inverter to a power bar at that point.

One note about using this "small" inverter to power phantom loads: if you load it up with all of your toys, it's possible to burn up nearly 100W of power. Over the course of 24 hours, this can add up to a lot of energy:

Phantom loads are any electrical loads that are not doing immediate work for you.

100 watts x 24 hours per day
= 2,400 watt-hours or 2.4kWh

A load of 2.4 kWh is almost half of the daily total energy production of most off-grid systems. Tread lightly and limit the use of this power to truly essential items!

Metering and Such

At this point you are probably wondering if I run around the house with a notepad and calculator, chastising Lorraine for using her hair dryer too long or making the toast too dark while recording every volt and who knows Watt. Actually, we hardly look at the system at all. Once you install

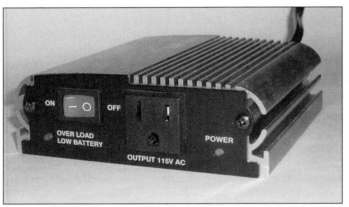

Figure 3-18. For those very small "always-on" loads, wire a small inverter such as this one to a power bar or have your electrician install a small "whole house" inverter to specially marked outlets strategically placed throughout the house.

your equipment and load the house with all the required electrical goodies, within your average production limits the system will almost take care of itself. Lorraine says that our system is magic, almost invisible.

For grid-interconnected designs, the system *is* invisible. The only notification you get that all is well is a statement from the utility advising you that you have a credit of $500 on your electrical bill. (Of course some people just have to keep running outside to watch the electrical meter running backwards or switching on the computer to check the metering and account data.)

As with any piece of complex machinery, a bit of care and management is required (chapter 17 will deal with this issue). Status metering, battery

hydrometer checks, and other system parameters will provide a comprehensive snapshot of the health of your own power station.

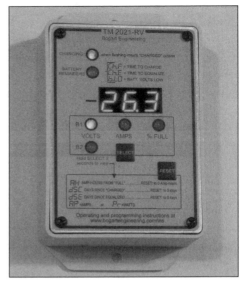

Figure 3-19. Every off-grid power system should have metering capabilities. A proper meter will lead to increased load control knowledge and system reliability. (Courtesy Bogart Engineering)

Is It Economical?

Solar thermal and biomass heating systems will usually provide a satisfactory return on investment provided the equipment is of good quality and can be depended on to work over the amortization period. Because of the relatively low capital cost for these systems, it always makes sense to consider heating technologies before electricity generation.

Grid-tie system valuation requires a bit of fancy mathematics to ensure that the income stream meets your investment requirements. When considering the value of a grid-tie PV system, a dealer will be able to identify the total sun hours per year and take into account system derating factors such as buildings or trees blocking the sun in order to arrive at a gross energy production figure. Multiplying the gross energy figure by the electricity tariff will give you a gross yearly income figure. Capital costs less government incentives will then allow you to arrive at the ROI figure.

Of course, many people do not bother calculating the ROI or worrying about cash flows because they want to "do their bit for the environment" or release themselves from the penury caused by excessive electrical bills. In most instances, reputable dealers will be able to either give you a good rule-of-thumb investment value or assist you in working the numbers.

Off-grid homeowners determine the break-even point based on how far the installation is from the electrical grid. If your electrical utility is more than ½ mile (0.6 km) from your house, the system may pay for itself from the second you turn on the first light. Moreover, you can add to this the benefits of no electricity bill, minimal environmental pollution, and the feeling of self-sufficiency you get the next time your neighbor's lights go out during that dark and stormy night

If you are planning to construct a new home and the building lot has not been purchased you may have another savings angle. Property values are determined by location, location, and of course, location. If that location happens to be beside a beautiful lake or mountain setting with the grid two miles away, you can be sure the land will not be as desirable as it might otherwise be. Use this to your advantage. Many people have never heard of an off-grid system or are unwilling to use one. Construction costs for a lengthy grid extension drive property prices down, possibly to the point where the decrease exceeds what the renewable energy system will cost to install, making the economic case better still.

As with any rule-of-thumb, there is always a group of people who will blow holes in the theory; folks living off-grid seem to be one of them. Over the years, I have noticed more and more people installing off-grid systems where the electrical utility sails past the entrance of their property. There can be no sound economic reason for doing this, and I have noted that the people who break the rules tend to be those "who exhibit rugged individualism and self- sufficiency," as we learned earlier.

Chapter 4 will highlight a number of on- and off-grid renewable energy installations and outline the costs for each system. By analyzing each of these lifestyles and comparing it to your own you will be able to determine a reasonable "back of the napkin" budget that matches your personal requirements.

Selecting a Dealer

Education and ambition are the engines that get the renewable energy wagon rolling; but you still need a dealership to sell you the

Figures 3-20 and 3-21. Environmental stores offer a wide array of ecologically responsible products including energy-efficiency devices and renewable energy products. Arbour Environmental Shoppe in Ottawa, Canada, has a perfect mix of useful products to help lessen your environmental impact.

Figure 3-22. Solar-power radios, energy-auditing software, photovoltaic-powered toys, rainwater collection tanks, and eco-friendly household products are but a few of the resources and educational products carried by many environmental outlets. Some of these products are kits that make great gifts and can even distract kids from the television set for a few hours—an interesting way of reducing energy consumption.

wagon in the first place. There are thousands of technologies designed to save or generate energy that "guarantee" they are excellent investments. Sadly, the guarantee isn't always valid. Although most dealers are honest, reputable people who are an integral part of the community and recognize that fleecing their clients is not in their long-term

best interests, others aren't quite as principled. And even if the dealer is doing a respectable job, competing technologies can make the purchasing process a complex one requiring expert and unbiased advice.

When faced with the prospect of spending money, whether a few dollars or a few thousand dollars, seek the advice of a reputable dealer. Local retailers who have been in business for a number of years and provide a clean, neat storefront are likely to provide honest and unbiased service. As you would with any sizable purchase, consider obtaining references from past clients as well as visiting installations to see exactly how well the product operates in the real world.

Energy-efficiency products (such as compact fluorescent lamps and low-flow shower heads) tend to be relatively inexpensive and may be purchased through local environmental or hardware stores. Renewable-energy-generation equipment is often "co-marketed" with complementary products such as wood stoves, swimming pool and spa products, or heating and air-conditioning equipment.

Just like every other retail product in North America, renewable energy systems may be purchased via the Internet from reputable dealers and used-equipment traders or through auctions on Web sites such as eBay. Unless you are an experienced home handyman (or woman) it may be wise to stick with local dealers you can trust. While it may be possible to have an inverter serviced over the phone by someone at a call center a thousand miles away, having a friendly supplier just down the street is very comforting. This is especially true if a critical piece of equipment goes "poof" and you are left in the dark. Remember, you can't put the smoke back in the box and expect it to work.

Many dealers not only sell

Figure 3-23. Renewable-energy-generation equipment is often "co-marketed" with complementary products such as wood stoves, or heating and air-conditioning equipment.

the equipment but offer *qualified* installation and repair service. Warming hot water with the sun sounds like child's play until you examine the technology required to harvest the sun's energy and transfer it safely and reliably into your water tank. Differential temperature controls, expansion tanks, vacuum solar collectors, and a myriad of other bits and pieces must be carefully orchestrated and installed, requiring quality workmanship to ensure a long and trouble-free operating life. Many jurisdictions require licensed contractors to perform the installation work, demanding building permits and insurance certificates prior to completion. Many dealers will not have such skilled staff in-house but will generally have a co-operative working arrangement with appropriate contractors.

Dealers are more than warehouses for the products they sell. They are experienced, professional, and educated people who want to be sure that you are satisfied with their products. While you may be able to purchase goods at a lower price, quality dealers are worth their weight in gold for the installation, operation, and lifelong servicing of their products.

Figure 3-24. When shopping for a renewable-energy products dealer, look for a store that is well maintained and stocked. Get references from current and past customers and try to arrange a site visit with a previous client. A dealer that is hesitant to offer this service may not be one you wish to deal with.

Endnotes

1   Although it is possible to physically run your electrical meter forwards and backwards as it credits and debits your electricity account, many utilities use an electronic metering system that keeps track of generated and consumed energy separately. Banked energy is not actually stored but rather flows into the distribution system for consumption by others. When you require your banked energy at some future time, you are simply drawing power from the grid while simultaneously debiting your account balance.

2   FITs are also known as Renewable Energy Standard Offer Contracts or Advanced Renewable Energy Tariffs. To learn more about FITs and their benefit to society read the report Smart Generation at http://www.davidsuzuki.org/Publications/smart_generation.asp

3   One unit of the primary fuel energy contained in coal goes to making electricity, while more than two units of thermal energy and electrical transmission system resistance are wasted. I can think of no other industry that would consider throwing away two-thirds of its raw material inputs.

4   The Ontario Power Authority is reviewing pricing as of April 5, 2009.

5   The process of "resynchronization" requires the inverter to match the voltage, frequency, and phase angle of the grid before reconnecting and trading energy. As this process is fully automatic in all grid-tie inverters, I draw the reader's attention to glossary and reference sections for further reading and research.

6   Battery usable capacity relates to how much energy can be withdrawn from a given battery bank without causing damage or reducing its life. This issue will be discussed further in Chapter 9, although a maximum "depth-of-discharge" of 50% is typical.

Chapter 4

A SHOWCASE OF RENEWABLE-ENERGY-POWERED-HOMES

Introduction

Before delving into the complexities and components of renewable energy electrical and heat generating systems, a review of real-world configurations both on and off the electrical grid would be prudent.

One of the first questions people ask regarding the installation of solar water heating or electrical generating equipment is, "So what did that set you back?" or "What is the payback for all that equipment?" (When was the last time anyone wondered what the payback was on a new car purchase, a trip to Disneyland, or a big screen TV?)

A similar situation occurs with folks who install off-grid electrical systems, with well-meaning people asking, "How can you live without "electricity"?"

The media are not much better. I have seen and heard media reports that a "normal" lifestyle can be supported only by spending upwards of a quarter of a million dollars if you're disconnected from the ubiquitous electrical grid.

As with most things in life, the truth requires a bit of digging to unearth. Looking at several examples of how people actually live with renewable energy may help shed some light on how to deal with pesky comments and questions and make good purchasing decisions.

In order to dispel some of the hype, this chapter will investigate a number of renewable energy systems, both on and off grid, ranging all the way from humble portable power units that can be taken camping to a luxuriously equipped home suitable for even the most discerning lifestyle.

In order to find the truth it is necessary to understand that the issue of system cost is

Figure 4-1. The size and cost of your renewable energy system is directly related to your needs and wants. Whether you want to add some TV time to those rainy days out camping or outfit a year-round home, energy demand will dictate final cost. Bonnie Klein and her partner Dave McDonald use this small PV array on their camping trailer to charge 6V golf cart batteries. An 1,800W inverter provides power to a stereo, laptop computer, microwave oven, cordless tool battery charger, and television for about the same cost as a gas generator but with none of the noise or pollution.

directly related to needs and wants: people living in mansions pay higher energy bills than retirees living in a condo. It stands to reason that a large home with numerous occupants (especially teenagers) and high energy demands will translate directly into a larger and more expensive power-generating system. Conversely, become more energy efficient and you will save money. This is true whether you are living downtown connected to the electrical grid or in a cabin miles away from society.

There is no "standard" system price, but rather a range of prices according to your needs and wants. Each of the selected dwellings in this chapter has been chosen to illustrate the type of lifestyle you can expect for a given capital expenditure. At the same time, the reader is cautioned that these are late 2009 prices and will vary depending on dealer markup, the "do-it-yourself" factor, and market trends in your area. In all cases, new equipment was installed and third-party installation labor costs have been factored in.

This chapter looks at renewable energy systems in the following order:

- Portable Equipment for Camping and Emergency Power
- Off-Grid RVs, Trailers, and Cottages
- Full-Time Off-Grid Residences
- Grid-Connected Homes Powered by Renewable Energy

Portable Equipment for Camping and Emergency Power

Portable RE systems can be designed to fit many applications. Starting with the ubiquitous solar-powered calculator at one dollar, renewable energy technologies are now available for just about any item. For fewer than ten dollars, you can now purchase flashlights and radios that require nothing more than a bit of sunlight or human muscle power to wind or shake the device into operation.

Figure 4-2. Using the caloric renewable energy contained in one raisin will provide enough muscle power to shake this LED flashlight, generating electricity and providing approximately 20 minutes of light.

Figure 4-3. This 150W power box can be used to power small home appliances during a power outage, charge cell phones or laptop computers, and keep small modern grid-powered devices running during a camping trip.

With the recent development of low-cost, high-power electronics, emergency backup power sources can be purchased at most hardware and department stores. Small units such as the 150W portable power box shown in Figure 4-3 are perfectly sized to power a small radio or energy-efficient desk lamp, recharge cell phone or laptop batteries, and even jump-start cars in a pinch. This model is currently available for $75 or less.

The 1200W power eliminator shown in Figure 4-4 is available for under $200 and is literally a small off-grid system on wheels. With its large internal battery and alternating current inverter, the unit is capable of powering small tools such as a drill, house lights, or a television during power outages. People with small cottages or hunting camps often use these portable units to avoid having to install a more expensive, permanently installed off-grid system or continuously run a generator just to keep a few lights on at night.

Photovoltaic panels like those shown in Figure 4-5 can be coupled to many models of battery eliminator boxes in order to recharge the internal battery. The manufacturer will recommend a specific size or power rating of PV panel to match the charging requirements of the unit. When large PV panels are connected to a battery bank, a device known as a charge controller may be required to prevent overcharging. Make sure you consult the power eliminator data manual for correct operation and connection information.

Figure 4-4. This 1200W power eliminator is literally an off-grid system on wheels. A unit of this size can provide a reasonable amount of energy at the cottage or hunting camp. Because of the limited amount of energy storage in the battery bank, it is wise to ration electricity carefully to prevent appliance shutdown. The larger the electrical load wattage (volts x amps), the shorter the running time.

Figure 4-5. Recharging battery eliminator units is easy if they are coupled to an appropriately sized photovoltaic (PV) panel. After connecting the panel to the battery inputs, place the panel in direct sun for several hours to recharge the power unit's battery bank. A voltage regulator (as discussed in Chapter 10) may be required. Consult the battery eliminator manual for correct PV panel and regulator requirements.

Taking your RV "Off-Grid"

The battery charger charges the RV's batteries when grid power is available

The charge controller prevents the RV's batteries from overcharging.

Figure 4-6. If you are tired of the side-by-side parking spaces provided by many RV parks, why not consider "dry RVing." No matter what size of RV or camping trailer you own, it is possible to outfit the unit with a small off-grid system as shown here and take the convenience of the electrical grid anywhere.

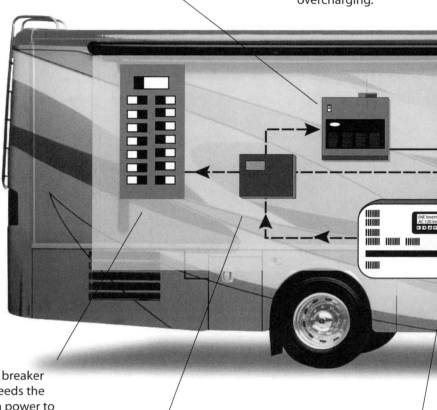

The AC breaker panel feeds the RV with power to run all the RV's AC loads.

The prewired transfer switch has an input for the grid power and inverter. The output power goes to the AC breaker panel.

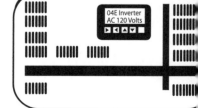

The inverter converts the DC battery power to AC power.

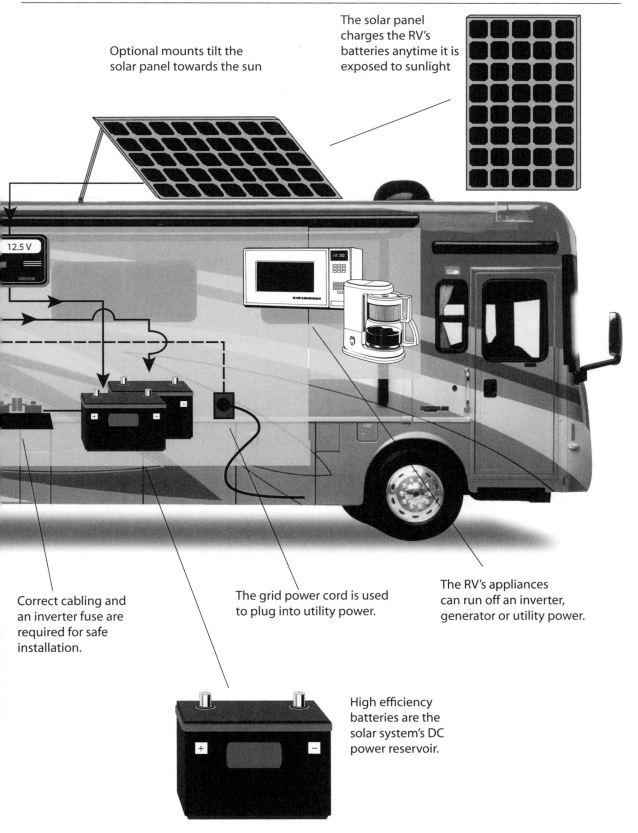

The solar panel charges the RV's batteries anytime it is exposed to sunlight

Optional mounts tilt the solar panel towards the sun

12.5 V

VXR3500

Correct cabling and an inverter fuse are required for safe installation.

The grid power cord is used to plug into utility power.

The RV's appliances can run off an inverter, generator or utility power.

High efficiency batteries are the solar system's DC power reservoir.

Over-discharging or overloading these devices will greatly reduce the life expectancy of the batteries.

With a portable satellite Internet transceiver powered by the off-grid system, it is possible to stay in touch with friends using Skype Voice over Internet Protocol (VoIP), read the paper, and even watch television from just about anywhere on earth. Similar systems can easily be adapted to sail and power boats.

As with any off-grid system, it is necessary to balance the electrical load consumption with the system's ability to supply energy. The worksheet in Appendix 7 will assist you with these calculations.

Figure 4-8. Even if your home or cottage isn't off grid, sometimes it just doesn't make sense to extend electrical wires under paved parking lots or over extended distances. A completely self-contained lighting system incorporates PV panels, a battery bank, and a charge controller, often for less cost than a line extension. (Courtesy Phantom Electron Corporation, www. phantomelectron.com)

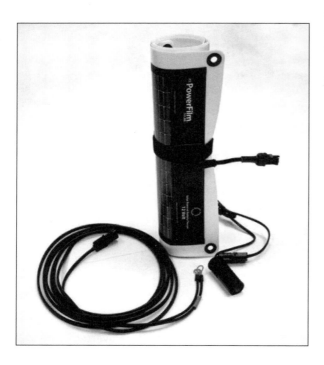

Figure 4-7. If you just can't leave your cell phone or Blackberry at home when you go camping, perhaps you need to take a stress management course. Alternatively, these lightweight photovoltaic panels can be used to charge the batteries of a GPS, an MP3 player, and other small electronic devices. Flexible PV panels are so light and pliable that they can be rolled right into your sleeping bag. (Courtesy PowerFilm Solar, www. powerfilmsolar.com)

4.1

The Kemp Off-Grid Camper

Seasonal and Weekend Off-Grid Solar System

System Cost:
Electrical	US$1,500.00
Portable Generator	US$1,400.00
Solar Thermal	n/a

Figure 4.1-1. A small camping trailer connected to an equally small off-grid system provides many of the comforts of home without the frequent and annoying buzz of a gasoline-powered generator.

An evening walk around a local camping park was interrupted by the cacophony of what at first appeared to be massive cicadas in the woods. It didn't take long for us to realize that the actual culprits were dozens of gasoline generators buzzing away, each powering a handful of lights and little else.

Although the noise was annoying enough, my brain started running "the numbers" and I came to the conclusion that one generator could handle the work of all the machines running simultaneously. The irony of someone operating a 5,000W generator just to run a couple of 12W compact fluorescent lamps bothered me to no end. Surely people must know about small off-grid systems for camping use? Then again, maybe not.

When Lorraine and I decided to purchase a "lightly used" camping trailer and place it at a secluded spot on a lake near our house, I jumped at the opportunity to give us some of the comforts of home in a quiet, off-grid setting.

The 23-foot travel trailer came equipped with a propane cookstove, oven, water heater, and refrigerator. These appliances consume too much energy to be operated by electricity generated by a small off-grid system and therefore run on propane. This left a number of devices that could be operated on either battery voltage (direct

Figure 4.1-2. The author enjoys the tranquility of a secluded camping site which has limited facilities. Biological power operates the kayak, and a couple of solar patio lights illuminate the way back from those late night paddles.

current) or 120V alternating current. I decided to operate the majority of the remaining appliances and devices on 12VDC supplied directly from a deep-cycle battery bank. The batteries are charged primarily with a photovoltaic (PV) panel that converts sunlight to electricity and a small backup generator is available to do the job during periods of high electrical demand or during inclement weather.

Figure 4.1-3. A 23-foot travel trailer is equipped with a propane stove, oven, and water heater, which consume too much energy to be operated from a small off-grid electrical system. The cookstove, range hood fan, operate from 12VDC and all of the room lights were converted to high-efficiency direct-current fluorescent models.

Figure 4.1-4. As a friend recently commented, "the camper provides just enough conveniences to keep it comfortable, yet simple."

Figure 4.1-5. The propane-powered refrigerator/freezer requires too much energy to operate from a small off-grid electrical system.

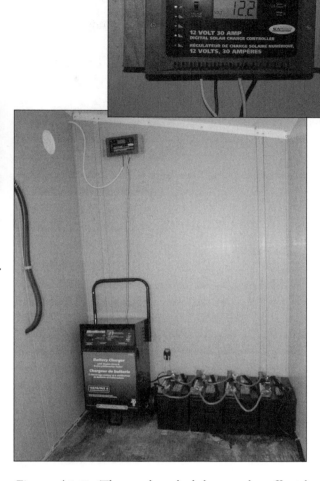

Figure 4.1-6. The off-grid system was built into a standard garden shed, allowing the two 120W, 12V photovoltaic panels to be mounted to the sloped roof. The panels face "solar south" and are steeply angled, allowing them to catch the maximum amount of summer sunshine.

This type of "battery direct" off-grid system is not only simple to design and install but is also lower cost than its larger alternating current cousin that is used in the majority of full-time residences.

The list of items connected to the 12V electrical system includes:

- Fluorescent room lights
- Stove-top ventilation fan and light
- Water pump
- Stereo radio and iPod speaker system
- DC to AC inverter to operate TV/DVD player
- Central furnace (electric controls/fan; propane heating)
- Camper ventilation fan
- Electric razor
- Automotive-style vacuum cleaner

The camper's central rooftop air conditioner and microwave oven are devices that we did not really feel were necessary for our enjoyment and given their very high electrical demand they would greatly increase the cost and complexity of the off-grid system. However, they could be operated in a pinch using the gasoline generator.

Figure 4.1-7. The garden shed houses the off-grid electrical and water supply system for the camper, while the balance of the room is used for storage of camping gear. The battery bank consists of four deep cycle marine batteries that are used to power electric trolling motors for fishing boats. These batteries are relatively inexpensive, available from many retailers, and easily recycled. The battery bank is also connected to the camper battery terminals using large-gauge (# 8 AWG) electrical wire. A fuse (top left of the battery bank) protects the circuit in the event of a cable fault. A large residential battery charger (lower left) is used to quickly charge the batteries from the gasoline generator. The photovoltaic charge controller and meter (top left) connects the PV panel to the batteries, ensures proper charging, and displays battery state-of-charge information.

Figure 4.1-8. Most RV campers come equipped with built-in battery chargers and electrical load distribution panels such as the one shown lower right. This battery-charging unit is far too small to charge the off-grid battery bank and must therefore be augmented with the larger model shown in Figure 4.1-7. The distribution panel also contains circuit breakers to protect both the AC and DC circuits that run throughout the camper. Mounted to the left of the distribution panel is a 120W micro-inverter that converts 12VDC to 120VAC. The output of the inverter powers all of the outlets in the camper, albeit with very limited capacity. It is used to charge the iPod battery and operate an electric razor and LCD TV.

Figure 4.1-10. The iPod and FM stereo unit is battery operated, making it portable for use outdoors. The 120V micro-inverter-powered outlets can charge the internal batteries of the iPod unit. The battery charger draws approximately 20W during the early stages of charging and next to zero after an hour or two.

Figure 4.1-9. This LCD television set with built-in DVD player is the perfect distraction for rainy days. The unit plugs into the 120V outlets (powered by the micro-inverter) in the camper and draws a miserly 45W of power during operation.

Figure 4.1-11. The camper is situated without access to potable water, which is brought to the site using Jerry cans. For dishes, showers, and toilet flushing, filtered and lightly chlorinated lake water is used. A heavy-duty garden hose is connected to a coarse filter and submerged into the lake. The other end of the hose is connected to a 10-micron water filter unit the outlet of which is in turn connected to the inlet of a 12VDC water diaphragm water pump such as this model by Shurflo. An on/off switch allows the pump to be activated as required.

Figure 4.1-12. The author is shown here placing the outlet end of the water pressure pipe into the camper water tank fitting. Adding approximately one ounce (30 ml.) of household bleach to the tank will provide enough chlorination to prevent algae growth and subsequent water odor.

Figure 4.1-13. The camper contains several fixtures that can be supplied from the non-potable lake-water system, including a flush toilet, sink, and shower.

Figure 4.1-14. Black water discharged from the toilet is stored in a small holding tank, which is periodically pumped out. Gray water from the sinks and shower are fed to an underground "dry well" which receives the water and discharges it into the ground. Be sure to use all-natural biodegradable cleaning supplies and soaps with this type of system.

Figure 4.1-15. The Honda EU2000i inverter is extremely quiet, has very low emissions, and is economical on fuel, making it a good match for this application.

Campers who are currently using a generator as their only power source would be advised to consider the simple off-grid system design described here. A system that includes only a battery bank and charger will allow lights and other low-power devices to be operated without the constant running of a generator. Whenever the batteries become discharged, the generator can be started, quickly replenishing them.

4.2
The Beevor Family Cottage Seasonal and Weekend Solar System

System Cost:
Electrical	US$1,400.00
Solar Thermal	n/a

The Beevor family enjoyed "camping" in their 600-square-foot (56-square-meter) off-grid cottage but found the tedium and inconvenience of oil lights and the lack of TV for the rainy days a bit tiresome. Paul decided to do something about it and installed a small 12VDC system in the spring of 1999.

"Even with such a small system, everything about the cottage became a pleasure," enthuses Margaret. "We have gone from prehistoric times to living the life of luxury!" With children underfoot, having a TV and VCR and doing away with the outhouse can certainly change how life at the cottage feels.

"Our current system comprises two Siemens SP65 photovoltaic cells installed with a roof mounting system," Paul explains. "The power output of the two panels is 130W, which is sufficient for our small and occasional loads. We use the cottage for three seasons, so the lack of winter sun is not a concern."

The output from the PV panels is regulated using a ProStar 30A voltage regulator to prevent battery-overcharging conditions. The energy is stored in gold-cart-style deep cycle batteries. Electrical power is routed to a 12V fuse box which supplies the cottage with DC power suitable for the small loads. 12V recreational vehicle lights and appliances are used.

"We also wired the cottage for standard 120V AC supply in case we decide to add an inverter at

Figure 4.2-1. The Beevor family enjoyed "camping" in their 600-square-foot (56-square-meter) off-grid cottage but found the tedium and inconvenience of oil lights and the lack of TV for the rainy days a bit tiresome.

Figure 4.2-2. Paul Beevor decided to do something about the lack of electricity and installed a small 12VDC system in the spring of 1999. The family's off-grid system comprises two Siemens SP65 photovoltaic cells installed with a roof mounting system. The cottage is used for three seasons per year, so the lack of winter sun and resulting energy is not a concern.

Figure 4.2-3. "Even with such a small off-grid system, everything about the cottage became a pleasure," enthuses Margaret. "We have gone from prehistoric times to living the life of luxury!"

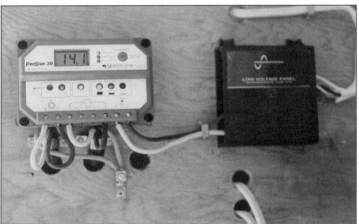

Figure 4.2-4. The output from the PV panels is regulated using a ProStar 30A regulator to prevent battery-overcharging conditions. Voltage is fed to the cottage wiring as 12VDC, necessitating the use of low-voltage RV or camping-style appliances and lights.

Figure 4.2-5. Electrical energy is stored in golf-cart-style deep cycle batteries.

some future time," says Paul. "In the meantime, I plan to add a Shureflo 12VDC water pressure pump and accumulator tank to provide pressurized water for the toilet and two sinks." As with most small cottages, the Beevors use a propane cookstove and a wood stove for heating.

The Beevors' low-cost energy system proves that good things can come in small packages.

4.3

Fairmont Hotels' "Kenauk Cabins"

A Luxury Resort Company Goes Off Grid

System Cost:	
Electrical	US$7,500.00
Solar Thermal	n/a

Figure 4.3-2. Located north of Montebello, Quebec, Canada these off-grid eco-cabins are accessible only by traveling down a 20-mile (32 km) gravel road.

Anyone who is familiar with the Fairmont hotel chain (www.fairmont.com) will know that it is the largest luxury resort company in North America. One of its properties, the famous Château Montebello, is a stunning red cedar castle which is the largest log building on the planet. Located in the town of Montebello, Quebec, Canada the resort attracts a variety of people, many of whom come to enjoy the natural beauty along the Ottawa River.

For those people who still want topnotch service but prefer to "rough it," the people at Fairmont have started to include eco-destinations in their array of product offerings. One of these is a series of off-grid eco-cabins (Figure 4.3-2) located north of Montebello that are accessible only by traveling down a 20-mile (32 km) gravel

Figure 4.3-1. Fairmont has started including eco-destinations and cabins in its array of product offerings. The cabins located on this private lake offer the ultimate in peace and quiet that the countryside has to offer.

road. These cabins are located on a private lake, offering the ultimate in peace and quiet that the countryside has to offer.

System installer Mike McGahern of Ottawa Solar explains: "The cost to run the electrical grid to these cabins would be enormous, well beyond the value of the property. Running a generator 24/7 to run a few lights and power a stereo was out of the question from an environmental standpoint. The only realistic option was solar."

Each cabin was equipped with a non-tracking PV array (Figure 4.3-4) with a nominal rating of 660W. Maximum power point trackers (MPPTs) were installed to increase the energy available to charge a small 24V, 530-amp-hour battery bank. A 2500W inverter provides the cabin with standard 120V electricity to power the various appliances.

Propane gas does the high-energy work of space and water heating (Figure 4.3-9) as well as powering the cookstove and refrigerator, while the off-grid system takes care of the rest. "The electrical appliances in the cottage are limited to necessities, taking the cabins to just one step above camping," Mike continues. "There are approximately six compact fluorescent lamps inside and on the exterior of the building. In addition, a deep-well 120V water pump supplies pressurized drinking and washing water. There is also a two-way radio that is powered by the off-grid system to call for room service should the need arise."

This is, after all, a Fairmont property.

Figure 4.3-3. Each cabin is equipped with a non-tracking PV array mounted in a traditional pole-mount configuration.

Figure 4.3-4. The array is constructed using six 110W PV panels interconnected to operate at 24VDC.

Figure 4.3-5. The inverter, MPPT voltage regulator, circuit breaker, and battery bank are located in a crawlspace area under the cabin.

Figure 4.3-6. The completed PV array provides a rated output of 660W which in turn charges a 530-amp-hour battery bank. Low-voltage DC is stepped up to 120VAC using a Magnum Energy inverter/battery charger.

Figure 4.3-7. There is insufficient electrical energy produced by this system to operate an air conditioning unit, although in the backwoods a ceiling fan can be equally appreciated.

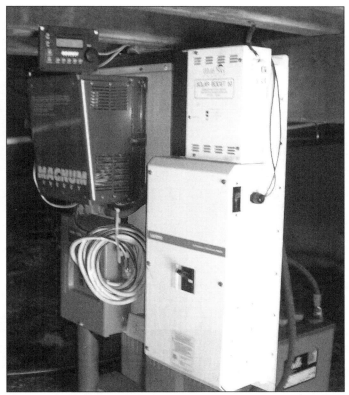

Figure 4.3.-9. All of the electrical components are mounted on a single "power board," which makes installation and wiring interconnection a snap for the installing electrician.

Figure 4.3-8. Energy-efficient compact fluorescent lamps provide all of the interior and exterior lighting for the cabins.

Figure 4.3-10. During the heating season, a propane-powered direct-vent heater provides the cabin with warmth. The off-grid system powers a self-contained distribution fan housed within the heater.

4.4

The Houston/ Lefebvre Earthship 3

An Affordable Entry into Solar-Powered Energy Independence

System Cost:	
Electrical	**US$13,250.00**
Solar Thermal	**n/a**

You wouldn't be inclined to think of a pile of worn out old car tires as part of a sustainable lifestyle or as a particularly energy-efficient building material, but this is exactly what Hillary Houston and her partner Raymond Lefebvre used to build their earth-bermed home.

Raymond explains that old tires can be carefully filled with an earthen mixture and rammed or compressed to form a solid building wall. Stacks of tire upon tire form the walls of the house and provide a large thermal mass, which is a prerequisite for solar thermal heating.

Most tire-based, earth-bermed homes are constructed in a bungalow-like design (Figure 4.4-2) and feature large amounts of south-facing glazing. Because of the radiused shape of the tires, wall construction does not have to follow standard 90° corners. This allows interior rooms to feature a more organic, curvy look.

Raymond explains: "It took approximately 800 tires to build our 1200-square-foot home, and you can't use just

Figure 4.4-1. You wouldn't think of a pile of worn out old car tires as part of a sustainable lifestyle or as particularly energy-efficient building material, but this is exactly what Hillary Houston and her partner Raymond Lefebvre used to build their earth-bermed home.

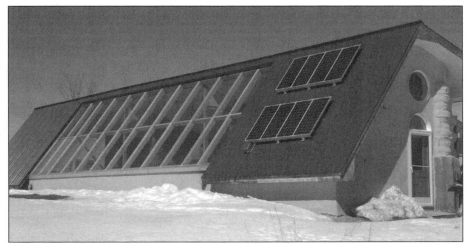

Figure 4.4-2. The south-facing wall of Hillary and Raymond's earthship home collects both solar thermal energy through the vast glazing area and electricity from 600W of photovoltaic panels.

Figure 4.4-4. In this view, the "lumpy" surface of the tires can be seen beneath the final parged coating of the exterior wall. Large roof overhangs are used to protect the tires and finished wall surface from the ravages of excessive moisture in this location. A protective barrier for the thermal insulation can be seen at the ground wall barrier. Also note the satellite television dish, proof of a no-compromise lifestyle.

Figure 4.4-3. The exterior walls of the earthship are parged with a colored concrete that gives the home its adobe-finished look.

any tire. We learned fairly quickly that you can't just have the junkyard drop off a load of tires and expect to use them, as uniformity is extremely important."

"Once we got the hang of selecting the proper tires for our project, we made absolutely sure that we didn't bring in any tires that were not going to end up inside a wall," interjects Hillary. "Early in the process we received 150 tires we couldn't use. It was a steep learning curve."

In northern cold-weather climates, the exterior walls are clad with a suitable insulating material and then parged with natural earth or concrete stucco materials (Figure 4.4-4). The tremendous mass of the interior walls (Raymond

Figure 4.4-5. The north-facing wall is completely bermed by the surrounding earth, providing mechanical support for the building as well as additional thermal buffering from the elements. The roof is punctured by skylights which can be opened during the summer season (or on sunny winter days) to provide additional cooling for the house. Screen doors and windows in the east and west walls are the air intake vents.

Figure 4.4-6. Hillary and Raymond welcome you into their home on a blustery and cold late-winter day. The outside temperature is approximately 26°F (-3°C), while a toasty 73°F (23°C) greeted us on the inside without the benefit of any heating source except the sun

Figure 4.4-7. Living inside a greenhouse might seem a bit odd, but nothing beats picking fresh salad greens in the middle of winter.

estimates them at a quarter of a million pounds) provides a combination heat sink and energy radiating system. During the daylight hours sunlight is admitted into the home through the south-facing glazing, brightly illuminating the interior (Figure 4.4-9) while the wall structure absorbs the heat energy, preventing the interior space from overheating. During the dark hours the opposite condition occurs and heat energy is reradiated from the walls into the living space.

As a result, the heat demand of this home is relatively low, with the majority of the energy being provided by the building's passive solar envelope and an airtight woodstove (visible at the rear in Figure 4.4-8). Strategically placed propane wall heaters (Figure 4.4-17) provide backup heat and are used to take away any chill on overcast days.

Electrical power for the house is provided by a 600W photovoltaic array (Figures 4.4-2 and 4.4-13) mounted next to the solar glazing on the metal-clad roof. The 24VDC output is wired directly into a 1300-amp-hour battery bank (Figure 4.4-14) which in turn supplies a Xantrex 2500W modified square wave inverter (Figure 4.4-15). The entire home is wired for 120V operation, including the one-half horsepower deep-well pump. A two-stroke gasoline-powered 4000W generator provides battery-charging power during the dark months of November and December.

A tour through the photo gallery makes it very clear that Hillary and Raymond are satisfied with their lifestyle, and they have shown that full-time off-grid living does not have to break the bank.

Figure 4.4-8. The open-concept design and high sloping roofline make the home appear much larger than its 1200-square-foot area suggests.

Figure 4.4-9. The south-facing view forms a magnificent vista overlooking the forest and fields full of deer.

Figure 4.4-10. Raymond shows off their high-efficiency refrigerator and freezer unit. Because the couple wanted to stay within a very tight budget for their off-grid system, choosing very energy-efficient appliances was critical.

Figure 4.4-11. An electric dishwasher was not included in the financial or electrical plans for this home. Hillary demonstrates the most efficient way of getting dishes clean.

Figure 4.4-12. The peaked ceilings and large expanse of glass contained in this home provide a light and majestic feel that belies the fact that the entire south-facing wall is below grade level. Perhaps Tolkien considered this design for his subterranean hobbits.

Figure 4.4-13. A 600W photovoltaic array connected in a 24V series circuit provides approximately 90% of the home's yearly energy consumption. A 4000W gasoline generator provides the balance of energy required during the dark days of November and December.

Figure 4.4-14. Electrical energy is stored in this 1300-amp-hour deep cycle battery bank. A well-sealed, ventilated walk-in closet provides a suitable storage area.

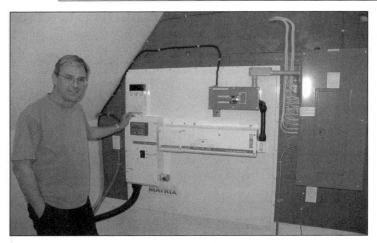

Figure 4.4-15. Raymond checks the electrical power status on his Xantrex inverter board. This prewired system contains all of the necessary components to operate a renewable-energy-based home. The installing electrician simply mounts the panel to the wall and connects it to the battery bank, photovoltaic panel, backup generator, and standard house electrical circuit breaker panel shown at the right of the photograph.

Figure 4.4-16. Electrical wiring within an earthship must be run inside conduit for protection. In this view, the electrical outlet and supply are mounted to the rough-coat masonry covering the tire wall. The finish parging will be applied directly over the electrical wiring conduit. Note the pop cans embedded in the scratch-coat parging. They take up space and reduce the quantity of finish-parging materials required.

Figure 4.4-17. Two propane heating units are installed, one at either end of the house. They are used primarily for backup heating when the couple travels or to take the chill off on those cool fall nights.

4.5

Anderson-Von Mertens Home

Off-Grid in New Hampshire

System Cost:	
Electrical	US$84,000
Thermal Collectors	US$ 7,100
Wood Pellet Boiler	US$ 5,500
Design Charges	US$13,750
Heating System Components	US$27,750
Heating System Plumbing	US$24,300
Gypcrete/Concrete Flooring	US$ 6,678

Figure 4.5-1. Given the difficulties and costs involved in extending the electrical supply across conservation-protected lands, Chris Anderson decided to invest in an off-grid system that was large enough for his family's home and had sufficient capacity to connect a future neighboring house as well. (Courtesy Chris Anderson)

I have to admit to having one of the best research assistants I could ask for. During the planning stages for revising *The Renewable Energy Handbook*, I suggested to Michelle Mather that we should be on the lookout for a very large off-grid system, something almost over the top. Well, she didn't disappoint me when the material for the Anderson-Von Mertens home appeared in my inbox.

A quick bit of addition will tell you that the Anderson-Von Mertens family has invested the price of a nice condo in their generously sized off-grid electrical and heating system. I will admit that as I started to review the data I was puzzled about why the system was so large given the energy-efficient appliances and the small small family size. Clearly there was more than met the eye.

"Anna and I were living in California and had been in discussions with my company, Borrego Solar Systems Inc., to open a branch office in the northeast," Chris Anderson explained. "Multiple stars aligned that made the move a perfect opportunity. Being able to move back to where my wife's family lives helped, as did the opportunity to continue working for my company. The fact that my mother-in-law happened to own property on lands protected by a conservation easement and was willing to give us the property as an incitement to move back was the icing on the cake."

As Chris explained to me, the property was large enough to allow two houses to be built, and one of their friends from California would eventually build a home next door. Given the difficulties and costs involved in extending the electrical supply across the protected lands, Chris decided to invest in an off-grid system that was large enough for his home and would have sufficient capacity to connect the friend's future home as well. "Our site has good solar exposure for part of the day, while the other lot provides a different sun window and might be better suited to a small wind turbine," Chris said. "Connecting the two houses together would make a stronger off-grid system and would probably eliminate the need for any generator operation to augment our energy supply."

Figure 4.5-2. The building lot was cleared of pine trees by an all-woman-owned and operated wood processing company.

Figure 4.5-4. The house is hydronically (hot water) heated using an in-floor heating system comprising plastic (PEX) tubes which are embedded in polished gypcrete, a spray-on concrete-like material. Floors are warm to the touch, with the large mass absorbing heat from the passive solar arrangement and buffering air temperature.

"Being the chief technical officer for Borrego Solar I am always looking at testing and using the technologies we sell," Chris continued. "Building this off-grid home allowed me to use a variety of solar and related equipment while allowing me to work with our marketing team to see if Borrego should enter the off-grid market as an addition to our residential grid-tie and commercial core business."

Once I was made aware of these details, Chris' decisions regarding the size of his energy infrastructure made much more sense.

The lot where the house was to be built was heavily treed with pine, so before construction could begin, a woman-owned and operated milling company was hired to clear the property and harvest and mill the wood which would be used in the construction.

Following frame and rough construction, the energy subsystems were installed. The house is heated using in-floor hydronic systems which require a plastic (PEX) tubing to be installed on the subfloor. A gypcrete layer is then sprayed on

Figure 4.5-3. The trees that were harvested during the clearing operation were cut and milled on-site and used in the building of the home.

top of the piping, encasing the tubing. The mass of the gypcrete acts as the energy storage system for the passive solar properties of the home, working in tandem with the active hydronic heating.

Energy from the seven-panel solar thermal system is collected and stored in a primary 1,200-gallon (4,500 liter) storage tank (Figure 4.5-6). This tank is equipped with a series of copper tubing heat exchangers that allow energy from several sources to be transferred into or out of the water-filled tank. Water is one of the best storage mediums for heat, having a very high specific heat capacity compared to other substances. For example, water will store approximately 4.5 times more thermal energy than concrete (960 joules per kilogram • °C for concrete versus 4,200 joules per kilogram • °C for water)[1].

Solar thermal energy is directed to one of the submerged heat exchange coils and energy from a wood pellet boiler uses a second coil. A third coil takes heat from the storage tank and directs it to the heat distribution plumbing system located directly behind the water storage tank. Hydronic heating is used on all three levels of the house.

The maximum operating temperature of the 1200-gallon (4543 liter) storage tank is 165°F (74°C) which is the limit imposed by the plastic lining. When the main storage tank reaches its operating temperature limit, hot water is pumped to a second "overflow" tank with a capacity of 738 gallons (2800 liters).

Chris mentioned to me that "placing the wood pellet boiler in the basement was a bit of a mistake as we had to move approximately three tons of pellets down the stairs. The following year I added a hopper and auger system which moves pellets from the garage to the basement, so make sure that readers interested in installing one of the units understands this." Point taken.

Although the home is 2,850 square feet (265 m²) in size, the energy-efficient design and solar thermal system help to offset heating costs. Chris indicated that they burn approximately three tons (3,000 kg) of pellets per year at a cost of $220 per ton.

Figure 4.5-5. The installation crew has just put the finishing touches on the seven-panel solar thermal collectors located on the south-facing side of the roof.

Figure 4.5-6. The energy from the solar thermal collectors is captured in an antifreeze mixture and pumped into one of the heat exchange coils submerged in the insulated water storage tank.

Electricity is generated using a photovoltaic array of 42 panels mounted on the south-facing roof of the house. The array comprises 27 Sharp 160W and 15 Sharp 167W panels, for a whopping total of 6,825W rated capacity.

A 12kW Kohler propane generator is also available for battery charging and to supply power directly to the home in the event of a PV system failure.

The electricity produced by the PV system is stored in a Rolls brand (Surrette in Canada) battery bank comprising 24 model S4-60 batteries rated 350Ah each. The battery bank gross rating is 1,500Ah at 48V and provides between 22 and 23kWh of usable electricity.

Figure 4.5-7. The heat storage tank is shown filled with water with the insulated cover in place. Antifreeze fluid is pumped from the third heat exchanger submerged in the tank and directed to the heat distribution system located just beside Chris' right shoulder. From here, heated fluid is pumped throughout the house hydronic heating system as required.

Two Sunny Boy model 4248U inverters provide the home with grid-standard 120/240VAC supply, with a maximum continuous rating of 4,200W each.

Figure 4.5-8. Electricity is stored in a battery bank comprising 24 individual batteries rated 350Ah each, for a total capacity of 1,500Ah at 48V and providing approximately 22kWh of usable electricity. The Sunny Boy model 4248U inverters seen to the right of the battery bank generate industry-standard 120/240VAC to power the home.

This photo montage of the home and grounds demonstrates how well Chris has designed his system to provide a no-compromise lifestyle.

Figure 4.5-9. A total of 42 PV panels providing a total of 6,825W are installed on the south-facing roof of the house. The arrangement neatly threads the solar thermal and electric systems together.

Endnotes

[1] Specific Heat is the amount of energy that is required to raise the temperature of a given mass of substance by a given temperature interval. For example, it requires 4186 joules of energy to raise the mass of 1 kilogram of water 1 degree Kelvin. Note that 1 degree Kelvin is equal to 1 degree Celsius. The correct syntax for this measure is: $c = 4186 \ J/(kg \cdot K)$

4.6

The Kemps' Zero-Carbon Home

Sustainable Off-Grid Living

System Cost:
Electrical   US$31,000.00
Thermal   US$ 4,600.00

Figure 4.6-1. Designing a traditional country-style farmhouse while integrating passive solar thermal technology can be a challenge. The large roof overhang (also visible in Figure 4.6-2) forms the front porch and provides a means of shading the house from direct sunlight in the summertime while allowing the low-angle winter sun to penetrate into the building envelope.

There is no point in arguing with a woman who has made up her mind. When the time came for Lorraine and me to talk about selling our old Victorian house we had pretty much made up our minds that the new home would be sustainably built and energy efficient and would look as if it had been built a hundred years ago. That much we could agree upon. Where the difficulty lay was in choosing the building lot.

I was aiming for a nice quiet cottage-style lot on a river. Lorraine had another plan, and she dragged me kicking and screaming to a lot she already owned. Now, a normal person would have jumped at the opportunity to avoid purchasing land. My holdup was the fact that Lorraine is the "baby" of eleven in her family and the lot in question was on the back end of the family farm.

I have one sister. As Lorraine and I discussed matters with the building inspector I realized that this location would have me within a stone's throw of 50+ nieces and nephews and assorted in-laws who would want to visit the horses, borrow a cup of sugar, and stay for the afternoon.

Fifteen years later, I have come to enjoy this sort of community living arrangement. This was reinforced during the ice storm of 1993 (grid power was down for 13 days) and the blackout of 2003 (we were eating ice cream at the time and didn't find out until someone phoned). On both occasions, it was nice to be able to offer a hot meal or coffee while acting as a hub in the community.

But I digress. Our home was built to look like a traditional country farmhouse from the early 1900s. We had a pretty good idea of the general layout and, working with our architect, we were able to develop a home that required the least amount of building material while giving us the greatest square footage. We decided on a two-story design with a completely finished basement. This made it possible for the house to be built around a catalytic wood stove which permits heat to circulate naturally throughout the home without the need for energy-robbing circulating fans.

Although many of the building techniques we used in 1993 were cutting edge at the time, several of these materials and designs have worked their way into the mainstream. For example, ultra-strong composite flooring trusses were used to span the 28-ft (8.5 m) width of the living room without the need for support posts or beams. Although the "particle board" trusses were unusual at the time (and caused a lot of consternation with our building inspector) they're quite common

Figure 4.6-2. Lorraine and Bill Kemp along with Shadow and Cedar enjoy a pretty standard lifestyle while powering their home with renewable energy from the sun and wind.

now. (Come to think of it, we caused our poor building inspector a lot of anxiety as we seemed to keep bending the rule book like an Indian yoga instructor).

Designing a traditional country-farmhouse-style of home while integrating passive solar thermal technology was a challenge. The large roof overhangs visible in Figures 4.6-1 and 4.6-2 form the front porch and provide a means of shading the house from direct sunlight in the summertime while allowing the low-angle winter sun to penetrate into the building envelope.

Living life off grid is definitely not the same as living life without power.

In keeping with our sustainability requirements, we opted for a siding material made from wood scraps which were formed into an artificial clapboard siding. The house is insulated primarily with blown-in cellulose which is manufactured from recycled waste-paper products. Areas that were difficult to insulate were tackled with spray foam (icynene) insulation, which also forms its own vapor barrier.

Interior wood trim and flooring were harvested from locally grown forests and processed at local sawmills. In addition, local tradesmen and subcontractors were used in an effort to keep our money in the local economy.

The domestic water system is based on a standard drilled well with a deep-well submersible pump and a large water pressure accumulator tank to minimize pump cycling. Low-flow and ultra-low flow appliances were selected to keep our water consumption well below half the national average of 91 gallons (345 liters) per person per day. Using natural and phosphate-free cleaners and a septic tank with an effluent filter and leaching bed allows waste water to percolate through the earth and right back into the water table to become usable again.

Rainwater is collected and automatically fed through drip irrigation to perennial flower beds. When grass trimming is called for, a solar-charged, battery-powered 24V electric mower is used. This mower works symbiotically with the seasons: more sun in the summer permits charging without concern about energy consumption. When we have heavy cutting work or fertilizing to do, we put a call out to the horses and they're more than happy to oblige. (Surprisingly, the horses trim around the wild and planted flowers, which saves me a whole lot of time trimming.)

Although energy efficiency and being off grid were front and center in our list of wishes, we did not want to sacrifice our lifestyle. We carefully assessed the electrical load of all the appliances required in our day-to-day life. We then investigated the equipment necessary to generate the electrical power required and prepared a financial budget.

The electrical generating equipment originally comprised a photovoltaic array (Figure 4.6-20) with a peak electrical rating of 1200W made from an array of 16 individual panels rated 75W each. As we moved away from propane-powered appliances, our electrical requirements increased somewhat, necessitating the addition of two 165W panels, mounted along the top. The manufacturer's rating is based on an optimum amount of light falling on the panels, but in reality the sun's energy and thus electrical output always fall below this ideal rating. On a clear, bright sunny day our 1530W array will output approximately 1200W, and a bit more in winter

The array is mounted on a solar tracking device which follows the sun on its daily path from east to west. The tracker is also equipped with an automatic elevation device which causes the panels to change angle as a function of the seasons. During the winter months, the photovoltaic array is oriented almost perpendicular to the ground (vertical), while during the summer months the panel will lie almost flat as the sun approaches solar noon.

In addition to the photovoltaic array, we have a Bergey 1500W wind turbine (Figure 4.6-24) mounted on a 100-ft (30 m) guyed lattice tower. Completing the electrical generating mix is a 10kW diesel generator (in the building shown in Figure 4.6-23) which is fueled with a mixture of between 30% and 100% biodiesel, depending on ambient temperature. We would have preferred to use 100% biodiesel at all times, but in this area temperatures can reach as low as -22°F (-30°C), making such a choice impossible. (For more information regarding biodiesel fuel see Chapter 16.)

On a yearly basis our photovoltaic panels provide approximately 80% of our total electricity requirements, while the wind turbine provides 15% and the backup generator provides the remaining 5%. Because of local weather conditions, neither the wind turbine nor the photovoltaic panels provide significant output during the month of November and part of December. It is during this approximately six-week period that the backup generator becomes a necessity.

The relatively low energy output of the wind turbine is due to the area having a fairly low annual wind energy rating. We knew that the wind turbine would not supply a significant percentage of our yearly energy requirements, but it does reduce the generator run time considerably. We estimate

Figure 4.6-3. Bill and Cedar offer a tour of their home. Anyone knowing Bill realizes that the first stop on any tour is in the kitchen for a ubiquitous cappuccino.

Figure 4.6-4. Although energy efficiency is doubly important in off-grid home designs, living a spartan lifestyle isn't necessary. Balancing energy production and demand is important in developing a successful off-grid system. And yes, there is even room for a cappuccino machine.

Figure 4.6-5. The house plan calls for plenty of open space, allowing both air and light to circulate freely. An efficient floor plan reduces construction materials and cost.

Figure 4.6-6. Major heat-producing appliances such as this cookstove are normally powered by propane gas. The small microwave oven (on top of the cookstove) is used occasionally for reheating and defrosting. A larger model could have been chosen had there been a desire to use one more regularly.

that generator operating time has been reduced by approximately 300 hours per year owing to the installation of the wind turbine, resulting in reduced fuel consumption, lower maintenance costs, and decreased wear.

Electrical energy from all of the electrical sources is fed into a very large battery bank (Figure 4.6-22) with a gross capacity of approximately 3,500 amp-hours, providing approximately six days of power without any input from the renewable resources. Our battery capacity was increased from 1000 Ah to 3,500 Ah during the conversion from propane to high-efficiency electrical appliances, which will be discussed in the zero-carbon section following. The control system is designed to automatically start the generator and recharge the batteries should they reach a predetermined low-power level of approximately 50% depth of discharge. Low-voltage power from the batteries is fed to an inverter bank with a total output capacity of 6kW, which in turn supplies household electrical needs.

Figure 4.6-7. Baking bread in the old cookstove may be reminiscent of grandma, but with free electricity and the relative efficiency of automatic bread makers, along with the requirement for less elbow grease, Lorraine prefers the modern approach. A food processor, blender, and standard Sears refrigerator complete the mix of kitchen appliances.

Figure 4.6-8. African violets love diffuse, indirect light. Thanks to the layout of the roof overhangs, late spring, summer, and early fall sunlight never comes directly into the house. Conversely, the sun's low winter angle shines deeply into the house interior, providing welcome warmth.

Figure 4.6-9. The living room is the center of the audio and visual system. A home theater system with large screen TV, CD, DVD, tape, surround sound amplifier, powered subwoofer, and satellite system works perfectly with today's sine wave inverter power.

Figure 4.6-10. A satellite dish is one way to stay connected in the woods, providing TV as well as high-speed Internet links to the urban world and beyond.

Figure 4.6-11. Household chores need not require muscle power. A standard central vacuum system with powered "beater bar" forms part of the household mix of appliances.

Figure 4.6-12. The Staber washing machine is the most efficient model in North America, although many front-loading machines are quickly catching up.

Figure 4.6-14. The horses require pressurized water, supplied by the house pressure system, as well as lights and even their own stereo.

Figure 4.6-13. Commuting 20 minutes each way to the "local" gym is no way to live sustainably. It's far better to have one in the house and invite friends to stop by. A mix of equipment such as a treadmill, elliptical trainer, stereo, and TV are all supported by the off-grid system. (No, running on the treadmill does not charge the batteries. Sorry!)

Figure 4.6-15. With Lorraine doing all that work in the previous pictures, she has earned the right to relax in the hot tub, in water which is heated and circulated using the power of the sun.

Figure 4.6-16. Bill's office has all of the devices one would expect. A computer, printer, fax machine, and photocopier are fairly standard fare. In addition, some not-so-normal equipment such as electronic test and measurement equipment and large-scale drafting machines all operate off grid.

Figure 4.6-17. We expected the library sitting area to be a bit dark, so a SunPipe (see Chapter 2.4) was installed to brighten things up a bit. This unit provides the equivalent light output of a 400W lamp without the heat loss of a typical skylight and with no need for electrical power.

Figure 4.6-18. This view of the living room and staircase shows how the heat from the catalytic wood stove can circulate through the upper and lower floors. Heat rising up the wide staircase displaces cooler air, which flows down the stairs, forming a continuous loop. Air will continue to loop until thermal equilibrium is reached between the two floors.

Figure 4.6-19. This propane-powered, freestanding stove provides a cozy setting for dinner parties. It is also the backup heat source (furnace) when the wood stove is not used or during extended traveling. Many people cannot believe that this little unit is capable of heating the entire house, but it is! Over the next year or so we plan to rebuild this unit to operate on biodiesel fuel.

Figure 4.6-20. Eighty percent of the home's electrical energy is produced by this photovoltaic (PV) array and sun-tracking device. The panel generates approximately 1,200W of power for as long as the sun shines.

Figure 4.6-21. With Bill leaning on the PV array, you get a pretty good idea of the size of these units. This one is approximately 8 ft high by 16 ft long (2.4 m x 4.8 m).

Figure 4.6-22. The electricity provided by the deep cycle battery bank is what keeps the house running when the sun sets and the wind stops. There is sufficient energy storage in this battery to power the house for approximately six full days without any input from the PV array or wind turbine.

Figure 4.6-23. When the battery bank shown in Figure 4.6-22 is discharged to a certain level, the diesel generator housed in this outbuilding will start, quickly charging it back up. This generator is operated on a mixture containing biodiesel, which is a renewable, low-emission fuel.

Figure 4.6-24. This wind turbine is rated 1,500W and will generate power in non-linear relationship with wind speed. Simply stated, wind turbines are not for every location, and considerable study of Chapter 7 is recommended before you purchase and install one. This unit provides 15% of the total energy requirements of the house.

Developing a Zero-Carbon Lifestyle

Approximately one-third of a home's total energy requirement is used to provide domestic hot water. In our case, the requirement for hot water is increased because Lorraine and I love nothing more than to soak in the hot tub while listening to the loons calling on the lake. The most common fuel of choice (or necessity) for the off-grid home in North America is propane, but not wanting the resulting greenhouse gas emissions to kill off these lovely birds or the rest of the planet, we have worked aggressively to reduce our home's fossil fuel consumption to zero.

Although I am often chastised for saying this, *most people don't live off grid, they live on propane.* If you think about it for a moment, the reason is obvious. Electrical or plug load energy and wood-fired space heating are only part of the mix of energy that is required in a home. Most off-grid homes rely on propane gas for appliances with large heat demands such as a hot water heater, clothes dryer, cooking stove, and backup space heat source.

Relatively scarce and expensive-to-produce electricity is reserved for the lighter loads such as water pumping, refrigeration, lighting, and optional goodies such as audio/video equipment and computers.

Although propane is one of the lowest-greenhouse-gas-emitting fossil fuels, it is nevertheless a non-renewable resource. Always up for the challenge, Lorraine and I were determined to reduce our consumption of propane to zero and have worked diligently at developing both the lifestyle and technology to make it happen. (Before I go much further, I will admit that we

Current Propane Appliance or Function	Proposed Alternate Appliance	New-Renewable Energy Source
Teakettle	Electric Kettle Teakettle on Woodstove	Electricity and Wood
Stove	Induction Cooktop Woodstove Cooktop	Electricity and Wood
Oven	Crock Pot	Electricity
Oven	Dutch Oven on Woodstove	Wood
Hot Water	Solar Thermal / Wood Heat Recovery Unit	Solar and Wood
Clothes Dryer	Hydronic System Clothesline	Solar and Wood
Basement Heating	Hydronic System	Solar and Wood
Bread Baking	Bread Machine Wood Oven	Electricity and Wood
Hot Tub	Super-Insulated Hot Tub	Solar Thermal and Wood
Back-Up Heating	Liquid-Fueled Heater	Biodiesel (not yet completed)

Table 4.6-1. This table shows all of the appliances that off-grid homeowners typically power with propane or other fossil fuel. The author has successfully moved all of these items with the exception of backup heating to alternate, renewable fuel sources as noted.

have been lucky and have not had to travel during the winter since we switched to a zero-carbon-fuel diet. However, we still maintain a propane backup space and water heating system for travel times or if friends use our home while we are away. Stay tuned. I have every expectation of being able to use biodiesel for these items and hope to explain my results in the next edition or on the website at www.aztext.com.)

In order to reduce our propane fuel load, I assessed our energy consumption and made a cross-reference list of alternative appliances that could minimize our fossil-fuel consumption. The list can be found in Table 4.6-1.

Although it may seem contradictory to state on the one hand that big heat-producing items must be moved to propane and then to immediately move them back to electricity, there is a method to my madness. A quick look at the propane devices reveals them to be fairly standard items with very high-energy demand. For example, the teakettle is heated on the gas stove; the oven, which is large enough for a 20-pound turkey, must be pre-heated even if you are baking a dozen muffins. The hot water tank, clothes dryer, and space heater are obviously all big energy consumers.

This list shows what alternative energy sources can be used and if there is an energy-efficient way of doing the *required* job. Replacing the propane-heated teakettle with an electric model was easy. For fifteen dollars, we purchased a nice stainless steel model that should last a lifetime while using a reasonable amount of electrical energy. How reasonable? To find out, I measured both the time and the power required to complete the task in order to determine the energy demand and drain on our electrical system:

1350W x 6 minutes per liter to boil
= 135 watt hours or 0.135 kWh

This is equal to approximately 2% of our total daily energy consumption and does not seriously contribute to total electrical power demand.

There are some very neat ways of attacking the rest of the items on the list. For example,

two appliances, using two fuel sources, electricity primarily in the sunny summer and wood in the colder months of the year, replaced the propane stovetop.

Common in Europe, the induction cooktop unit (Figure 4.6-25) is a very energy-efficient

> *"Most people don't live off grid, they live on propane."*

appliance. Looking more like a piece of high-end stereo equipment from Bang & Olufsen® than a cooking appliance, its top surface is nothing more than a tempered glass plate. Turn the unit on high and it will not get hot to the touch. However, as soon as a pot or frying pan is set in place, electromagnetic energy causes the pan to immediately begin heating.

The beauty of this unit is that the majority of the thermal energy remains in the food, while a similar gas stove emits over 70% of the generated heat into the air. The induction unit is very handy during the summer months when there is lots of solar energy to burn and little impetus to heat your house. The "two-burner" unit pictured here costs less than $200.

Figure 4.6-25. Common in Europe, the induction cooktop unit is very energy efficient, looking more like a piece of high-end stereo equipment from Bang & Olufsen® than a cooking appliance.

Figure 4.6-26. This wood-fired cookstove unit is manufactured by Heartland Appliances (www.heartlandapp.com) and is equipped with a water heat recovery unit. The cookstove can therefore do triple duty during the cooler months: heat the house, cook food, and provide hot water. This cookstove replaced the propane model shown in Figure 4.6-6.

Figure 4.6-27. From left to right, a countertop convection oven, microwave oven, and induction stovetop provide several highly efficient electrically powered cooking options. The 18.5 cubic-foot refrigerator/freezer unit is one of the most energy-efficient models on the market, consuming less than 1 kWh per day, yet costs less than $350 at many retailers.

The old-fashioned wood cookstove shown in Figure 4.6-26 is not a restored piece from a junkyard but is in fact an updated and airtight replica of grandma's traditional model. This unit is manufactured by Heartland Appliances and can be equipped with a water heat-recovery unit that I have tied into the solar thermal water heating system. The cookstove can therefore do triple duty during the cooler months: heat the house, cook our food, and provide hot water. (See Chapter 5.6 for more details on hot water heat recovery systems.)

Other often-overlooked energy-efficient appliances can be found in storage sheds, basements, and attics—remnants of wedding and Christmas presents long since forgotten. The stovetop Dutch oven, rice steamer, crock pot, and breadmaker are all application-specific appliances that consume relatively little electrical energy and can easily replace propane-fueled appliances.

The propane oven was easily replaced with a combination of options. During the cooler months, the wood cookstove is always ready, nicely preheated to accept whatever you're preparing. Ready for duty during the summer months or to provide more cooking capacity in the winter are the tabletop microwave and the convection oven, both of which are efficient enough for off-grid homes (Figure 4.6-27).

You may be interested to know that I volunteered to cook the Christmas turkey using the wood stove unit last year. Everyone thought I was nuts except for Lorraine's Mom, whom I affectionately refer to as "Mrs. T." I told Lorraine that there was no chance I could screw this up and reminded her that her Mom must have cooked a hundred turkeys this way. Sure enough, the turkey was cooked perfectly, and with six "burners," a warming oven above, a convection stove, and a microwave all doing their duty, we finished cooking Christmas dinner about an hour earlier than usual.

Domestic hot water heating consumes a large amount of energy and I decided to develop a hybrid system that would use the ample summer sun when it was available while drawing additional energy from our two wood stoves (one being the space-heating stove shown in Figure 4.6-18 and the other our kitchen cookstove described above). This system works beautifully, providing plenty of guilt-free hot water for almost endless showers and the hot tub.

I chose to install a vacuum-tube-based solar thermal collection system (Figure 4.6-28) rather than a flat plate system after calculating that we would need to store a large volume of very high-temperature hot water whereas more typical systems deliver water close to the desired usage temperature.

The solar thermal system produces approximately 60% of our yearly hot water production, with the wood-fired system providing the balance. In most regions of North America it is possible to reduce domestic hot water heating fuel requirements by at least 60% through the installation of a solar thermal system.

Because the wintertime wood-based heat recovery system produces so much energy, I determined that it would be possible to direct some of this heat to other functions using a hydronic system. Hydronic heating is simply the transfer of thermal energy using water. By installing a small electrically operated control valve and pump to a hydronic loop, it is possible to transfer heat from the storage tanks to water baseboard heaters located in areas that need additional energy. In our house, the basement is super-insulated and remains at approximately 60°F (16°C) all year round. Not bad for summer but a bit too cool in winter. Allowing a thermostat to activate the pump and "zone valve" causes hot water to flow through the baseboard-heating units as required.

Likewise, the propane dryer was modified with a similar pump and zone valve arrangement, with the hot water being pumped through a small radiator installed in the air intake of the dryer. The pump and valve are connected in parallel with the drum drive motor of the dryer, causing hot water to flow whenever the drum is active. Selecting "air dry cycle" causes hydronically heated air to blow

Figure 4.6-28. Solar thermal hot water heating systems such as this vacuum-tube-based system are discussed in Chapter 5. The solar thermal array provides approximately 60% of the Kemps' domestic hot water requirements as well as providing energy to heat their hot tub.

Figure 4.6-29. A solar cooking unit such as this model completely eliminates the need for fossil-fuel energy for cooking, replacing it with free sun energy. There is nothing to be afraid of; these models work in much the same manner as a tabletop convention oven or slow cooker.

through the clothes, drying them in about twice the time of propane. Although the drying cycle is longer the energy used is free, and if you are only doing a load or two a day, the extra time makes no difference whatsoever. Of course, the clothesline is also available whenever the weather cooperates.

For more detailed information regarding this system see Chapter 5.6 Heating With Renewable Fuels.

On the advice of a friend, we have just begun to add solar cooking into the mix (see Figure 4.6-29). Later this summer I plan to add a rack to the PV tracker to hold the solar cooker. The PV tracker will ensure that the solar cooker remains perfectly aligned with the sun. Although Lorraine's desire is to actually do serious cooking with the unit, I am excited about the prospect of not overheating the house and running our air conditioning no longer than necessary.

Over the years we have had many visitors, some of whom come over for the first time at night, when the wind turbine and PV panels are hidden in the darkness. The comments are always the same. "I thought you didn't have any power out here," people say, as they wonder how the automatic yard lights and electric garage door openers can possibly function. Of course I enjoy this little game, showing off a pretty standard array of appliances crammed into our typical country home. Living life off grid is definitely not the same as living life without power.

Figure 4.6-30. PV-powered electric fencing protects approximately 25 acres of horse paddock area. The combination of PV- and rechargeable-battery-powered electric fencers makes great environmental and economic sense, eliminating the cost and waste of disposable batteries.

4.7

The Miller-Cameron Home

Upgrading an Urban Century Home

System Cost:

Solar Electrical	**US$14,000**
Solar Thermal	**US$5,000**
Grey Water Purifier	**US$2,800**
Drain-Powerpipe	**US$900**

The Miller Cameron family live in the beautiful lakeside city of Kingston, Ontario, where they made the decision to upgrade their turn-of-the-century brick home and create a comfortable and sustainable showcase.

Rob and Cindy made a list of projects they wished to undertake during a needed renovation to ensure that the house could accommodate them and two growing children. Rob explains that, "the house required renovation and with various government rebates and incentives available, this was the time to get the job done."

Rob's list was quite extensive, but he was quick to point out that updating the home was more financially viable than purchasing another and that many of the updates he was planning would provide mid- and long-term return on investment. It did not hurt matters any that the

Figure 4.7-1. The Miller 1½-storey 1,500-square-foot home was gutted to complete the required updates, which included the addition of a 600-square-foot insulated concrete form (ICF) addition.

Figure 4.7-2. ICF construction provides an insulation value of between R-45 and R-50, which is double that required for conventional home construction. Long-term thinking provided the Miller-Camerons with a healthy return on investment as a result of lower heating and air conditioning costs, and it also gave them a more comfortable home.

Figure 4.7-3. The home was completely gutted to allow for improved insulation as well as new doors and windows.

home had been in the extended family for three generations and was available at the right time.

The 1½-storey 1,500-square-foot home was gutted to complete the required updates, which included the addition of a 600-square-foot insulated concrete form (ICF) addition with an insulating level of between R-45 and R-50, which is unheard of in traditional home construction. In addition, the family completed the following list of updates to their home:
- new energy-efficient windows and doors;
- urethane spray foam insulation in old 2" x 4" (substandard) walls to increase insulation rating to R-30;
- high-efficiency natural gas furnace;
- 1,360W grid-connected photovoltaic array;
- two-panel solar hot water heating system;

- instantaneous modulating hot water heater;
- Power-Pipe waste water heat exchanger;
- sun tube natural light pipe;
- compact fluorescent lamps;
- high-efficiency Pacific Energy wood-burning stove;
- Brac Greywater recycling system.

Rob is quick to explain that even at a tariff rate of 42 cents per kWh he is not going to make a quick buck. "The installed cost of a grid-connected system is approximately $1,000 per watt or about $14,000 for our mid-sized system," Rob says. "I expect to generate between 1,600 and 1,700 kWh per year of energy, so it will take approximately 18 years for a simple payback. However, this is about the same as receiving 5.5% on your investments, which after the debacle of 2008 is not looking too bad!"

Of course hindsight is 20/20, and if Rob had waited another two years he might have been able to enjoy a doubling of both the rate of income for his green energy and his return on investment. In any case, the ROI is positive, which is much more than can be said for granite countertops.

Another popular yet flawed green-energy plan is known as net metering. In this program, only one meter is required, which measures energy flow both into and out of the house. The concept was developed by well-meaning bureaucrats to create a simple green-energy program on the assumption that a small RE system could offset the energy consumed by the homeowner—in other words, the amount of energy generated by the PV system would equal the amount of energy the homeowner consumed.

Such a program assumes that the clean energy generated by the PV system is worth the same as dirty coal-generated electricity or nuclear power with its garbage stream that must be babysat and paid for for millions of years. This is another example of government logic—or lack thereof.

For those considering a grid-connected renewable energy system, stay away from net metering tariff programs unless you are given the

Figure 4.7-4a, b, c, and d. The home is shown going through its metamorphosis from a drafty century-old house to a modern, sustainable home.

power-generating equipment for free; otherwise your ROI will be zero.

The Miller-Camerons have done an amazing job of integrating a number of advanced renewable energy technologies into their urban home. Although the underlying house may be over 100 years old, the advanced technologies demonstrate how all modern homes should be constructed with an eye to the future. Underscoring this point is the fact that the family recently won a Liveable City Award in the environment and sustainability category.

Figure 4.7-7. Most jurisdictions require that you install two meters if you are selling renewable energy to the electrical distribution grid. One meter is required to purchase energy; the other is for the sale of energy to the grid. Having two meters allows homeowners to purchase energy at one rate while selling it at a higher, green-energy tariff. For example, Rob Miller pays approximately 12 cents CAD per kWh for electricity to run the house while selling his PV electricity to the grid for 42 cents CAD per kWh .

Figure 4.7-5. During the home renovation, the Miller Cameron family is seen in front of the house with the 1,360W photovoltaic electrical array in the background. Connecting micro-scale renewable energy systems to the distribution grid is a relatively new experience in North America. Homeowners in the province of Ontario can sell electricity to the distribution grid at a rate of 42 cents CAD per kWh. This rate is currently being revised to 83 cents CAD. This higher rate will place Ontario on a similar footing with world green-energy leader Germany.

Figure 4.7-6a and b. Many electrical distribution companies throughout North America are learning about renewable energy right along with their homeowner customers. A qualified electrical contractor and equipment dealer are worth their weight in gold. Above left, Rick Rooney of Quantum Energy installs the inverter.

Figure 4.7-8. A solar thermal heating system is installed on a south-facing roof. The two flat panel units supplied by EnerWorks provide approximately 60% of the family's hot water heating requirements.

Figure 4.7-9. The EnerWorks solar water heating system comprises a small pump and control system that mounts to the outside of the hot water storage tank. In this closed-loop active system, a mixture of propylene glycol antifreeze and distilled water is circulated between the solar collectors and the solar thermal tank water heat exchanger. Rob indicates that the ROI for this system is about 14%, making it an excellent investment.

Figure 4.7-10 Hot water from the solar thermal system is fed into this high-efficiency on-demand hot water heater. Because the Bosch Aquastar model 125BS uses modulating technology, it will automatically adjust burner demand based on water inlet temperature. Therefore, on sunny days the water heater may not even come on. ROI for this model is expected to be approximately 20%.

Figure 4.7-11. A Pacific Energy advanced combustion wood stove was installed in a central location, lending warmth and beauty to the home.

Figure 4.7-12. A Brac Greywater recycling system (www.bracsystems.com) receives (grey) waste water from sinks, washing machines, bathtubs, and even air conditioning condensate, clarifies it, and feeds it to bathroom toilets. As approximately one-third of the water used in houses is for toilet flushing, an immediate 30% reduction in water consumption and sewage treatment can be expected. Based on the marginal charges the Miller Camerons pay for water and sewer, they expect to receive a 13% ROI on this unit. Obviously these units should be mandatory for all new construction and wherever water systems are under strain.

4.8

Surya House

The Dunkley-Fisher Natural Living Home

System Cost:

Electrical	n/a
Solar Thermal	US$4,400
Masonry Stove	US$8,100
Stove Installation/Options	US$3,400

The first thing one notices upon entering Jill Dunkley and Andy Fisher's home is the light. The house is infused with the same feeling of light and color as if you were still outdoors, wandering in the meadows of wildflowers that surround the house. Perhaps this is intentional given the name—Surya House. "'Surya' means 'Sun' in Sanskrit, the ancient language of the Yogis," Jill explains.

Jill and Andy built Surya House in 2007, constructing the home with care and respect for the earth on which it sits. Using sustainable building materials and incorporating energy- and resource-saving technologies such as super-insulation, passive solar heating, water-saving technology, drain water heat recovery (Power-Pipe), and rainwater capture for the gardens, the couple has developed a home that truly meets its designer/builder's goals.

Jill and Andy work from their home, he as a psychotherapist and Jill as a Yoga therapist and instructor. Both believe that the physical and spiritual environment in which they practice augments their work and ongoing research. It is clear that every effort has been made to ensure that this sense of comfort and tranquillity radiates throughout the home, informing even the technologies that make the house so unique.

The house is constructed using super-insulation techniques to ensure the lowest possible energy consumption. This is evidenced by the fact

Figure 4.8-1. Jill Dunkley and Andy Fisher's Surya House is an oasis of tranquillity nestled in the forests of Eastern Ontario. (All photographs courtesy Beth Girdler)

Figure 4.8-2. The ultra-high-efficiency house is infused with the same feeling of light and color as if you were still outdoors.

Figure 4.8-3.

Figure 4.8-4.

Figure 4.8-5.

that the house requires only one cord of firewood per year to maintain a comfortable temperature. Perhaps the most visible piece of "technology" in the home is the massive masonry stove located in the open-concept central living area. Andy says: "These beautiful heaters operate very differently from typical woodstoves. Rather than attempting to control the rate of combustion to extend firing time, the masonry stove requires fast, hot fires to heat the thermal mass of the stove. A complex flue system slows the passage of smoke through the stove, ensuring excellent heat transfer to the rock. After the fire has burned out, the stove continues to radiate heat for hours. You can even cook with this unit after the fire has extinguished."

Figure 4.8-6. The masonry stove is the centrepiece of the home and provides a division between the eating, sleeping, and living areas. Masonry stoves provide hours of even heating long after the fire has burned down. They are also environmentally friendly, as the oxygen-rich fire creates little smoke and particulate matter.

Figure 4.8-7. A traditional airtight parlor stove located in the lower living area provides a touch of heat for yoga classes.

Figure 4.8-8. This view of the south-facing side of Surya House shows the windows that act as solar collectors and the large roof overhangs that act as sunshades during the late spring, summer, and early fall periods. The vacuum tube solar thermal collector can also be seen in this view.

Although the home can be heated with wood, the combination of super-insulation and passive solar heating are the main reasons that the house thermal systems work so well. "We oriented the house with the long side facing solar south to capture the heat of the sun," said Andy. "Roof overhangs have been designed to shade the windows during the summer months yet provide maximum solar gain during the fall, winter, and spring. The acid-stained concrete floors absorb this solar energy, preventing the house from overheating and allowing the sun's warmth to radiate slowly during the day and evening."

This last point is a very important one and concerns a critical component of any passive solar design. Without the ability to store and "buffer" the sun's energy, the living space will quickly overheat on sunny days and become cold on cloudy days and during night hours. The innovative use of stained concrete (Figure 4.8-7) provides a tough yet cozy flooring material.

A solar thermal system, water-saving appliances, a rainwater-harvesting system, a root cellar, a large vegetable garden, and an adherence to permaculture principles all serve to provide additional sustainable amenities for the house. Future plans include nurturing honeybees and creating an edible forest garden. Jill and Andy are rightfully proud of their achievement in living lightly on our delicate planet.

Figure 4.8-10. Jill and Andy are rightfully proud of their achievement in living lightly on our delicate planet. Surya House is a testament to their commitment.

4.9

Earth Energy Case Study

The Goodman-Leblanc Home

System Cost:
Geothermal System:	US$23,076
Actual Cost:	US$10,877
	(see text)

Figure 4.9-1. The Goodman-Leblanc family lives in a fairly typical 2,200-square-foot, ten-year-old house in an area where winters are cold and summers are hot and muggy. In short, this is one of the best places to test the performance and financial claims of a geothermal system. (Photographs courtesy Jeff Goodman and Lynne Leblanc)

Jeff Goodman and Lynne Leblanc are probably the very best people I know for showcasing geothermal technology to a skeptical public. Their home is of fairly typical construction dating back about ten years, covers 2,200 square feet of living space, and is located in Eastern Ontario, an area that is known for its bone-chilling winters and hot, muggy summers. If you were to ask where to put a geothermal system through its paces, this is the place.

But that is only part of the story. Jeff is, ummm, how to put this politely, a bit "anal" when it comes to record-keeping and calculating the investment benefits of just about anything he buys. I suppose this is not surprising, considering that he is an environmental scientist. Lynne, on the other hand, is involved in marketing, and after seeing the results of their geothermal investment she is now working for the geothermal firm and selling these units to other families. You might say they have a bias, but one they acquired only after the system proved both its capabilities and its return on investment.

Shortly after Jeff and Lynne moved into the new house, they (naturally) started keeping track of the cost of running their oil furnace ($1,700+ for the first year) and maintaining it. The home did not have air conditioning at the time and an oil-fired water-heating tank provided domestic hot water.

"It was pretty easy to calculate the mid-term operating and maintenance expenses for space heating and hot water," Jeff explains. "I did some quick calculations and realized that we should be able to get a positive return on investment by installing a closed-loop geothermal system. Not to mention that we would receive the added benefit of having air conditioning and reducing our environmental impact by removing oil from our energy diet. The total greenhouse gas emission reduction was four tons a year or the equivalent or removing four cars from the roads."

I asked Jeff to provide a "few notes of fact" after our numerous discussions and, true to form, I received an entire story line including all the financial mathematics for the project from start to finish. Here's Jeff's story:

"As first-time homeowners we had little equity in the home and the interest rate on our mortgage reflected what the banks typically offer in this situation. We made a choice to pay an extra $100 per month against the mortgage so we could get into a better rate in a shorter time.

At the end of the first year we totaled the

receipts and realized that we'd spent $1,707.79 for oil to heat our home and our water, including hot water tank rental fees. With various low-cost energy-efficient upgrades to the home (CF bulbs, water-saving faucets) and careful attention to electricity use (elimination of phantom loads, etc.) we kept our cost of electricity for the year to $688.82 for a total of $2,396.61 for utilities. All figures include taxes and represent the out-of-pocket total expense for the year. The cost of oil was far better than for our previous rental property and this pleased us, although we already had plans for a far more efficient system to look after climate and hot water in our home. After sheer luck prevented catastrophe as a result of multiple cases of furnace failure during the winter, we felt we needed to replace the oil furnace right away and would need to borrow the money to do it.

To install a typical oil combustion system, we would have been looking at the following costs:
- High-efficiency oil furnace: C$3,600
- Oil hot water tank, 60 gal: C$2,350
- Indoor oil tank: C$1,600
- Central air: C$3,000

(All figures include taxes and installation)

Figure 4.9-2. After multiple oil furnace failures and burning $1,700 worth of oil, Jeff and Lynne decided to replace the furnace system and install a geothermal unit.

The oil tank is included because our house insurance requires that we change the tank after approximately ten years, and it was time. Air conditioning would have been an upgrade to our home and it is included for reference.

The installation of all of the above components would have improved our oil consumption bills because of the increased efficiency of the new furnace and hot water tank, though the cost of the system as estimated above was $10,550—a nice figure for comparison, but not something Lynne and I ever considered.

Our choice was a geothermal heating and cooling system for our home. This system uses the ground temperature to heat our home in the winter and cool it in the summer and looks after 50% to 60% of our hot water demand. The technology is known by several names, including geoexchange and ground-source heat pump (GSHP). It represents the single most energy-efficient method of heating and cooling a home."

An important consideration for space heating/ cooling is energy efficiency. The best combustion furnaces on the market have 94% efficiency (94% of the energy consumed in the furnace is captured to warm the space). A geothermal system is considered to be 250% to 350% efficient (for each unit of electricity used to power the heat pump, 2.5 to 3.5 times that energy is extracted from the ground and used for space heating). Although electricity is required to operate the geothermal system, it is only used to move heat, not to create it. The efficiency rating is known as the coefficient of performance (COP) and may be expressed either as a percentage or as a decimal; for example, 275% may be expressed as a COP of 2.75.

Geothermal systems are available in different variations, and Lynne and I chose a "closed-loop" system. In this configuration 1500 feet (457 meters) of tubing is buried in our backyard. A biodegradable ethanol and water blend (an *antifreeze*) is circulated through the tubing, which is buried six feet below ground. A small, energy-efficient pump inside the home circulates the liquid through the tubing. By the time the liquid re-enters the home it has equilibrated with ground temperature, roughly 4°C (39°F) in our area. The fluid then passes through the heat pump. In the winter months, the heat pump warms the house by extracting energy from the antifreeze. During this process the antifreeze is cooled from 4°C to approximately 3°C—a small change, but the thermal differential provides sufficient energy to warm our house, even when the outside temperature reaches -25°C (-13°F). Below this outside temperature, a 10kW supplementary heater may 'flash on' briefly to help maintain indoor air temperature.

In the summer months, the heat pump works in reverse, warming the antifreeze by extracting heat from the house and providing cool air to circulate through the ducts. The warmed antifreeze is sent back through the 1500-foot (457 meter) loop and is returned to ground temperature.

The system also warms our hot water by running water from the hot water tank through a heat exchanger located on the refrigerant coils inside the heat pump and then back to the water

Figure 4.9-4. The underground heat transfer pipe is unrolled and placed in the trench. At a depth of 6 feet (2 meters) the ground temperature maintains a relatively constant temperature, allowing heat to be moved into the house and "harvested" by the geothermal system. Although it may be hard to believe that 4°C (39°F) ground can provide any heat, the "amplification" effect of the geothermal compressor and refrigerant system is highly effective.

Figure 4.9-3. The Goodman-Leblancs chose to install a closed-loop geothermal system which required a 1500-foot loop of in-ground piping to collect earth energy. A contractor begins the digging process using a high-hoe to excavate the 6-foot (2 meter) deep by 750-foot (230 meter) long trench. For those who have smaller urban lots, do not despair, as vertically drilled systems can also be installed (see Figure 5.8-7).

Figure 4.9-5. The pipe is rolled out forming a supply and return loop. A hole is drilled (cored) through the foundation wall allowing the pipe loop to connect to the geothermal system. The hole is sealed and insulated and the trench is backfilled. Note that the trench depth and length are determined by the size and efficiency of your house as well as by local conditions.

Figure 4.9-6. After the pipe is laid in the trench, workers apply compressed air to the system and check for leaks. Once the pipe is installed, it should last a lifetime without failure.

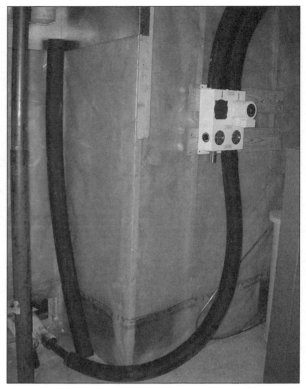

Figure 4.9-7. A biodegradable antifreeze solution of water and alcohol is pumped into the piping system. This picture shows the supply and return side of the ground loop exiting the top of the antifreeze circulation pump assembly (white), with the supply and return loop to the geothermal system shown at bottom.

Figure 4.9-8. Condensation from the air conditioning cycle is collected and sent to an in-floor drain via the white drainpipe, center. The ground loop supply/return pipes are insulated using a closed-cell insulation sleeve.

heater. The hot refrigerant coils transfer heat through the heat exchanger to the water heater. This is known as a desuperheater option and further improves the efficiency of the system, providing between 50% and 60% of our hot water demand during the winter months. During the summer air-conditioning season, there is very little extra heat developed by the refrigerant system, so we are considering adding a solar thermal heating unit to provide the balance of our hot water needs.

The system was quoted at $21,105 installed and included increasing the size of our ducts to accommodate the extra air flow needed for these systems. There was the additional cost of hiring an operator and equipment to dig a trench

Figure 4.9-9. The geothermal system including air handler is shown in this picture. All forced-air geothermal systems use larger-than-average air ducts to ensure even heat flow throughout the house. Unlike a standard furnace that is oversized and designed to blast very hot air for a short period of time, the geothermal system may run continuously, supplying air at the exact temperature required by the home.

Figure 4.9-10. Geothermal systems may include an optional backup electric heating unit that can provide supplementary heat during abnormal cold spells. Provided the house is well insulated and draft free, this device should remain idle the majority of the time.

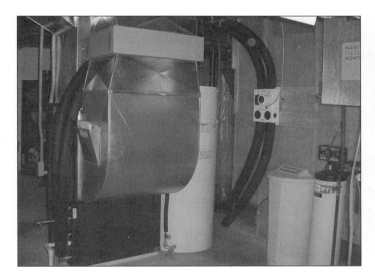

Figure 4.9-11. This view shows the geothermal unit, air handler, and ground loop pipe entering the circulation pump and leaving the house. Note that the water heater is located next to the geothermal system and is connected to the desuperheater unit.

Figure 4.9-12. This view shows the large air handling ducts required by the geothermal system.

5-6 feet deep, 2 feet wide, and 750 feet long to accommodate 1500 feet of tubing (750 feet out from the house, 750 feet return). We had a logging trail on our property and chose to bury the loop under this trail. We were told to budget $2,000 and the final cost was $1971.48 for a total installed cost of $23,076.48.

The cost of the geothermal system was projected at $12,526.48 more than the oil system capable of looking after all of the same climate and hot water needs for our home. So why choose the geothermal system over the oil system? There were other factors to consider beyond the initial cost of the system:

- Lifetime costs – The geothermal system is projected to have a much longer life than any combustion furnace.
- Energy costs – The high efficiency of the geothermal system was projected to provide as much heat as our old furnace for $600/year, air conditioning all summer for $39 (that is not a typo) and hot water for $300/year. Quick math shows this well below the $1700+ spent on oil without the benefit of air conditioning in the summer.
- Home value – The geothermal system, according to the bank, resulted in a dollar-for-dollar ($23,000) increase in the value of our home, resulting in…
- A better mortgage rate – Because of the bank's incentives, we were offered the best mortgage rate if we installed the geothermal system
- Greenhouse gas emissions – A geothermal system reduces our emissions by four tons.
- Renewable energy heating system – Heat can be taken from the ground in the winter months and the sun will warm it again the following summer. Likewise, summer heat can be drawn from the house and fed back into the ground, a process that can be repeated endlessly.

Summary:

- The total cost in oil for the preceding 12 months was $1,707.79. The total *increase* in our electrical bills after the installation of the geothermal system was $915.64. This gives a net savings on utility payments of $792.15 for the year (46%). The cost of oil has increased 57% since these initial calculations were completed and if I were to re-run the number now, our savings would be in excess of $1600.
- The new mortgage rate offered by our bank meant that a total of **$231.12 was saved in interest for the year** (more of our mortgage payment went to principal as a result of the new rate, despite the increased amount owing). An important note here is that the bank does not set the monthly payment for our mortgage against the new rate but calculates it against the prime rate for a higher monthly payment and increased contribution towards the principal.
- Our total mortgage payment for the year increased by $2,329.20. So our total utility savings plus an extra $128 per month left our pockets towards our mortgage. From the point above, it should be noted that there was an additional contribution towards principal this year of $2,329.20 + 231.12 = $2,560.32.
- The increased mortgage payment and better rate result in **our mortgage-free date coming 21 months sooner**.
- Twenty-one months of mortgage payments under our pre-geothermal scenario would mean $25,932.48 in additional mortgage payments. **The time we spend mortgage-free will pay back the investment on the system and more**.
- While the installed cost for the system was $23,076.48, we received a tax rebate of $692.29 and a grant under the EnerGuide for Houses program of $957.00 for a total installed cost of $21,427.19.

- If we were to build a new house we would definitely equip it with a geothermal system instead of a conventional combustion system (oil, gas, etc.). **It is important to note** that builders shouldn't try to justify the expense of installing a geothermal system by telling owners that they will recoup the $25,000 cost in utility bill savings. **Rather, the difference between this system and a conventional system (in our case $21,427.19 minus $10,550 = $10,877.19) is the appropriate target.** Energy savings alone, at today's rates (Fall 2009), would see this paid back in approximately six to seven years, resulting in an after-tax return on investment of approximately 15%. However, this ignores lifetime and maintenance costs which offer further savings with the geothermal system.

- The indoor temperature selected on the geothermal system was 21°C and it was not turned down at night to avoid using the 10kW heater in the mornings to 'fast heat' the house. (i.e. a geothermal system is more energy efficient if it is *not* turned down at night, unlike combustion furnaces which save money when used with programmable thermostats).

Conclusion:

The geothermal system was an excellent investment. The increased value to our home is not something we focused on because we intend to keep this house for many years, although it is comforting to know we could make this investment back. The utility savings are in line with the model predictions and amounted to a 46% improvement during the first year of operation. The extra money we put towards our mortgage each month is offset by the utility savings and will see our mortgage paid off almost two years earlier. While we had been told that it would take years to pay for itself, we feel that this system paid for itself before it was turned on for the first time. Our bank is essentially paying for the system and giving us a profit thanks to the lower interest rate;

the utility savings are extra gravy in the deal. In any event, a similar system installed today would yield an after-tax ROI of approximately 15%. Not bad considering the drubbing most investments have taken in recent memory.

A geothermal system is a worthwhile consideration for anybody upgrading their old furnace, building a new home, or renovating a home for resale purposes. All factors considered, it was the most economical choice we could have made. As for the drastic reduction in greenhouse gas emissions—well, it's hard to put a dollar amount on feeling this good about something so important to today's children and our collective future."

Figure 4.9-13. Jeff and Lynne are the proud parents of their own energy-, money-, and environment-saving geothermal system.

Chapter 5
HEATING AND COOLING WITH RENEWABLE ENERGY

P rice signals capture everyone's attention. When oil was a couple of dollars a barrel very few people worried about how much of it they consumed to operate cars and heat homes. In the early 1970s automobiles weren't much smaller than the tanker ships used to flood North America with Saudi Arabian crude, and home design and construction gave little attention to energy efficiency and related fuel consumption.

In 1973 an organization known as OPEC unilaterally raised oil prices, providing the wake-up call that shook Americans out of their fuel consumption complacency. Prices soared and those who heated their homes with oil went from being energy pacifists to being scared silly almost overnight. As if that wasn't bad enough, all Arab oil producers stopped shipping oil to the United States as a way of protesting American policy regarding Israel. Price increases of over 500% coupled with uncertain supplies exposed the vulnerability of the United States economy, which depended then as it does now on foreign oil supplies.

Following considerable political posturing by the Nixon administration, the embargo was lifted and life returned to an uneasy state of normal. In the years following the energy crisis considerable work was done to improve the efficiency standards of homes and automobiles. Unfortunately politicians have very short memories, and the crisis of 1973 was quickly forgotten as fuel prices and supplies stabilized, causing most North Americans to return to their gluttonous ways.

The beginning of the 21st century has brought about a new energy crisis fostered not by embargoes but rather by concerns about the environment, stability of supply, and terrorism. Although the economy is still completely addicted to oil and demand continues to rise unabated, OPEC remains relatively friendly to the West. It should; after all, as of 2003, American consumers have drained the astounding sum of $7 trillion from the economy simply to burn Arab oil and watch it go up in smoke. According to a report in the October 25, 2003 issue of *The Economist*, this estimate does not include subsidies of staggering cost, such as the ongoing military presence in the Middle East as well as development subsidies and tax breaks to the large oil companies.

Environmental degradation also continues unabated, as neither the government nor oil companies consider it a cost to be added to the balance sheet. Canada exports nearly 8% more oil to the United States than does Saudi Arabia. The province of Alberta, through its phenomenally dirty tar sands oil extraction program, reaps the short-term economic windfall even though the process makes it one of the largest greenhouse gas emitters in the world. With governments addicted to increasing gross domestic product and job creation, energy efficiency and the environment are of no more concern than they are to the Saudi Arabian oil sheiks with whom they compete for market share.

The long-term view of energy is to move the world economy away from nonrenewable resources to clean-burning, carbon-neutral renewable fuels such as clean-electricity, ethanol, biodiesel, and perhaps hydrogen for use in transportation infrastructure. Wind turbines, photovoltaic panels, and clean-burning biomass provide complementary energy sources for the production of electricity. The transition has already started. In 2003 Germany announced that it had already decommissioned one nuclear power plant as a result of increased use of wind turbine and photovoltaic electricity generation technologies. On the residential home front many renewable energy technologies such as solar domestic hot water systems are currently available and economically viable.

Renewable Fuels

Burning any fuel has some negative impact on the environment. Fossil fuels are nonrenewable and contribute smog-producing chemicals and greenhouse gases—mainly carbon dioxide—into the atmosphere. Some nonrenewable fuels are better than others. Natural gas and propane are the best, while low-grade coal is the worst. Consider that a natural gas or propane cook stove allows the burning fuel to vent *into* the house. Try *that* with coal or oil.

Wood and scrap wood pellets, switchgrass, biodiesel, and ethanol are different from fossil fuels in that they are renewable, easily replaced energy storage units. A growing plant absorbs nutrients and water from the ground as well as carbon dioxide from the atmosphere. Photosynthetic processes within the plant convert carbon dioxide gas into carbon, which is stored in the structure of the wood or oil seed. When dry wood or biofuels are burned, carbon is consumed, releasing heat and carbon dioxide gas. The good news is that burning the plant products properly exhausts no more greenhouse gases into the atmosphere than simply letting the tree or plant rot on the forest floor or farmer's field. However, allowing even the best wood stove to burn smoldering, smoky fires releases ash and other pollutants into the atmosphere.

Some nonrenewable fuels are better than others.

Solar Energy – Something for Everyone

If you are designing a new home or retrofitting an old one, an even better energy source to consider is the sun. A properly designed and oriented home absorbs free energy on sunny days, helping reduce your reliance on burning anything (See the Fisher-Dunkely home tour in Chapter 4.) Both passive and active solar heating systems can be used. Passive solar heating is just that: passive use of properly oriented walls, windows, and architectural house features. Active solar heating involves a series of solar collection and storage units designed to increase the "density" of the captured energy. Within the broad range of renewable fuels and "solar thermal" technologies are a number of options for the rural and urban homeowner to consider, each with its own pros and cons. While the number of configurations is virtually infinite, let's assess those technologies that follow the *eco-nomic* path:

- 5.1 Passive solar heating
- 5.2 Active solar air heating
- 5.3 Solar water heating
- 5.4 Solar pool heating
- 5.5 Active solar space heating
- 5.6 Heating with renewable fuels
- 5.7 Space cooling systems
- 5.8 Earth Energy with Geoexchange

5.1

Passive Solar Heating

Solar energy is free, non-polluting, and renewable. Why don't more people use it? My guess is that most people think that houses using solar energy have to look like something George Jetson would live in. Well, don't worry. Geodesic dome designs aside, you won't have to live in a house that upsets your local town hall planning committee or causes you to be barred from block parties for life in order to live with a passive solar system. Another possible reason for being afraid of solar energy is its variability and the fact that your building contractor isn't able to quantify it. If you require heat, just install a big furnace and then forget about it, right? Everyone knows it's not quite that easy with solar energy.

Not so fast. Living with solar energy does not require a house full of complicated electronics and miles of glass. What passive solar energy does require is following some fairly simple guidelines to allow your home to take full advantage of the sun's energy.

Step 1 – House Orientation

Orienting the house to collect solar energy may seem obvious, but it's not. Most people (let alone city planners) give no thought to ensuring that the house is oriented in such a way as to capture as much sun as possible in the winter and provide proper shading in the summer. This is how it's done:

- Using a compass at the proposed building site, locate magnetic south. Consult the magnetic declination chart in Appendix 4 and determine the compass correction for true or solar north and south. Using this heading, place the long axis of the home at a right angle to it.

- Place deciduous trees in a path between the summer sun and the house. This provides shading for the house and glazing. When the autumn winds blow, the leaves fall from the trees, allowing the welcome winter sun indoors.

- Plant pine, cedar, or other evergreen trees to the north and east to provide a windbreak. In treeless areas, wind and storm blocks can be made using rock outcroppings, hills, or even your neighbor's house.

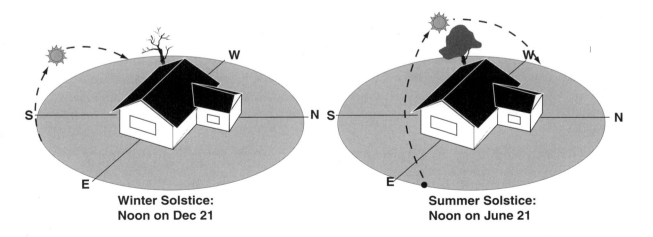

Winter Solstice:
Noon on Dec 21

Summer Solstice:
Noon on June 21

Figure 5.1-1. Orient the house to take advantage of winter solar gain and summer shading.

These orientation strategies apply whether you are in the country or the city. If you are designing a home in a city, try to evaluate the sun's path over your lot and take into account any shading effects caused by future buildings. The Solar Pathfinder shown in Figure 6-10 works wonders if you are concerned about your home's view of the sun, or solar window.

Figure 5.1-2. This home, located in Eastern Ontario, is designed to capture the maximum amount of winter sun energy, while avoiding the high-angle summer rays.

Figure 5.1-3. This lovely veranda also acts as the south window summer shading system. The roof overhang was designed to allow sunlight to enter the house structure from mid-September through mid-March.

Step 2 – Insulate, Insulate, Insulate

If you haven't taken the time to review Chapter 2, "Energy Efficiency," go back and read it. A house that is under-insulated and prone to air leakage will quickly put an end to your quest for comfort, energy efficiency, and low-cost operation. **Remember that it is far less expensive to conserve energy than to acquire more energy to heat a substandard house, even if the energy is free!**

If you live in an area where summer air conditioning loads are high, consider adding radiant barrier insulation in the attic, as discussed in Chapter 2.

Step 3 – Window Design

Windows are the passive solar collectors of the house. Vertically oriented glass will catch the low winter sun and miss the high summer sun angle. This means you can create a "normal" looking home and still get maximum solar efficiency.

- Place the majority of the desired glazing on the south side (long axis) of the house.
- Choose windows that are of the highest quality you can afford. Refer to Chapter 2 for information on window construction. Remember that triple-glazed glass with Krypton gas fill is the most efficient available, reducing nighttime energy loss.
- Do not use low-E glass on the south side of the house or on any window that you wish to use as a solar collector.
- Ensure that window units that can be opened are installed on the side of the house facing the prevailing winds as well as on the opposite side of the house. You can open these windows to provide nighttime cross-flow cooling of the home.
- Limit the amount of glazing on the north, northeast, and northwest sides of the house.
- Do not over-glaze the south side. Excess window area equates to excess heat.

Step 4 – The Finishing Details

These finishing touches will assist in making the job work "just right":

- Provide proper home ventilation using an air-to-air heat exchanger unit as outlined in Chapter 2.

- Provide sun shading over the east-, west-, and south-facing windows as required, to prevent the summer sun from heating the structure.

- Use summer sunscreen blinds similar to the units shown in Figure 2-10.

- Install a thermal absorber and buffer system in the living area. The sun will quickly overheat a properly insulated house, making it unbearable. A solar mass such as cement/tile flooring or walls, finished in dark colors, will absorb heat, helping to moderate temperatures. During the evening hours, this stored energy is radiated back into the room, helping to keep temperatures constant.

The key to ensuring that your passive solar design does not cause extreme cycling of interior temperatures is to balance the solar collector area (windows) with the thermal storage and buffering mass of the floor and wall systems. When these components are not in balance the results are very similar to a greenhouse: hot during the day and cold at night.

An educated architect or heating contractor will be able to assist you in developing a heat-loss calculation for your home. Alternatively, a home inspection auditor who is capable of performing an air leakage test may be able to assist with this calculation. Armed with this information, you will be able to calculate what percentage of the yearly heating load can be expected from solar energy. With the sun providing approximately 1000W per square meter of window area, it's a shame not to take advantage of this free energy source.

Figure 5.1-4. This recently built off-grid "earthship" design is very heavily glazed by northeastern temperate area standards. It will require plenty of summer shading and cross-flow ventilation to keep it from over-heating in the summer. Sunshades or greenhouse whitewash paint can be used to reduce heat gain. (Note the photovoltaic panels mounted to the right of the windows.)

For readers who would like more detailed design information on the subject, an excellent book is available that teaches construction techniques, theory, and mathematical modeling of heat gain and loss. *The Passive Solar House* by James Kachadorian (ISBN 978-1933392035, Chelsea Green Publishing Company) is labeled as "a building book for the next century" and I highly recommend it to anyone who wishes to push the building design envelope.

5.2
Active Solar Air Heating

A simple, unique solar heating concept is produced under the trade name Solarwall® by the firm Conserval Engineering Ltd. This cladding material is mounted on the south, east, or west walls. Tiny perforations allow outdoor air to travel through the exterior surface. During the day, as outside air passes through the panel and along its inside surface, it absorbs the sun's energy, warming the air which in turn rises. The preheated, warm air is then drawn into the building's ventilation system, reducing the load on the heating system.

During sunny days the Solarwall® preheats intake air by between 30°F and 50°F (17°C and 28° C), depending on the desired flow rate. Snow can reflect up to 70% of solar radiation into the Solarwall®, increasing the system's performance further. The system also works on cloudy days, as

Figure 5.2-1. The Solarwall® system capitalizes on the fact that a building's exterior cladding will be exposed to the energy of the sun and captures that energy for fresh air heating and ventilation. (Courtesy Conserval Engineering Ltd.)

diffused light can provide up to 25% of the total radiation available on a sunny day.

The Solarwall® can even help cool your building in the summer. The cladding structure shades the interior wall and any hot air that accumulates at the top of the unit escapes through the perforations, keeping the wall surface cooler than it would be if it were exposed directly to sunlight. Electrically (or manually) controlled dampers allow unheated air into the building, maintaining air quality. The dampers can also switch to the wintertime position, providing preheated air intake.

A slightly modified version of the vertically installed Solarwall® is the roof-mounted system shown in Figure 5.2-4. If you are building a new home or planning to re-roof an existing house, the incremental capital cost may be negligible.

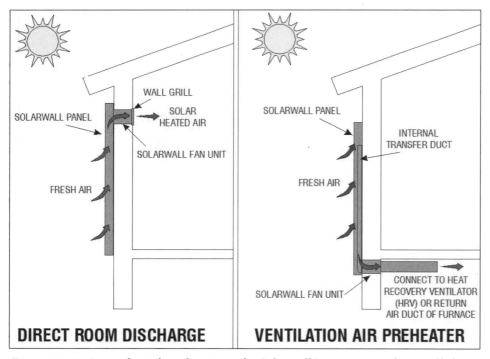

Figure 5.2-2. In residential applications, the Solarwall® system may be installed as a direct room discharge unit or as the preheater for the intake of a furnace or heat recovery ventilator system.

Figure 5.2-3. This view of a partially completed home shows the Solarwall® prior to the installation of the fieldstone siding desired by the homeowner. The design will work well with the stone facing, providing a very aesthetic finish.

An alternative to the Solarwall® system is a retrofit product known as the Cansolair unit. The Cansolair is a cross between the Solarwall® system and a greenhouse, contained within a package of the same size as a sheet of plywood. The heated air may be delivered to the home's heat recovery ventilator or via a thermostatically controlled room ventilation fan as shown in Figure 5.2-5.

The economic case for solar thermal heating systems depends on many factors including the age of your home, its heating demand, and its thermal efficiency. There is little point in installing a system that may only provide a small percent of total heat demand because of a thermally leaky house. On the other hand, an energy-efficient home that has a correspondingly lower heat demand may receive a significant portion of its total yearly thermal energy from such a system. It will be necessary to compare the heat loss of your home with the heat production of the solar thermal unit to determine how much purchased energy, in the form of oil or gas for example, the system will offset. This is a complex calculation and is best left to professional heating and cooling contractors and their computers. If you are willing to give the calculations a shot, then consider purchasing the book *The Passive Solar House* discussed above.

Figure 5.2-4. The Solarwall® system has been adapted to act as a metal roof structure as well as an air solar thermal collector. (Courtesy Solar Unlimited Inc., Utah)

Figure 5.2-5. The Cansolair solar heating unit may be installed as a retrofit device on the south-facing wall or roof. Using a thermostatically controlled fan unit (shown above right), the system can provide preheated intake air to the building ventilation system or directly to a room The unit produces approximately 2,900 kWh of energy per year based on 2,200 hours of sunlight. At a hydro cost of 10 cents per kWh, this equates to almost $300 per year of free energy.

5.3

Solar Water Heating

When you ask people about solar systems almost everyone will talk about heating for domestic hot water supply, and with good reason. Approximately one-third of the energy bill for a home is related to hot water heating costs. According to a Florida electrical utility, for every $1000 you pay in energy costs, $300 goes toward water heating. Stated another way, after heating and cooling costs, domestic hot water heating is the next most expensive home energy cost (see pie chart in Figure 5.3-2).

While everyone expects solar energy to make economic sense in the tropical or desert areas of the southern United States, what about the areas to the north, including Canada and its Arctic regions? The maps shown in Figure 5.3-3 indicate the percentage of hot water heating energy that

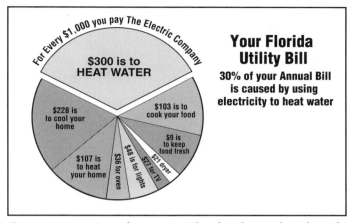

Figure 5.3-2. According to a Florida electrical utility, for every $1000 you pay in energy costs, $300 goes towards water heating.

can be achieved in different geographic locations. Throughout the densely populated northeastern section of the United States and Canada, homeowners can achieve an average of 55% of their water heating demand from solar. If you happen to live in Anchorage, Alaska or even Inuvik in Canada's Arctic region, you can still achieve an impressive 35% average savings in annual water heating costs.

You may be wondering if your house will have hot water only in the amount indicated in the maps in Figure 5.3-3. Not to worry. All solar thermal systems work in conjunction with your existing hot water heating system, providing "pre-heating" of the water supply. For example, if the desired hot water temperature is 140°F (60°C), the solar water heating system may preheat the incoming cold water supply to 110°F (43°C), reducing the fuel demand of your hot water heater while maintaining adequate supplies of hot water. Over the course of one year, your hot water fuel bill will be reduced by an amount approximately equal to the percentages indicated. The capital cost for the solar water heating system divided by the yearly fuel savings will provide the simple return on investment, which will almost certainly exceed the ROI that can be expected from most financial investments.

Figure 5.3-1. Solar thermal water heating systems are used throughout the world and with good reason. Approximately one-third of the energy bill for a home is related to hot water heating costs. Almost everyone is aware that Arizona and Florida can achieve huge savings with solar systems, but what about the majority of people who live in more northerly climates? It may come as a surprise, but if you live in Anchorage, Alaska or Inuvik in Canada's Arctic you can still achieve a 35% reduction in hot water heating costs by using a solar thermal water heating system. (Courtesy EnerWorks Inc.)

Figure 5.3-3. These maps of Canada and the United States show the average annual solar energy that can be captured to reduce home hot water heating costs. Solar thermal hot water systems are one of the most economical means of putting the sun's energy to work. A typical family of four can expect to receive a return on investment of approximately 10% to 14% if they are located in an area with a minimum of 50% solar heating penetration. For those lucky enough to live in the Arizona desert or sunny Florida, the return on investment can exceed 20% according to industry figures. (Courtesy EnerWorks Inc.)

Before Installing a Solar Thermal System...

Most people practice their Luciano Pavarotti while taking a shower; I expect that the engineering minds at Renewability Energy Inc. focus not on singing but on just how much hot water is going down the drain. In fact the hot water used for household tasks such as laundry, dishes, showers, and baths requires enormous amounts of energy, but much of the heat contained in the water is not extracted before the job is completed. If the drain water were cold after the shower or laundry, the energy efficiency of the task would be 100%. However, this is never the case, and a considerable amount of energy actually goes down the drain in the form of warm waste water.

The Power-Pipe™ heat recovery system is an amazingly obvious device which transfers energy

Figure 5.3-4. A close-up view of the top section of the drain and spiral-wound supply pipes of the Power-Pipe system. After installation the entire assembly is wrapped in insulation, preventing heat loss and increasing thermal efficiency. (Courtesy Renewability Energy Inc.)

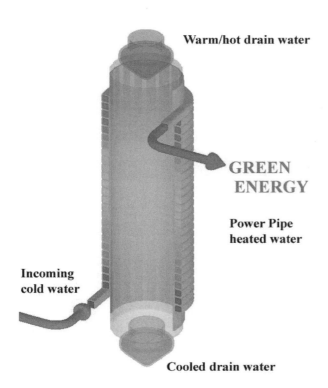

Figure 5.3-5. As warm waste water pours down the large vertical pipe, heat is transferred to the spiral-wound water supply pipe which in turn feeds a hot water heater or the cold side of plumbing fixtures. (Courtesy Renewability Energy Inc.)

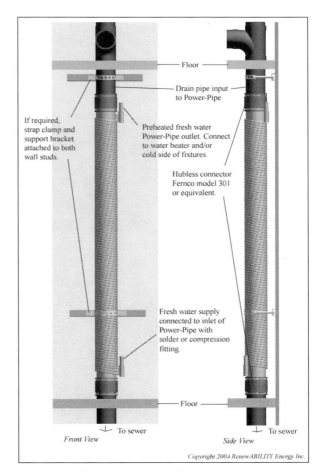

Figure 5.3-6. Although it is possible to retrofit your home with a Power-Pipe heat recovery system, it is far easier to do the installation during construction. (Courtesy Renewability Energy Inc.)

from warm waste water to incoming potable cold water which in turn supplies the home's hot water heater. As shown in Figure 5.3-4, a copper drainpipe is mounted in a vertical orientation, allowing waste water to pass through on its way to the sewer. The vertical orientation is important in order to ensure that drain water "skins" along the inside diameter of the pipe as it makes its way to the sewer, simultaneously increasing the surface temperature of the pipe. A second smaller pipe is wrapped in a tight spiral around the vertical drainpipe. The bottom or inlet side of the spiral pipe is connected to the incoming cold water supply. As the warm drain water pours down the larger vertical pipe, heat is transferred to the spiral cold water supply pipe. Potable water exiting from the top of the spiral assembly is thus pre-warmed and may be sent to the solar water heater or fossil-fuel hot water heater, reducing energy requirements further still.

Using less hot water to begin with will also reduce energy costs. Review Chapter 2 regarding energy efficiency to reduce water heating fuel bills further.

Understanding the Basics of Solar Domestic Hot Water Systems

Solar hot water heaters work in conjunction with your existing fossil-fuel or electrically operated water heating tank, supplying a percentage of your total hot water heating requirements. They will reduce your hot water heating costs, greenhouse gas emissions, and smog and atmospheric pollutants.

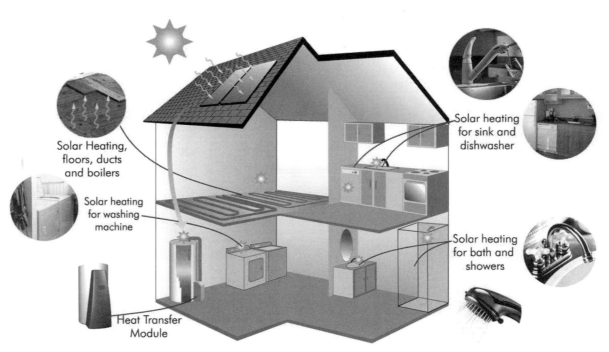

Figure 5.3-7 is a simplified view of a typical SDHW system. The typical urban solar water heater consists of roof-mounted panels which capture the sun's energy as heat. A circulator pump causes a transfer fluid to absorb this energy and pass it to a heat exchanger which in turn heats water in a storage tank. Solar-heated water is most often used in the kitchen, laundry, or bathroom, although it can also provide heat for space, pool, and spa applications as discussed later in this chapter. (Courtesy EnerWorks Inc.)

Energy Collection

Solar collectors are best mounted on the roof, facing solar south and mounted at an angle approximately equal to your geographic latitude in much the same manner as electricity-producing photovoltaic panels as discussed in Chapter 6. Solar collectors are available in a number of different configurations, although flat plate designs (see Figures 5.3-1 and 5.3-8) are the most common. Evacuated tube collectors (see Figures 5.3-9 and 5.3-10) make up the balance, although alternative collector configurations such as the concentrating, active sun-tracking unit from Menova Engineering shown In Figure 5.3-11 show up on the market from time to time.

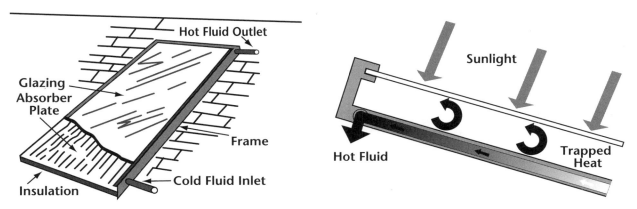

Figure 5.3-8. Flat plate solar collectors are miniature greenhouses which capture sunlight, trapping heat between a glazing material and an insulated rectangular box. A series of thin pipes attached to a dark-colored heat absorbing plate is placed inside the collector box. A circulating fluid such as water or antifreeze passes through the pipe assembly and heat collector plate, absorbing the trapped heat. If the heated fluid is potable water, it is supplied directly to the water heating unit. In severely cold climates an antifreeze solution is used instead of water, requiring the use of a heat exchanger to transfer heat energy to the hot water system.

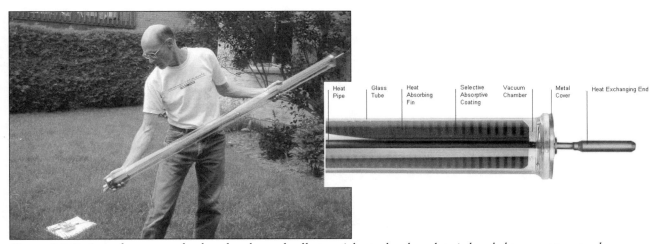

Figure 5.3-9. A single evacuated tube solar thermal collector. A heat absorber plate is bonded to an apparatus known as a heat pipe. The assembly is placed inside a glass tube which is sealed and evacuated, creating a vacuum. The collector plate absorbs the sun's energy, causing a fluid inside the heat pipe to boil. Heat is then transferred to a condenser bulb mounted at the top of the unit. If you have ever made coffee and placed it inside a thermos bottle you know how long the drink stays hot. In a similar manner, heat cannot escape from the evacuated tube collector, providing a system that is not affected by extremely cold weather or high water temperatures. (Courtesy Carearth Inc.)

Figure 5.3-10. Evacuated tube solar collectors are most commonly mounted to a "header" in groups of ten. The transfer fluid flows through the header, absorbing heat from the condenser bulb located at the top of the evacuated tube. The header assembly is well insulated, preventing heat loss.

Figure 5.3-11. Solar thermal collection systems can also go high-tech. This Power-Spar™ from Menova Engineering Inc. uses a spherical mirror to concentrate the sun's energy directly onto a fluid transfer header, which has the effect of amplifying the sun's energy. Additionally, the Power-Spar™ contains a unique solar tracking unit which ensures that the mirror is correctly aimed at all times. The Power-Spar™ may also be fitted with photovoltaic panels to produce electricity at the same time as heat energy.

(For the purposes of this discussion, I am not considering unglazed solar thermal collectors, which are generally used for spa and pool heating only. See Chapter 5.4 for further details on this technology.)

I am often asked which type of solar collector is the best choice. Opinions vary greatly, but deciding which model is "best" requires looking at the facts.

Although one can compare efficiencies, construction materials, absorber coatings, and a host of other "esoteric" data, what really matters at the end of the day is whether the system will perform as required and provide the best value for the capital expended.

Based on this non-technical assessment, I would suggest that potential purchasers review the Solar Rating and Certification Corporation website (www.solar-rating.org). The SRCC is a non-profit organization that provides the solar thermal water heating industry with third-party, non-biased evaluation of its products. The corporation provides rating documentation for registered products, allowing comparisons between models. On the website home page, follow the Ratings link and then click on Glazed Collector Rating Listings. Then scan the list of manufacturers until you locate the model you are interested in reviewing. Click on the SRCC Number link on the left and a new page will be displayed detailing the test results, as shown in the example model in Figure 5.3-12.

The Collector Thermal Performance Rating table indicates the amount of thermal energy generated by the collector during a clear, partly cloudy, or cloudy day, in both imperial and metric units (thousands of BTU per panel per day and megajoules per panel per day respectively). Select the weather pattern for your area and then select the category, which for the vast majority of applications will be C or D. The temperatures in the Category column represent the difference between the temperature of the water entering the solar collector and the temperature of the air surrounding it. This temperature differential is also known as delta-T (ΔT). Generally speaking, the greater the temperature differential, the lower the efficiency of the solar collector. Having said that, all types of collectors operate within a fairly narrow band of efficiencies in "normal" climatic conditions (Categories C and D) and vary greatly outside this standard set of conditions.

By way of example, the data in Figure 5.3-12 for a solar collector in the cloudy area of coastal Washington State indicates between 3.6 and 11.1 thousands of BTU of energy per panel per day. This rating can be compared to ratings for other panels, and system costs can be evaluated. Keep in mind that panels with differing surface areas will naturally capture more or less energy. To keep the playing field even, make sure you are comparing apples to apples.

Assuming that quality of installation, warranty, and similar factors are equal, the solar thermal system that provides the most energy for the lowest dollar cost in your area will be the "best" choice for your application.

Regardless of which solar thermal collector you choose, it must have an unobstructed view of the sun between the hours of 9 a.m. and 3 p.m. throughout the year. Before purchasing a system, ensure that your site provides the necessary solar exposure. In addition, the collectors must be mounted so that they are angled from the horizontal at an angle equal to your latitude and are facing as close to solar south as possible, bearing in mind that a phenomenon known as magnetic declination must be considered when a compass is used for alignment. Magnetic declination maps provide a correction to compass readings in order to locate true north and solar south. Consult Appendix 4 to determine the magnetic error in your location.

Remainder of System

There are numerous types of solar water heating systems available, including simple yet effective home-built designs such as the batch heater shown in Figure 5.3-13. In this design, a water heater stripped of its outer jacket and insulation blanket

SOLAR COLLECTOR CERTIFICATION AND RATING  SRCC OG-100	CERTIFIED SOLAR COLLECTOR

	SUPPLIER: Acme Solar
	                        123 Main Street,
	                        Anytown, Anystate USA
	MODEL: solar1
	COLLECTOR TYPE: Glazed Flat-Plate
	CERTIFICATION#: 2007099A

COLLECTOR THERMAL PERFORMANCE RATING

Megajoules Per Panel Per Day				Thousands of BTU Per Panel Per Day			
CATEGORY	CLEAR DAY	MILDLY CLOUDY	CLOUDY DAY	CATEGORY	CLEAR DAY	MILDLY CLOUDY	CLOUDY DAY
A (-5°C)	38.1	28.8	19.5	A (-9°F)	36.1	27.3	18.5
B (5°C)	35.0	25.6	16.3	B (9°F)	33.2	24.3	15.5
C (20°C)	30.0	20.8	11.7	C (36°F)	28.5	19.7	11.1
D (50°C)	20.4	11.9	3.8	D (90°F)	19.4	11.3	3.6
E (80°C)	11.6	4.1	0.0	E (144°F)	11.0	4.1	0.0

Original Certification Date: 31-Sep-08

COLLECTOR SPECIFICATIONS

Gross Area:	2.487 m^2	26.77 ft^2	**Net Aperature Area:**	2.297 m^2	24.72 ft^2
Dry Weight:	45.4 kg	100. lb	**Fluid Capacity:**	3 liter	0.8 gal
Test Pressure:	1103. KPa	160. psg			

COLLECTOR MATERIALS

Frame:	Aluminum Extrusion
Cover (Outer):	Low Iron Tempered Glass
Cover (Inner):	None
Absorber Material:	Tube - Copper / Plate - Aluminum
Absorber Coating:	Sputtered Selective
Insulation Side:	Isocyanurate Foam
Insulation back:	Isocyanurate Foam & Fiberglass

Pressure Drop

Flow		ΔP	
ml/s	gpm	gpm	in H$_2$O
20.00	0.32	28.00	0.11
50.00	0.79	115.0	0.5
80.00	1.27	256.00	1.03

TECHNICAL INFORMATION

Efficiency Equation [NOTE: Based on gross area and (P)=Ti-Taj]   Y INTERCEPT   SLOPE

S I UNITS: η= 0.7250   -3.23740 (P)/I   0.01038 (P)^2/I   0.6   -3.7 W/m^2.°C

I P UNITS: η= 0.7250   -0.5703 (P)/I   0.0010 (P)^2/I   0.6   -0.7 Btu/hr.ft^2.°F

Incident Angle Modifier [(S)=1/cosθ - 1, 0°<θ<=60°]   Model Tested: solar1

Kα = 1   0.058 (S)   -0.274 (S)^2 **Test Fluid:**   Water

Kα = 1   -0.23 (S)   Linear Fit   **Test Flow Rate:**   49.8 ml/s   0.79 gpm

REMARKS:

August, 2009

Certification must be renewed annually, For current status contact:

SOLAR RATING & CERTIFICATION CORPORATION

c/o FSEC • 1679 Clearlake Road • Cocoa, FL 32922 • (321) 638-1537 • Fax (321) 638-1010

Figure 5.3-12. The Solar Rating and Certification Corporation (www.solar-rating.org) provides rating documentation for registered solar thermal water heating products, allowing buyers to make informed comparisons of different models. This sample data sheet shows a typical test report. (Courtesy Solar Rating and Certification Corporation)

and painted flat black is placed in an insulated plywood box. The assembly is covered with a glass or plastic glazing material and angled to face solar south as discussed previously. Household pressurized water is supplied to the bottom of the tank, while heated water is removed from the top and fed to the hot water heater in the home.

The insulated box and glazing act in the same manner as a greenhouse, trapping thermal energy which is transferred from the outer surface of the tank to the water. As water heats its density decreases, causing it to float to the top of the tank where it can be drawn for household use. Batch water heaters are not freeze resistant and must be used in temperate climates or drained and bypassed during winter months.

The commercial batch storage unit shown in Figure 5.3-14 is common throughout the developing world, as these units are not considered aesthetically pleasing to the North American eye and can be susceptible to freezing. (Check with the manufacturer if you live in an area that is prone to freezing.) These units are, however, extremely simple, relatively inexpensive, and may be ideally suited for three-season cottages in northern environments or full-time residences in more temperate areas.

Direct or open-loop systems heat the same potable water that is circulated throughout the residential plumbing system. These units can be actively controlled with pumps and temperature-measuring devices or may operate using the thermosyphon or natural convention loop effect as noted above.

The most common configurations in use in North America today are active drain-back systems and closed-loop designs that use a pump, temperature sensors, and electronic controls to capture and move thermal energy. The control system monitors the temperature of the water in the storage tank and compares it to the solar collector temperature. Whenever the solar collector is a few degrees warmer than the water in the tank the pump will activate, causing water to flow and collecting thermal energy in the process.

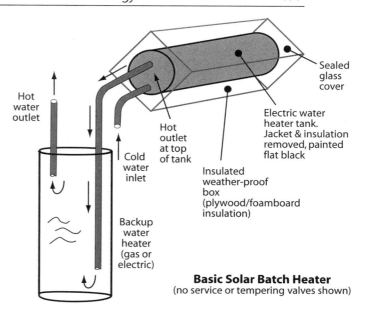

Basic Solar Batch Heater
(no service or tempering valves shown)

Figure 5.3-13. The basic solar batch water heater is a staple renewable energy feature of the do-it-yourself crowd. Although cheap and simple to construct, this design does not differ significantly from the more sophisticated commercial model shown in Figure 5.3-14.

Figure 5.3-14. Batch solar and passive open-loop systems use natural convection to circulate water through the collectors and water storage tank.

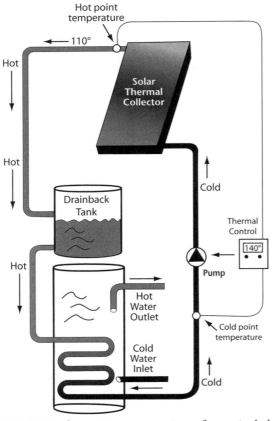

The control unit can also monitor the temperature of the water storage tank; if it exceeds a given set point level the pump will be prevented from operating, thus limiting maximum water temperature.

The drain-back system is shown in schematic view in Figure 5.3-15. This design allows potable water to be the primary heat transfer medium between the solar collector and the storage tank, with the water draining back into the small storage tank whenever the pump switches off. Draining the solar collector of water during cold, cloudy periods, on winter nights, or when the system is idle ensures that water will not freeze within the system.

Drain-back units may be equipped with a heat exchanger as shown in Figure 5.3-15. If a heat exchanger is provided, heat will be transferred from the hot water within the heat exchanger piping to the relatively cooler water contained in the storage tank. With this configuration, there is no possibility of cross-contamination between the solar collector water loop and the potable household system.

Figure 5.3-15. A schematic representation of a typical drain-back solar water heating system is detailed in this drawing.

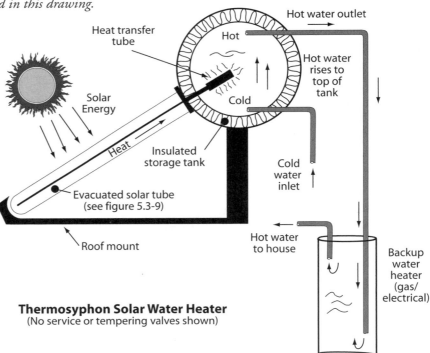

Figure 5.3-16. A schematic view of the commercial solar batch water heating unit shown in Figure 5.3-14.

Thermosyphon Solar Water Heater
(No service or tempering valves shown)

The closed-loop glycol system is the preferred configuration in areas where freezing is a concern. A mixture of food-grade propylene glycol antifreeze (it *must* be safe; soft ice cream contains 30% of the stuff) and water continuously remains in the solar collectors, plumbing lines, and heat exchanger. A differential electronic controller and temperature sensing devices are connected to an alternating current circulation pump in the same manner as in the drain-back system described above. It is also possible to replace the alternating current circulation pump with a direct current model connected to a small photovoltaic panel mounted in the vicinity of the solar thermal collectors. This configuration eliminates the need for temperature sensors and the differential electronic controller by activating the circulation pump whenever the sun shines.

Figure 5.3-17. The drain back system is relatively simple and theoretically immune to freezing, making it a popular choice for domestic solar thermal hot water systems in moderate climates. Water contained in a storage tank is circulated through the solar collector and the hot water tank, which is equipped with a heat exchanger unit. When an electronic controller stops the circulation pump, water will drain back into the storage tank, preventing freezing. (Courtesy Alternate Energy Technologies, LLC)

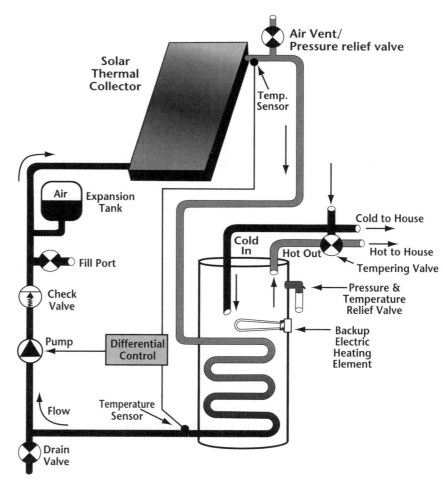

Figure 5.3-18. The active, closed-loop system shown above uses a heat-collecting antifreeze fluid which is circulated through the solar collectors and heat exchanger. Energy is stored in either a single-storage water heater tank (indicated) or in a two-tank configuration, according to the manufacturer's design.

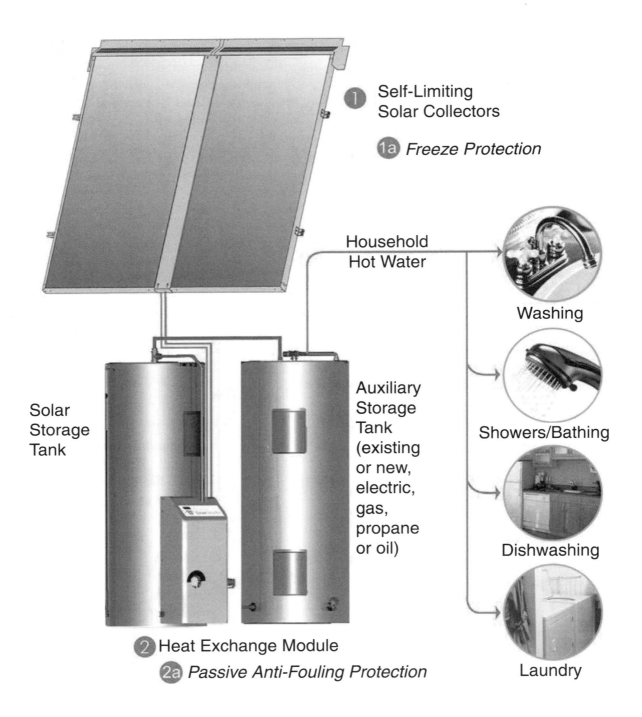

1 Self-Limiting
Solar Collectors

1a *Freeze Protection*

Household
Hot Water

Washing

Solar
Storage
Tank

Auxiliary
Storage
Tank
(existing
or new,
electric,
gas,
propane
or oil)

Showers/Bathing

Dishwashing

2 Heat Exchange Module
2a *Passive Anti-Fouling Protection*

Laundry

Figure 5.3-19. A mixture of food-grade propylene glycol antifreeze which remains in the solar collectors and plumbing lines at all times makes the closed-loop glycol system completely freeze-resistant in any climate. A differential electronic controller and temperature sensing devices activate a circulation pump, transferring energy to an existing hot water tank via a heat exchange module. (Courtesy EnerWorks Inc.)

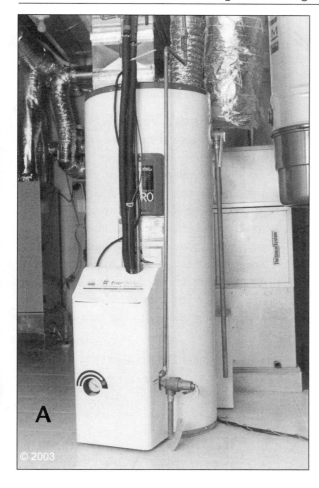

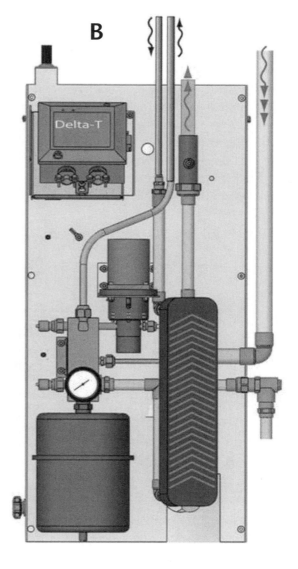

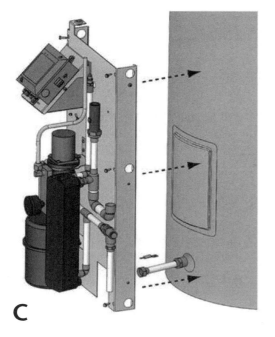

Figure 5.3-20. An EnerWorks Inc. integrated solar thermal system is shown mounted to an existing hot water heating unit (A). A detailed front view of the unit is shown (B) with the cover removed, while an isometric (angled side) view is shown (C). The electronic differential controller is mounted on the top left corner of the control panel with the circulator pump, heat exchange module, and pressure tank mounted below. (Courtesy EnerWorks Inc.)

Circulation Pump

Centrifugal-type hydronic heating pumps are used in solar hot water heating systems to circulate a heat-collecting fluid. They are chosen because of their low power consumption, high reliability, and low maintenance. For long life it is recommended that bronze or stainless steel pumps be used in this application even though they are slightly more expensive than cast-iron models.

These pumps are readily available from any plumbing supply store that specializes in hydronic heating systems.

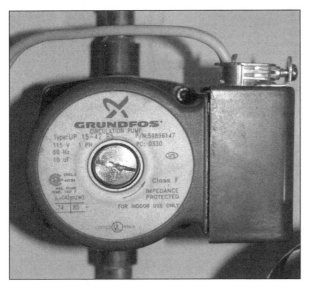

Figure 5.3-21. The Grundfos brand of centrifugal-type hydronic heating pump is commonly used in solar hot water systems because of its low power consumption, high reliability, and low maintenance. Other manufacturers such as Taco and Hartell provide similar models.

Plumbing

Solar thermal systems are generally plumbed using standard domestic copper plumbing pipe which is wrapped in an insulation sleeve material such as Rubatex-brand Insul-Tube. An alternative material is PEX Tubing manufactured for the hydronic (in-floor) heating industry.

To minimize heat loss it is recommended that plumbing lines be routed inside the building envelope as soon as possible after leaving the solar thermal collectors. If you are planning on building

a house, it is a good idea to pre-plumb lines from the location of prospective solar collectors to your mechanical room. The cost is insignificant and doing this will save a considerable amount of grief during a future installation process.

Check Valve

A one-way check valve is installed on the outlet port of the circulation pump to prevent the convection flow of warm fluid from the hot water storage tank to the cold solar collectors at night. Check valves may be spring or gravity operated. Gravity-operated check valves must be installed in the orientation shown in the installation manual. In applications using an AC circulator pump, the spring-operated check valve is preferred, as it is less susceptible to internal leakage caused by minute particles circulating in the heat-collecting fluid.

Fill and Drain Ports

One fill valve and one drain valve must be provided to allow the filling of the circulation system with water or glycol solution. In order to fill and pressurize the circulation system, a positive-displacement high-pressure pump is required to overcome the vertical "head pressure" related to the force of gravity and the expansion tank air pressure.

Expansion Tank

An expansion tank allows the water or glycol solution in the closed-loop system to expand and contract as a function of its temperature. Without an expansion tank the buildup of pressure would cause the plumbing lines to burst when the circulation fluid is heated by the sun.

Hydronic expansion tanks such as the model shown in Figure 5.3-22 contain a rubber bladder which is in turn connected to the plumbing circulation lines receiving the circulation fluid. With the tank empty the air chamber is pre-charged to approximately 15 psi (103 kPa) using a bicycle pump or compressor. As the circulation fluid expands and contracts as a result of the changing temperature, the pressurized air bladder maintains a relatively constant pressure in the plumbing system.

Your hydronic heating component supplier will be able to determine the optimum size of expansion tank based on the quantity of circulation fluid in the system.

Pressure Relief Valve

Pressure relief valves are designed to prevent the catastrophic explosion of plumbing lines in the event of excessively high pressure buildup. A setting of 50 psi (345 kPa) is common for most hydronic heating systems. All pressure relief valves including the model shown in Figure 5.3-23 are equipped with a drain port which must be connected to a capture bucket, floor, or sanitary drain.

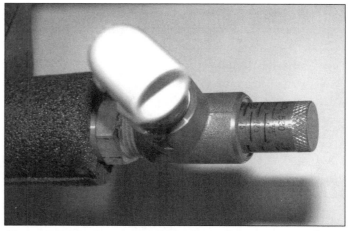

Figure 5.3-23. A pressure relief valve will prevent an explosion of the plumbing lines in the event of an excessively high pressure buildup.

Pressure Gauge

A pressure gauge indicates that the circulation system is operating within acceptable pressure limits. The normal operating pressure is approximately 15 psi (103 kPa) greater than static (pump off) pressure of the system. The static pressure is determined by the vertical height of circulation fluid and is based on a pressure increase of 1 psi (6.9 kPa) for every 29 inches (0.74 meters) of plumbing rise above the gauge height. For example, the static pressure for a plumbing line that rises 14 feet (4.3 meters) is 5.8 psi (40 kPa).

Static Pressure
= (14 feet rise x 12 inches/foot) / 29 inches per psi
= (168 inches) / 29 inches per psi
= 5.8 psi

The operating pressure will vary somewhat depending on the temperature of the circulation fluid: the warmer the temperature the higher the pressure reading; the cooler the fluid the lower the pressure.

Also note that the static pressure will increase as soon as the heat transfer fluid is pressurized against the expansion tank bladder.

The pressure gauge may also be used to detect leaks in the circulation system. A slow, continuous drop in static pressure over time indicates a fluid leak.

Figure 5.3-22. An expansion tank allows the water or glycol solution in the closed-loop system to expand and contract as a function of its temperature.

Figure 5.3-24. An air vent is installed at high points in the circulation system and is used to remove trapped air (left). A pressure gauge (right) indicates that the circulation system is operating within acceptable pressure limits.

Air Relief Valve

Air trapped in the fluid circulation lines will lower the heat exchange efficiency or stop the circulation of fluid altogether. To remove trapped air, relief valves (Figure 5.3-24,) are installed at high points in the fluid circulation lines and at the solar collector panels. During commissioning these valves are opened, allowing the trapped air to escape to the atmosphere. Automatic air relief valves may be installed as an option.

Atmospheric gases are often dissolved in the circulation fluid and make their presence known through a gurgling sound in the plumbing lines. If the circulation system is completely free of air there should be no sound at all when the pump is running. During the yearly maintenance review of the system it may be necessary to activate the relief valves to eliminate any trapped air.

Heat Exchanger

The heat exchanger transfers energy from the solar collectors to domestic water in the storage tank. Heat exchangers may be external to the storage tank, such as those shown in Figure 5.3-20, or internal, as shown in Figures 5.3-18 and 5.3-25.

Heat exchangers are available in single- or double-wall designs, indicating the number of mechanical barriers between the heat-absorbing

circulation fluid and the potable water in the storage tank. Local plumbing codes may require the use of double-wall heat exchangers to prevent the contamination of potable water in the event of a leak. Check with your local municipality to determine what codes may be in effect prior to purchasing you solar thermal system.

Heat exchangers work when the circulation fluid is at a temperature higher than that of the water contained in the storage tank. Heat transfer is also affected by the surface area (size), thermal conductivity, and flow rating of the heat exchanger. If you are designing your own solar thermal system, contact a hydronic heating contractor to determine the appropriate size for your installation.

Figure 5.3-25. The heat exchanger transfers energy from the solar collectors to domestic potable water in the storage tank.

Antifreeze

To prevent the collector circulation fluid from freezing, an antifreeze solution of 50% propylene glycol mixed with 50% demineralized or distilled water is used. Common automotive antifreeze containing ethylene glycol must not be used because of its toxic nature.

The collector loop is filled with the antifreeze mixture using a positive displacement pressure pump. Sufficient fluid is introduced into the collector circulation loop to develop approximately 5 to 10 psi (34 to 69 kPa) of working pressure above the static pressure (static or pump-off pressure plus 5 to 10 psi = working pressure) as indicated on the pressure gauge. The additional pressure is caused by the pressure of the heat transfer fluid working against the pressure of the accumulator tank bladder.

Temperature Sensors

Temperature sensors are located on the hot fluid outlet of the solar collector and the bottom or cold heat transfer pipe of the water storage tank. The sensors, known as "thermistors," change their electrical resistance as a function of temperature, thereby feeding information to the electronic controller module.

The temperature sensor shown in Figure 5.3-26 is applied to the cold side of the thermal storage tank plumbing as indicated in Figure 5.3-18. Prior to attaching the temperature sensor to the plumbing lines, thermally conductive grease supplied with the controller module is added to ensure a proper temperature reading. The sensors are secured in place with a screw-type hose clamp as indicated in Figure 5.3-26 and wrapped with pipe insulation.

Electronic Control Module

The brain of the solar thermal system is a device known as a differential temperature controller. This unit monitors the circulating fluid temperature at the outlet of the solar thermal collector and on the cold side of the water storage tank. When the temperature of the solar thermal collector outlet exceeds the cold-side temperature of the storage

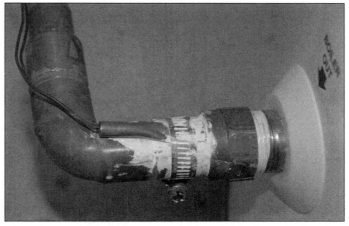

Figure 5.3-26. Temperature sensors are located on the hot fluid outlet line of the solar collector and the bottom or cold line of the water storage tank. They feed temperature information to the electronic controller module.

tank by a predetermined number of degrees, the circulation pump is started and the heat transfer process begins.

A differential temperature in the order of 10°F to 25°F (5°C to 14°C) is required in order for adequate heat transfer to begin. Electronic controls such as those shown in Figure 5.3-27 also contain a high-temperature safety limit or cutoff circuit which will shut the circulation system down once the water storage tank reaches a predetermined setting. The upper limit is adjustable and is normally set to 170°F (77°C).

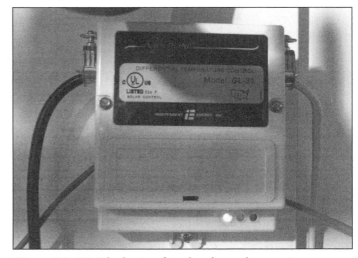

Figure 5.3-27. The brain of a solar thermal system is a device known as a differential temperature controller.

Setting the turn-on differential too low will cause the pump to "short cycle," turning on and off the circulation pump. Setting the differential point too high will waste thermal energy.

The correct setting will cause the controller to short cycle only in the early morning, late afternoon, or during a cloudy day. During the remainder of a sunny day, the pump should operate continuously during the peak solar window from approximately 9 a.m. until 3 p.m.

Temperature Display

Two temperature gauges are recommended in order to measure the circulation fluid temperature at the inlet and the outlet of the heat exchanger, with an optional third gauge installed in the hot water storage tank as shown in Figure 5.3-28.

A temperature differential of approximately 20°F (11°C) across the heat exchanger indicates effective operation of the system. Monitoring the maximum temperature in the water storage tank provides an indication of the effectiveness of the high-limit cut-out circuit.

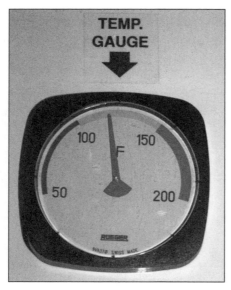

Figure 5.3-28. Two temperature gauges are recommended in order to measure the circulation fluid temperature at the inlet and the outlet of the heat exchanger, with a third gauge installed in the hot water storage tank as shown.

The House Side of the Plumbing System

So far we've concerned ourselves with the heat-absorbing side of the circulation system. The output of the solar thermal storage tank may be fed to a fossil-fuel or electrically powered water heater or directly to the house hot water supply.

Regardless of which installation is chosen, a temperature-limiting tempering valve should be added to the hot water outlet prior to feeding the household hot water circuits. The reason for this is that the solar thermal system will produce hot water of varying temperatures depending on the amount of energy captured by the solar collectors. If the solar thermal system is incapable of meeting the desired temperature set point, a backup heating element installed in the solar tank or a regular secondary hot water heater will heat the water to the desired temperature. On the other hand, if there has been a significant amount of sunny weather the water temperature in the storage tank may exceed the desired set point temperature, potentially scalding anyone who is not paying attention. A tempering valve will "mix in" an amount of cold water to ensure that the hot water supplied to the house is at the desired temperature, improving system safety and comfort.

Choosing and Installing a System

A qualified, experienced dealer will be able to help you select the water heating system that best meets your specific needs and can assist you in determining what state, municipal, or local utility tax incentives and rebates may apply. (You can also review the Database of State Incentives for Renewables & Efficiency Energy—DSIRE—at www.dsireusa.org.) If you have unpleasant recollections of some fly-by-night organizations selling solar thermal systems during the peak of the oil crisis, today's licensed professionals should be able to alleviate your concern.

It is wise to select dealers who are able to provide warranty service for both parts and labor and who have a satisfactory track record of

Return-assembly includes female quick-connect coupling and ball-valve

Supply-assembly includes female quick-connect coupling, ball-valve and Schrader valve for leak testing

Figure 5.3-29. Temporary service fittings are installed on the solar thermal plumbing system to facilitate flushing and charging. (Courtesy EnerWorks Inc.)

installing systems. Whenever possible, ask to see a previously installed system and try to discover if the owner is satisfied with both the installation and after-sales service.

Installing a solar thermal system is not the same as going to the hardware store to purchase a regular water heater. The dealer will have to perform a site evaluation, taking measurements and determining pipe routing and other details required to complete the installation. You will become an active part of this process.

Because you are making modifications to the structural and plumbing systems of your home, it will be necessary to install components that are certified by Underwriters Laboratories in the United States or the Canadian Standards Association to ensure fire and electrical safety. Purchasing approved products also lets you sleep at night with the knowledge that product designs have met construction-quality standards. In addition, you must obtain a building permit from your local inspection authority.

For local installation contractors and equipment suppliers in the U.S. contact the American Solar Energy Society at www.ases.org. In Canada contact the Canadian Solar Industries Association (CanSIA) at www.cansia.ca.

Charging the Heat Transfer System

At the completion of the solar thermal system installation, the technician will flush the plumbing lines to remove any chemicals and foreign materials. Following this step, a 50/50 mixture of demineralized or distilled water and propylene glycol antifreeze is pumped into the system as illustrated in Figures 5.3-29, 5.3-30 and 5.3-31.

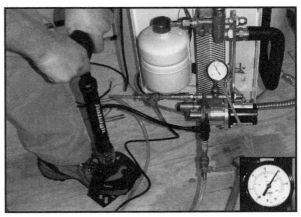

Figure 5.3-30. After flushing and cleaning the solar thermal plumbing, the installation technician attaches an air pump to the plumbing fittings and pressurizes the system. The service pressure gauge should remain fixed for a period of one hour to ensure a leak-free system. (Courtesy EnerWorks Inc.)

Return-assembly quick-connect and ball valve

Supply assembly quick-connect, Schrader valve and ball valve

Supply line (from reservoir to pump and appliance)

Charge pump

Return line from appliance

Reservoir with heat-transfer fluid

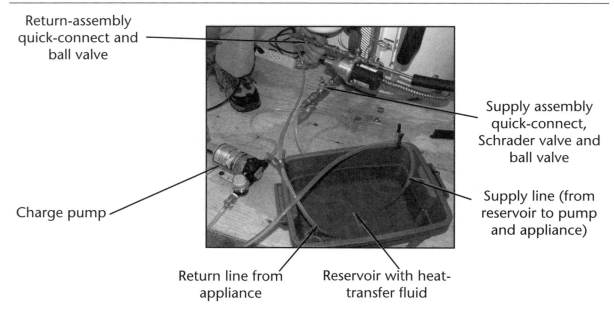

Figure 5.3-31. A positive displacement pressurizing pump draws heat transfer fluid from a container, pumps it through the solar system, including the rooftop collectors, and returns the fluid to the container. During the filling process, air trapped in the system will bubble in the return tank. When the fluid has circulated for five minutes without the presence of air, the system is ready for final pressurization. Closing the "return to tank" valve will cause the system pressure to build to the static level, at which point the system circulation valves are closed and the unit is ready to begin heating. (Courtesy EnerWorks Inc.)

Operating a Solar System

Once the system is up and running you will be able to monitor its operation by observing the water temperature in the storage tank as well as the differential temperature at the inlet and outlet ports of the heat exchanger, as discussed above.

Homeowners can achieve at least 55% of their hot water from solar thermal units.

You can also monitor the system for slow, minute leaks by watching the pressure gauge over a period of time. System pressure will rise and fall with a change in circulation fluid temperature; however, the average "resting" pressure should remain constant over time.

In extremely sunny weather it is very likely that the system will produce more hot water than you can possibly use. When this situation occurs the electronic control module will stop the circulation pump, preventing the temperature of the water storage tank from rising above preset "over-temperature" limits. Circulation fluid trapped in the solar thermal collectors of antifreeze-based systems will continue to absorb energy, increasing their operating temperature. Drain-back systems are not subject to this condition.

In situations where the circulation pump is stopped, a phenomenon known as "collector stagnation" occurs. If this condition persists the propylene glycol solution may prematurely break down, leaving a sticky, sludgy deposit in the plumbing components. To prevent this from happening manufacturers have developed a number of novel solutions. For example, EnerWorks has developed a "smart metal thermal actuator" which is installed on the underside of the company's flat plate solar collectors. During normal operation heat is drawn away from the collector by the circulating fluid and the actuator remains closed. When stagnation occurs, excess heat in the collector assembly causes actuators located at the bottom and top of the rear side of the panel to open, allowing fresh ambient air to ventilate and cool the collector assembly. This operation is shown in the photograph in Figure 5.3-32.

An alternative to having this heat energy "wasted" when the water storage tank is filled is to include a bypass valve that will direct energy to other areas of the home that may be able to put it to work. For example, the hot fluid could be plumbed to a baseboard hot water radiator unit to provide auxiliary space heat in the home. Alternatively, a second heat exchanger could be plumbed into a hot tub or swimming pool, providing an excellent use for this energy. In either instance, your heating contractor will install the necessary valve and control equipment to ensure that domestic hot water is heated first with the auxiliary loads operated secondly.

An examination of the antifreeze circulation fluid is recommended at least every two years. There are two ways of testing this fluid. Samples may be sent for analysis to the Dow Chemical Company under its free annual fluid analysis program. The company will provide a complete summary report with a results table and a cover letter explaining the condition of the sample and outlining necessary maintenance procedures for its products. (Contact Dow through the company website at www.dow.com/heattrans/index.htm.)

ACUSTRIP® Company, Inc. offers a simplified fluid test kit for propylene glycol antifreeze solutions. A sample of the antifreeze solution is placed in a clean glass jar. A test strip (Figure 5.3-33) is immersed in the fluid and then compared to a color chart, indicating percentage propylene glycol concentration and pH (acidity) level. The ideal concentration level is 50% propylene glycol to 50% demineralized or distilled water. The pH of the solution should be in the range of 8 to 10. If the glycol concentration is too low or the pH level too acidic, the system must be partially drained and new glycol added.

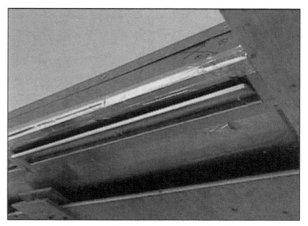

Figure 5.3-32. Manufacturers have developed numerous methods for preventing stagnation of the antifreeze circulation fluid. EnerWorks has developed a thermally activated vent which opens when stagnation conditions develop, admitting fresh ambient air. (Courtesy EnerWorks Inc.)

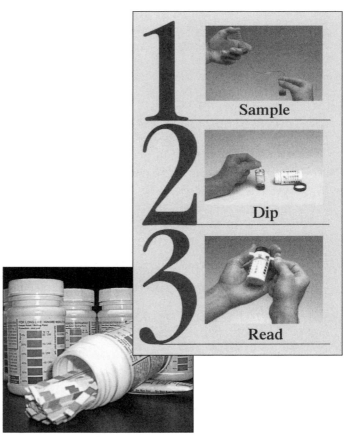

Figure 5.3-33. ACUSTRIP® Company, Inc. offers a simplified fluid testing program for propylene glycol antifreeze solutions. This test kit should be used at least once every two years.

5.4

Solar Pool Heating

Heating a swimming pool using the heat of the sun is probably one of the most obvious and cost-effective applications for solar energy. According to the Canadian Solar Industries Association there are more than 250,000 swimming pools in Canada, of which approximately 40% are heated. It is further estimated that homeowners spend more money to heat their pools than their homes. Switching from fossil-fuel or electric heating to a solar thermal system would recoup the initial capital in less than 2.6 years, yielding a return on investment of approximately 38%. According to the company Ameco Renewable Resources (www.amecosolar.com), using natural gas for pool heating costs in excess of $2000 for a typical swimming season even in sunny southern California!

The graph shown in Figure 5.4-1 indicates the increase in swimming pool water temperature as well as the extension of the swimming season when a solar thermal heater is used in conjunction with a solar blanket. A swimming pool that uses both of these items will be more comfortable and will save an enormous amount on heating bills. You will also eliminate tons of greenhouse gas emissions, literally! Figure 5.4-2 shows typical greenhouse gas emissions (in short tons: 1 t = 0.91 metric tonnes) during a single swimming season; consider this not only from an environmental perspective but from a financial one, as governments contemplate carbon-taxes and emission caps.

The Pool Heating System

To get the maximum benefit from your solar pool heating system, the collectors must be mounted on a south-facing roof or structure which is ideally angled to match your geographic latitude. It is also possible to place the collectors on the ground, although there will be a resulting drop in heating efficiency, requiring either the installation of additional collectors or an acceptance of reduced system performance.

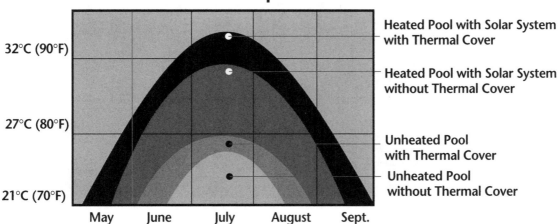

Figure 5.4-1. Using a solar thermal pool heating system and insulating solar blanket not only increases the pool temperature but can extend the swimming season by up to two months in the U.S. northeast and Canada.

Greenhouse Gas Emissions
From Residential Pool Heating Methods

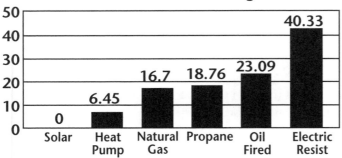

The required collector area will depend on the surface area of the swimming pool, the desired water temperature, and the amount of shading from trees and neighboring buildings. As a general rule of thumb the collector surface area should be approximately 60% to 70% of the surface area of the pool. For example, a swimming pool that measures 10' x 20' (3 m x 6.1 m) will require a minimum solar collector area of:

Minimum solar collector area
= 10 x 20 feet x 0.6
= 120 sq. ft. (11 m²)

If the site is partially shaded or the collectors are mounted on a flat roof, consider increasing the collector area to match that of the swimming pool surface area.

Figure 5.4-2. Heating a swimming pool with fossil fuel is similar to throwing dollar bills into a lake: very expensive and not a really bright thing to do. Heating a swimming pool with electricity in Southern California creates 40 tons (36.3 tonnes) of greenhouse gas emissions in a single swimming season while costing the homeowner in excess of $2000. A solar thermal pool heating system can reduce both of these figures to near zero.

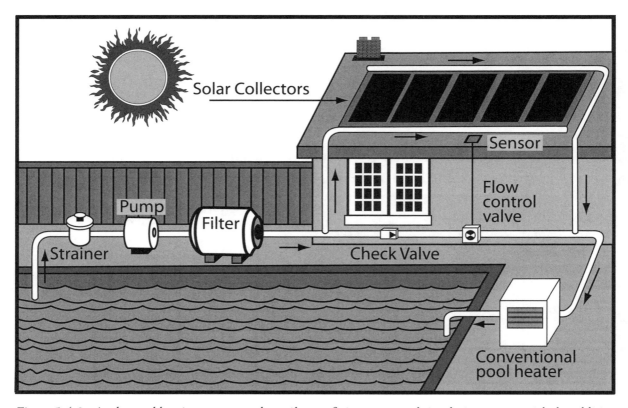

Figure 5.4-3. A solar pool heating system can be easily retrofit into most pool circulation systems with the addition of solar thermal collectors, thermostatic controller, and a check and flow control valve. With the addition of a pool cover, a conventional backup pool heater should not be required.

With few exceptions, the existing filter pump will be capable of circulating pool water through the solar panels. In areas where freezing is a concern, ensure that the system is equipped with drain spigots and properly sloped plumbing lines to ensure complete water drainage for winter storage.

Controlling the temperature of the swimming pool may be accomplished in one of two ways. The first method incorporates nothing more complex than a timer to turn the circulation pump on and off in conjunction with sun up and sun down times. The advantage of this configuration is simplicity and low cost. The disadvantage is reduced pool circulation time and lack of control over pool water temperature.

Figure 5.4-5. With few exceptions, the swimming pool pump should provide sufficient circulation of water through the solar collector panels. Temperature control may be as simple as a day/night timer turning off the pool water circulation at night.

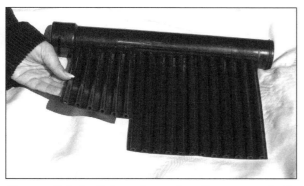

Figure 5.4-4. Solar thermal collectors used for swimming pools may be as simple as these flexible rubber pad units manufactured by Enersol or the high-efficiency flat plate collector design discussed above. A typical collector surface area is approximately 60% to 70% of the swimming pool surface area.

An automatic solar pool heating control consists of an automatic diverter valve which channels water to the solar collectors based on differential temperature control and desired pool temperature. When the pool reaches the desired operating temperature, the diverter valve redirects the water flow from the solar collectors to the filtration system.

In either design an auxiliary fossil-fuel or electric pool heater may be added to the system. In this configuration the solar collectors will preheat the water before it reaches the backup heating unit. Fossil fuel or electricity will be used only to make up the difference between the solar thermal output temperature and the desired pool water temperature set point. However, the judicious use of a pool thermal blanket should eliminate the need for any backup heat.

Figure 5.4-6. This three-season drain back solar pool heating unit extends the swimming season by two months and increases the water temperature when used in conjunction with a solar pool blanket.

5.5

Active Solar Space Heating

Before looking at how an active solar thermal system figures into the design, let's review the basic operation of a typical hydronic or hot water heating system, which can be adapted to solar thermal space heating with relative ease.

Hydronic in-floor heating systems are becoming increasingly popular, in part because of the continuous, even warmth they provide throughout the home. As the heat is developed in the flooring material, even hard-to-heat products like ceramic tile stay nice and toasty for cold toes. Many people who suffer from airborne allergens find that their symptoms are greatly reduced with hydronic heating, as there is no dusty air being blown about the house through the centralized fan-forced heating system. Lacking a fan, hydronic systems are also very quiet and energy efficient.

The concept behind hydronic heating is quite simple. Heated water (or an antifreeze mixture) is pumped through a series of flexible plastic pipes located under the flooring or embedded in the insulated concrete slab of your house. A zone thermostat turns a very tiny, energy-efficient pump on and off as heat is required in each zone. For areas where freezing is a problem, environmentally friendly (propylene glycol) antifreeze is added to the water. This is also required for outdoor boiler heating systems described in Chapter 5.6.

If you prefer to have a forced-air system or would like to add central air conditioning, a hot-water-to-air heat exchanger can be used. This unit installs in the blower/ductwork assembly and transfers heat from the boiler to the circulating air in the same manner as a conventional forced-air furnace. A second air exchanger can be installed to add air conditioning to the mix.

Another advantage of hydronic heating concerns the boiler unit. Most boilers provide two "heating loops," one for hydronic space heating and the second for domestic hot water heating. These boilers can provide 100% of your space and water heating requirements using renewable resources such as biodiesel, wood, wood pellets, and even dried corn. Dual-fuel boilers are also available, typically combining an efficient wood-burning stove with an oil or electric backup energy supply.

Figure 5.5-2 illustrates a dual-fuel fired boiler which provides the energy for both space heating and, optionally, a second plumbing circuit for domestic hot water. The dual-fuel fired boiler has a wood-burning section on the top and an oil-fired burner on the bottom. Provided the wood fire is maintained, the boiler is programmed not to start the oil burner unit. This allows you to

Figure 5.5-1. With hydronic heating, hot water or antifreeze solution is pumped through plastic tubing located in or under your home flooring.

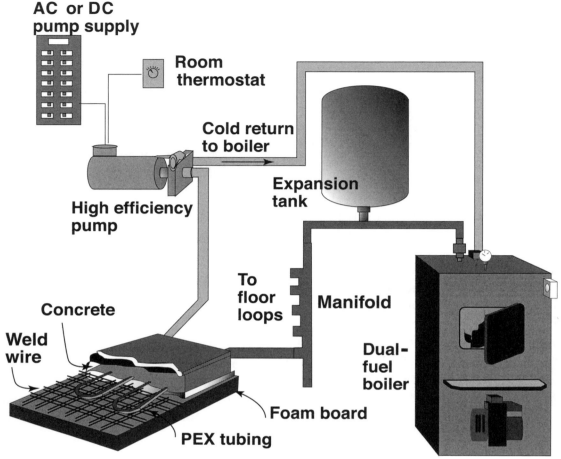

AC or DC pump supply

Room thermostat

Cold return to boiler

Expansion tank

High efficiency pump

To floor loops

Manifold

Concrete

Dual-fuel boiler

Weld wire

Foam board

PEX tubing

Figure 5-.5-2. A typical hydronic heating system is shown in schematic view.

supply up to 100% of your heating needs from a renewable source.

Hot water or a water/antifreeze mix leaves the boiler and enters a manifold line which contains an expansion tank that is placed in the system to allow the water to expand or contract in volume, depending on temperature. The manifold provides a number of outlets to feed the various zones or area heating loops, each equipped with its own circulation pump and thermostat.

The heating loops consist of, for example, a flexible plastic pipe known as PEX, manufactured by companies such as REHAU Inc. (www.rehau-na.com). PEX pipe may be encased in insulated cement flooring over grade or affixed directly to the bottom of interior subfloors. Heating an

insulated concrete slab provides a high level of thermal mass and helps to regulate the room temperature, thus providing much higher levels of comfort than central hot air furnace designs, where air temperature fluctuates as a result of the furnace cycling on and off.

PEX lines should be no longer than approximately 300 feet (91 meters), as the heat dissipates with distance. Cool fluid exiting the heating loop is drawn by a high-efficiency circulation pump and returned to the boiler water inlet. Hydronic heating contractors use circulation pumps from manufacturers such as Grundfos (www.grundfos.com) which are rated between approximately 40 and 80 watts at 120 volts.

For on-grid systems, Grundfos manufactures

one of the finest circulation pumps that has been proven to operate reliably over many years of service. Although a rating of 40 to 80 watts may appear acceptable, each pump can be expected to operate at a 50% duty cycle requiring 0.5 to 1 kWh of energy per day. It must also be remembered that each zone (area controlled by one thermostat) has at least one pump. For a system with six pumps, the energy requirement may exceed 3 kWh per day, a sizeable portion of wintertime energy production.

An alternative, low-energy selection is the El-Sid (Electronic Static Impeller Drive) pump from Ivan Labs. This pump does not have a conventional motor and spinning shaft but uses an electronic drive circuit to cause a magnet embedded in an impeller to spin. Being completely sealed, it will last "forever." The El-Sid pump is also very efficient. A 10W 12VDC pump can circulate water through two 300-foot (90 meter) heating loops. To put this in perspective, most people use 15W CF lamps for each table lamp. A direct-current-driven pump allows for either direct connection to the battery bank or connection through a converter module (see Chapter 11). Connection directly to the battery bank saves electrical energy by reducing inverter losses. It also ensures operation of the heating system in the event of voltage converter failure.

The controls for a hydronic heating system couldn't be easier. The pump is connected directly to a zone thermostat which in turn powers the pump on and off based on heat demand.

Heat loss calculations, PEX tubing layout, and other plumbing factors necessitate a discussion with your heating contractor to determine design and installation details.

Factoring in Solar Energy

As discussed in Chapter 5.3, a typical solar thermal hot water heating system will provide between 35% and 90% of your hot water requirements, depending on geographical location. What these figures do not tell you is when this energy is available and how it is dispersed over the year.

Figure 5.5-3. Active solar heating panels such as these evacuated tube thermal collectors can supplement your renewable energy hydronic heating system. The amount of solar energy developed during the winter and the heat loss of your home will determine the performance and cost effectiveness of the installation.

Appendix 5 and Appendix 6 illustrate the average sun hours per day for North America for the worst month and yearly average respectively. In the sun belt of the United States the average number of sun hours per day remains relatively constant throughout the year, meaning that solar energy is available during the summer and winter months. However, despite the availability of solar energy, home heating requirements in Arizona and Florida are minimal. In the northeastern United States and most of Canada solar energy is concentrated over the long summer hours. Once fall arrives and winter storms set in, sunlight hours per day drop dramatically, coincident with increasing heat demand.

A solar thermal hot water system design is balanced to provide an *annual average reduction* in purchased hot water heating fuel. Therefore, a solar thermal system will not meet 100% of the wintertime hot water requirement and will have little additional energy available for space heating.

Many people believe this is a problem that can be corrected by size: simply add more solar thermal panels and away you go. That's far too easy an answer. Firstly, a larger system may

provide ample hot water for space and other domestic requirements at the expense of having an enormous amount of excess energy that cannot be used in the summer. Secondly, the increase in capital cost for such a large configuration would almost certainly not be economical.

Perhaps storing some of this excess heat during the summer might solve the problem? Read on.

Thermal Storage and Heat Loss

It stands to reason that the amount of energy required for space heating will be dependent upon how much heat the home loses. Heat loss is a function of how well the home is constructed and includes such variables as building envelope volume, insulation quality, and air tightness.

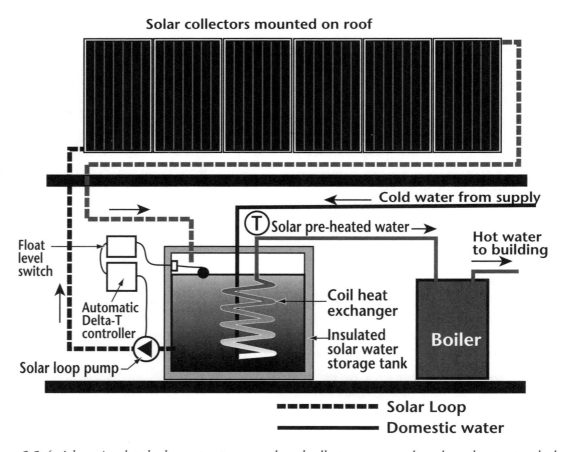

Figure 5.5-4. A large insulated solar water storage tank and collector array supply preheated water to a hydronic heating system. The practicality of such a system will depend to a large extent on the heat loss (energy requirement) of the building. It is not practical to store heat energy from one season to the next.

In addition, the difference between indoor and outdoor temperatures adds a further complication.

There is also a difference between temperature and heat that must be understood prior to tackling the challenge of thermal energy storage. If you heat 1 cup (237 ml) of water over a flame, its temperature will rise more rapidly than that of 1 quart (1 L) heated by the same flame. Therefore, as the mass of water increases, so does the amount of heat energy required to achieve the same temperature.

In order to store heat, a mass (or weight if you prefer) of a substance is required. Fortunately, water is an ideal substance as it has an unusual ability to store large amounts of heat energy, a phenomenon known as having a *high specific heat*. For example, the specific heat of lead is nearly 30 times lower than that of water, even though its density is higher. The measure of the specific heat of water is 1 BTU per pound per Fahrenheit degree (1 calorie per gram per Celsius degree). Expressed in English, this means that for every pound of water that is heated by 1°F, one BTU of energy is added. Conversely, for every pound of water that is cooled by 1°F, one BTU of energy is removed. (As the majority of heating contractors continue to work with British Thermal Units or

BTUs, let's stick with this system. For discussion purposes, 1 BTU = approximately 1 kilojoule or 239 calories.)

Consider that a tank of water weighing 1,000 pounds (120 gallons/ 455 L) which is heated to 50°F (28°C) above room or ambient temperature will have 50,000 BTUs of added energy. Assume for a moment that the tank is installed in a room which has a heat loss of 10,000 BTUs per hour. If we transfer 10,000 BTUs per hour from the storage tank to the room, it is possible to maintain a constant temperature in the room for approximately five hours. Once all of the heat energy is removed from the tank, the room temperature begins to fall.

Stored Energy (BTUs) ÷ Heat Loss (BTU/Hr) = Available Heating Time (Hours)
and:
Heat Loss (BTU/Hr) x Desired Heating Time (Hours) = Heat Storage Required (BTUs)

The simplest way to determine your home's heat loss is to have a heating contractor or home energy auditor run the calculations for you. It is not possible to generalize the heat loss for all homes and locations other than to say that the more extreme the winter temperature and the

Figure 5-5-5. A thermal storage tank can store an enormous amount of energy from intermittent sources such as solar thermal and wood boilers. When equipped with an optional electric heating element, the storage tank can provide domestic hot water and space heating energy. (Courtesy STSS Company Inc.)

larger and more poorly built (older) the house, the greater the heat loss.

A northern location provides a colder climate and fewer sun hours per day, resulting in lower heat production for storage.

A well-built home in wintry Montana may, for example, lose 400,000 BTUs of heat energy per day. In order to store this amount of energy for a one-month period the storage tank must be large enough to carry 12 million BTUs of energy:

400,000 BTU/day Heat Loss x 30 days' Desired Storage = 12,000,000 BTU

A storage tank containing water heated to 200°F (93°C) or 130°F (68°C) above ambient or room temperature would have to weigh approximately 92,000 pounds (42,000 kg) in order to store this much energy. To hold this mass of water the tank must have a capacity of 11,500 gallons (43,500 L). This is simply impractical.

On the other hand, if the heat loss is considerably lower, perhaps due to a more temperate climatic zone or better-built house, a smaller thermal storage unit may be practical. Furthermore, multiple energy sources can supply heat to one central storage tank. For example, the large storage tank shown in Figure 5.5-5 can be sized to hold up to 1,600 gallons (6000 L) of water with a weight of 13,000 pounds (5897 kg). Assuming a water temperature of 100°F above ambient, this equates to approximately 1.3 million BTUs of energy, enough to supply a family of four with a couple of weeks' worth of domestic hot water and perhaps a few days of space heat buffering.

The manufacturer STSS Company Inc. of Mechanicsburg, Pennsylvania has developed a variety of sizes of these unique storage tanks, each with the ability to be shipped flat and then expanded at the final destination. Water/glycol solution can be circulated in a closed loop through the spiral heat exchanger, supplying domestic hot water and space heating.

An integral electric heating element and thermostat provide backup energy to maintain tank water temperature. Heat exchanger coils may also be routed to a wood- or oil-burning boiler, heat pump, and solar thermal collectors for additional energy input and storage.

The complexity of thermal storage units dictates that a comprehensive heat loss analysis of your home be completed prior to sizing the energy input sources and possible thermal storage system. This applies to solar thermal heating as well as to traditional heating sources such as fossil-fuel and electric heating supplies.

Given the difficulty in cross-season energy storage (from summer to winter), solar thermal systems do not generally make *eco-nomic* sense in areas where winter heat demands are high.

Figure 5.5-6. Solar thermal systems are the most economically feasible means of harnessing the sun's energy provided they are installed professionally and for the correct application. These evacuated tube solar thermal collectors provide the homeowner with virtually free domestic hot water but would make little financial sense in a space heating system.

5.6

Heating with Renewable Fuels

If you ask people to name all the renewable home-heating fuels, chances are they will stall immediately after they mention "firewood." In many parts of the developing world wood is as scarce as the proverbial hen's teeth, leaving people to search for alternatives such as dried peat or cow dung. Fortunately we don't have to chase Bossy with a shovel in order to lengthen our list of renewable heating options.

Chapter 16 examines biodiesel as one renewable and clean-burning fuel for home heating and transportation requirements. For those heating with regular fossil furnace oil it may come as a surprise to learn that No.2 automotive diesel fuel heating oil and its renewable partner, biodiesel, are virtually interchangeable and may be blended at any desired ratio.

Biodiesel is a vegetable oil or animal fat derivative that is relatively clean burning as well as

Figure 5.6-1 Biodiesel is a vegetable oil or animal fat derivative that is clean burning and carbon neutral, which means that it contributes no net carbon dioxide greenhouse gas emissions to the atmosphere.

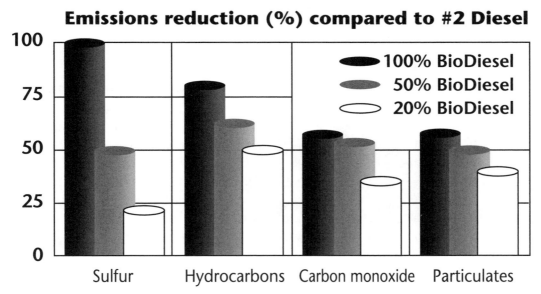

Figure 5.6-2. Biodiesel may be used in its pure form, designated B100 to indicate the percentage of biodiesel in the mix, or blended with fossil-fuel heating oil. This chart indicates the reduction in various atmospheric pollutants based on the blend concentration.

carbon neutral, which means that it contributes no net carbon dioxide greenhouse gas emissions to the atmosphere. With North American production of biodiesel running in excess of 30 million gallons per year (79 million liters per year) and numerous states offering it as a clean alternative to fossil-fuel heating oil, demand is starting to heat up.

Wood, Wood Pellets, and Corn

Heating with wood goes back thousands of years. Europeans who settled in the New World along the east coast and in Pennsylvania experienced winters in North America that had a ferociousness never seen in Europe. Rapid population growth coupled with poor energy efficiency of both wood-burning appliances and homes caused heavy deforestation and left towns with a grimy, sooty pall. It's no wonder that the majority of the population eventually jumped on the oil, natural gas, and electric heat bandwagon. Write a check, set the thermostat, and you have instant comfortable

Figure 5.6-3 Firewood stacked neatly in your garage or yard is like money in the bank. When properly burned in an EPA-certified wood heater it is a clean-burning energy-efficient heat source (Canada utilizes the EPA certification system). Wood pellets and dried corn also provide a clean-burning, renewable heat source with the added benefit of automatic stoking in many models of heating unit.

central heating—what could be better? Why would anyone take a step "backwards" into the wood-heating scene?

Since cavemen started using fire, there have been several advances in wood-heating technology. These advances have included new wood-burning stove designs that increase the amount of heat recovered and at the same time lower environmental pollution. Besides saving the environment, these technologies can also save your back; when you couple the energy-conservation techniques discussed in Chapter 2 with clean, high-efficiency wood-burning appliances, you reduce the amount of wood you have to cut, split, pile, and burn.

When you have decided on wood (for the sake of brevity "wood" includes wood pellets and dried corn and other biomass fuel sources) as your main or secondary source of home heating, check with your stove supplier or local building inspector for installation details. Since the installation of wood and pellet stove systems is complex and varies from region to region, no specific installation details are provided in *The Renewable Energy Handbook*. It is highly recommended that a contactor or professional installer do the installation work. Use the details provided here to increase your knowledge and understanding of wood-burning appliances and help you purchase the best unit to meet your needs.

Wood-Burning Options

Burning wood is not just about tossing a log into a fireplace or refurbished wood stove from the antique dealer and expecting to heat your house. It takes a high-tech wood-burning stove to heat cleanly and efficiently for a long period of time. There are several types of stove that fall into this category:

- Advanced combustion stoves
- Catalytic stoves
- Wood pellet and corn stoves
- Russian or masonry heaters
- Wood furnaces and boilers

Figure 5.6-4. This Vermont Castings (www. vermontcastings.com) catalytic wood stove easily heats this 3,300 square-foot, ultra-energy efficient home. Located in the very cold reaches of eastern Ontario, the home requires only two cords of firewood per heating season.

Figure 5.6-5. As this picture illustrates, there is an amazing array of wood-burning stoves and fireplaces available for purchase. A reputable dealer such as Embers located in Ontario, Canada is a good place to start your quest for home-heating energy self-sufficiency.

Designers and manufacturers of these units have as their primary aim the safe and clean extraction of every BTU of energy available from the wood you load into the stove. The reduction in frequency of wood loading lowers costs by reducing the amount of wood purchased or cut and processed.

The increase in efficiency comes first from creating an airtight burning chamber, causing controlled combustion, and then from developing a secondary combustion process, burning the smoke emitted from the fire before it reaches your chimney. (The masonry stove differs from this concept and will be discussed later in this chapter.) Although the EPA wood heater certification program was created to reduce air pollution, it resulted in added benefits like higher efficiency resulting in lower fuel consumption and increased safety. On average, EPA-certified stoves, fireplace inserts, and fireplaces are one-third more efficient than older conventional models. That means one-third less cost if you buy your wood and a lot less work if you process it yourself.

An airtight burning chamber is required to limit the amount of oxygen reaching the fire, lengthening the burning time and evening out the heat flow from the wood stove. This eliminates traditional fireplaces as efficient heat sources, as the energy they produce is negated by the massive amounts of cold makeup air, drawn from the outdoors, that is required to keep the fire burning. A slow, smoky fire also increases airborne pollutants and the risk of chimney fires as a result of unburned fuel coating the chimney as creosote.

The idea of burning the smoke before it reaches the chimney may sound like snake oil, but it's true. Modern wood-burning appliance technology is aimed at developing ways of capturing the unburned fuel present in the emission of smoke. Smoke is the result of burning wood decomposing into "clouds" of combustible gases and tar. Applying additional oxygen and heat or special catalyzing materials causes a secondary "burn." This results in increased heat output and lower atmospheric emissions as well as a decrease in fuel dollars and backache.

Advanced Combustion Units

Advanced combustion wood stoves expand on the concept of a simple box stove that has been in use for over a hundred years. Simple stoves allow the heat and smoke of the fire to travel in a direct path up the chimney. The advanced combustion stove places an "air injection" tube into the smoke path. Secondary inlet air is drawn in through the tube, increasing the oxygen content of the smoke and causing it to burn. The smoke then travels through a labyrinth path, radiating heat before it exits via the chimney. This approach is like adding a turbocharger to a car's engine: free horsepower from otherwise wasted exhaust.

Catalytic Units

Catalytic wood stoves operate on the same principal as an automotive catalytic converter. Unburned wood smoke is routed through a catalytic device, a ceramic disk made with a myriad of honeycomb holes running through it. The disk is coated with a special blend of rare earth metals that have

the unique feature of lowering the combustion temperature of the exhaust gases when mixed with a secondary supply of atmospheric oxygen.

As the smoke (or exhaust) passes through the catalytic device, it is mixed with oxygen from an air inlet and it ignites. If you watch the wood-burning stove's catalytic converter operating, you will see it glowing red hot and engulfed in flickering flames as the smoke is being consumed. A quick glance at the chimney on a crisp, cold January afternoon verifies the operation, with nothing but wisps of steam being emitted.

The downside of catalytic technology is the requirement to move a control handle from the bypass to operating position once the stove has reached operating temperature. Although this may sound like a minor issue, laziness or neglect in operating this control will negate the energy-efficiency and emission-reduction capabilities of the stove. (I will personally admit to leaving the converter turned off in the early fall and spring, when home heating demand is much lower.)

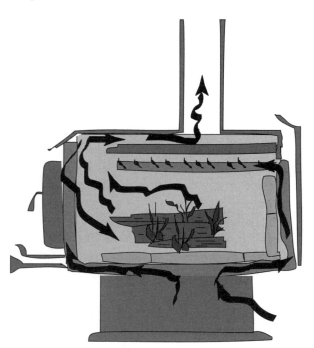

Figure 5.6-6. The advanced combustion stove relies on secondary oxygen intake and smoke burning to increase efficiency.

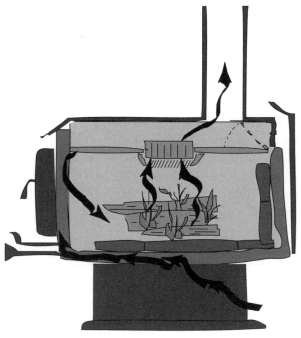

Figure 5.6-7. The catalytic wood stove uses a device made with special rare earth metals which allows wood smoke to burn in the presence of a secondary air supply.

Another consideration is the cost of replacing the catalytic converter after burning approximately 10 full cords (each full cord is 4' deep x 4' high x 8' long. / 1.2 m x 1.2 m x 2.4 m) of firewood.

Wood Pellet and Corn Stoves

Pellet stoves use dried grains such as corn as well as other biomass materials such as wood by-products from furniture, lumber, and pulp and paper industries. The waste material is ground and pressed together using naturally occurring resins and binders to hold the rabbit-food-like pellets (Figure 5.6-3 inset) together. As a waste product, biomass fuel pellets offer excellent synergy, heating your home while reducing landfill waste and needless greenhouse gas emissions at the same time. Pellets are convenient, as they are supplied in neat and compact dog food-size bags which can be stacked in your garage ready for use. Simply scoop a bunch into the hopper of the stove about once per day and the controlled feeding unit will automatically deliver the right amount of fuel to the burner.

As a small aside, many people mistakenly believe that hardwood pellets are a better choice than softwoods. This notion is incorrect; hard and softwood pellets have the same energy density per unit mass. Purchase pellet fuels with the lowest cost per pound or ton.

The only downside is: no electricity, no fire. A biomass stove draws electrical power 24 hours per day and may require too much energy for the off-grid system to supply during the dark, dreary days of winter. Use caution when evaluating the energy demands of these devices.

Most biomass stove manufacturers provide an automatic battery backup system to operate their units in the event of a power failure. This device operates in the same manner as an uninterruptible power supply (UPS) for a home computer. During normal operation the pellet stove operates from utility power and the UPS recharges simultaneously. When a power failure occurs, the UPS draws on a battery and inverter system to generate electricity, powering the stove. The UPS must be sized in accordance with the electrical load of the stove and the estimated number of hours before utility power returns.

Russian or Masonry Units

When we think of Russia, we think of winters in Siberia. In that part of the world, serious heat is needed, and there is no fooling around with wimpy stoves. The masonry heater is a serious unit. It is designed with the firebox and chimney lined with refractory brick and the flue routed through a labyrinth path designed to slow the smoke on its way to the chimney. An external facing of brick, stone, or adobe completes the design and increases the heat-absorbing mass of the unit.

When these units are operated, a fast, furious fire is ignited in the firebox. Smoke and its accompanying heat zigzag through the flue passages, giving up energy to the surrounding masonry work. Depending on the locale, one fast fire per day is all that is required to heat the

Figure 5.6-8. The wood pellet or corn stove uses a motor-driven auger to feed precise amounts of waste wood pellets into a fire pot.

masonry, with the stonework slowly radiating its stored heat into the house.

The masonry unit achieves its environmental passing grade by creating a hot, fast fire. Fast firing of a stove will generate the same amount of heat energy as a slow fire, but in a shorter amount of time. The resulting hot fire consumes more oxygen than a slow, smoldering fire, burning more completely and emitting fewer pollutants.

The downside of masonry heaters is their cost, the need to keep re-firing, and their size and weight. Assuming you have the space and the floor joists to support one of these big guys, they offer a warm and pleasing welcome to any home.

Wood Furnaces and Boilers

Wood furnaces and boilers come in two varieties: indoor and outdoor. Outdoor wood boilers are super-large wood stoves surrounded by a tank of water or antifreeze solution. Heat from the burning wood is transferred to the fluid, which is then pumped into the house for space and water heating. A control system located in the house regulates the temperature of the boiler by dampening the fire. Most units also provide an alarm signal to indicate when it's time to re-stock the unit. Furnaces, which provide warm air, are available as indoor units only.

Indoor wood boilers and furnaces are often dual-fuel fired, like the unit shown in Figure 5.6-11 from Benjamin Heating Products (www. benjaminheating.com). These units are very similar to outdoor models except for their size, emission of pollutants, and appetite for firewood. A major advantage of indoor, combination-fired boilers is the ability to supply heat even after the

Figure 5.6-10. Outdoor wood-fired boilers such as this model keep the wood and chips outside. Although these units are popular they are also inefficient, gobbling up vast quantities of wood while spewing out considerable amounts of atmospheric pollutants because of their inefficient burning chambers, which cause smoky, smoldering fires.

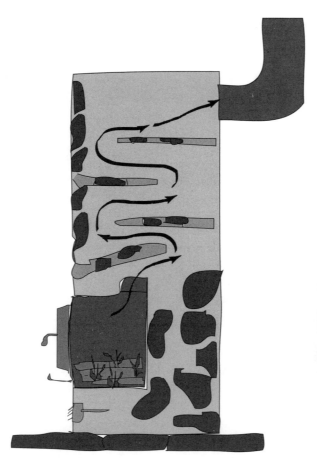

Figure 5.6-9. The size of a masonry stove makes it the centerpiece of the house. A popular design from the Old World, it provides heat long after the fire has burned down. (Courtesy Temp-Cast Enviroheat)

fire has burned down. The Benjamin unit contains both a high-efficiency wood-burning stove and an oil-fired backup burner which is used only when the wood-fired section cannot supply sufficient heat at the end of the burning cycle to maintain the desired boiler temperature. Units are also available with fan-forced heating which supplies a standard central air plenum or as hot-water boilers for use with hydronic heating.

The Specifics of Wood Heating

Wood heating systems can be adapted to almost any home heating system. The most popular style of heating system is the central warm air distribution design. A thermostat commands a centralized furnace to heat an air plenum or chamber. An electric fan circulates room air through the hot plenum to a series of ducts throughout the house. When the thermostat senses the room temperature is warm enough, the furnace turns off, causing the house to cool, whereupon the cycle repeats.

Hot water or steam boiler systems replace the hot air with water. These systems use either radiator units to transfer heat to the room or hydronic, in-floor hot water pipes as shown in Figure 5.5-1.

A wood stove or masonry unit uses direct radiation and convection to distribute its heat.

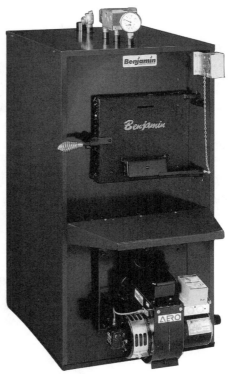

Figure 5.6-11. The combination boiler provides space and water heating in one compact unit. This model burns wood and oil efficiently and provides maximum flexibility of fueling options. (Courtesy Benjamin Heating Products)

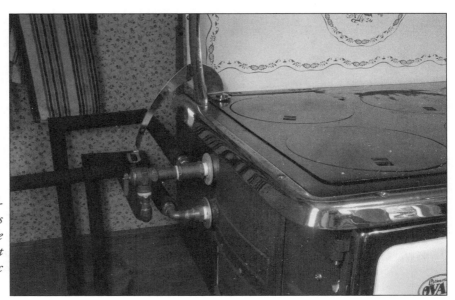

Figure 5.6-12. The Oval wood-cookstove from Heartland Appliances (www.heartlandapp.com) is available with an optional water-heating jacket that may be used for space or domestic hot water heating.

Figure 5.6-13. The Multi-Heat automatic stoker boiler is especially suitable for effective and environmentally friendly firing with biofuels such as wood pellets and corn. The unit's efficiency of approximately 90% rivals that of high-quality oil and gas furnaces. The large hopper must be filled approximately once or twice per week depending on heat requirements. (Courtesy Tarm USA Inc.)

You will have to decide if the heating system falls into one of the above designs based on several choices, including the following:

- Will the renewable heating system be your primary heat source or will it be supplementary? If it is to be your primary heat source, you must consider backup heating methods, particularly in colder areas subject to freezing.

- If you live in a more temperate area and travel little and renewable heating is to be your primary heat source, consider pellet or masonry heaters. These models can provide sufficient heat for up to one-and-a-half days. The Tarm USA Inc. Multi-Heat stoker boiler shown in Figure 5.6-13 can operate for up to one week without refueling. Heating for longer periods of time will require a friendly neighbor to stock the stove or an alternate heat source with automatic controls that activate when the fire dies.

- If you live off-grid, consideration must be given to the amount of electrical energy required to operate the heating system(s). Convection wood stoves require no electrical energy but must be tended frequently. On the other hand, the Tarm stoker boiler will operate independently for up to one week, albeit at a cost of electrical power that may be in short supply during the dark winter months. Propane heaters such as the model shown in Figure 5.6-17 require no electricity and integrate well as a backup heating source.

- Do you have ready access to wood or biomass pellets? Although purchasing wood and pellets can be less expensive than using fossil fuels, you will make up the difference in extra work carrying the wood and removing ashes. It may be cheaper still to harvest your own wood, bearing in mind the additional costs for tools, transportation, and time.

Anyone who has ever been to a cottage that is heated with an old Franklin or cookstove cannot imagine using wood as a primary source of heat. These old stoves gobble up armfuls of wood and never seem to get the place really warm. At other times it's the opposite problem: there is too much heat and belching smoke, making you wonder if your cottage has been transported straight into the depths of hell.

Many people use wood or wood pellet stoves as their primary heating source. Combining one of these stoves with proper energy-conserving techniques and passive or active solar heating will give you the best heating synergies. The answer to avoiding the cottage nightmare is good planning and the use of quality equipment.

As noted above, there are several systems for using wood as a primary heating source:

- convection space heaters
- wood-fired hot-air furnaces (single- or dual-fuel combinations)
- hydronic or hot water in-floor heating

Convection Space Heating

Convection space heating is the most common type of wood heating installation. No electricity is required to operate this type of unit, which is doubly good when the grid fails. Locate the wood stove in a central part of the house, with open passages to the upstairs and other rooms. An example of a well-placed unit is shown in Figure 5.6-14, where heat from the stove can easily move around the home and up the open stairwell, displacing cool air.

A common problem with space heating units occurs when people purchase a model too large for their needs. Make sure you discuss size with your dealer, and explain your requirements regarding primary or secondary heating as well as your floor plan. Most first-time wood stove users cannot believe that small units are capable of heating an entire house.

Figure 5.6-14. A wood stove placed in a central location near open areas of the home facilitates airflow, helping to ensure an even flow of heat. Remember to plan access for wood supplies coming in and ashes going out.

If your heating area is enclosed, or broken into many rooms, it is possible to use a small "computer fan" to blow air down hallways or between rooms. These units are available from wood stove dealers. Many building codes allow for grates to be cut between the ceiling of the heated room and the floor above, allowing natural air circulation.

Another option is to consider a pellet stove in tight locations. Because of their precise electronic controls, it is possible to lower burning rates below that of a typical wood space heater without degrading environmental operation and still provide a glowing hearth. An example of a space-heating pellet stove is shown in Figure 5.6-15.

A pellet stove hopper holds up to one-and-a-half days' worth of fuel. This offers the added advantage of not having to run back home to re-stock the stove or bugging your neighbor should you decide to stay out late at night.

Grid-connected homes with an existing central furnace provide an automatic means of supplying backup heat. When the fire goes out the thermostat turns on and the furnace kicks in. Using the furnace air distribution system also provides an alternate means of moving the heat from around the wood stove. Figure 5.6-16 illustrates such a design. Heat from the wood stove pools at the ceiling level. An existing or additional cold air return or "suction duct" is located near the ceiling. The furnace fan is left in "manual" or low-speed mode, drawing excess heat from the room and circulating it throughout the house.

An effective alternative for homes without centralized air distribution or constructed with cathedral ceilings is to place ceiling fans in the heated area, blowing warm ceiling air downward and mixing it with cooler room air.

Wood-Fired Hot Air Furnaces

Wood-fired furnaces operate in the same manner as fossil-fuel units. The main advantage of wood furnaces is the fan-forced, temperature-controlled system that allows even heat distribution. All models contain automatic dampers, thermostatic fan control, and safety shutdown dampers. Due

Figure 5.6-15. A wood pellet stove offers the look of a traditional wood-burning stove with the convenience of only having to add a few scoops of pellets into a hopper to provide equivalent heating capabilities. (Courtesy Harman Ltd.)

to the air-heating plenum design, central air conditioning may be added without difficulty.

Most models of wood-fired hot air furnaces are designed as dual-fuel units. These systems switch from wood to a fossil/biodiesel fuel or electric resistance heat source (not allowed off-grid) automatically as the fire dies down.

Before purchasing one of these models ensure that it carries the EPA certification label. Older designs of wood/electric furnaces that were sold during the height of the 1970s oil crisis are woefully inefficient and will eat you out of house and home.

Fan-forced hot-air furnaces are generally not recommended for off-grid applications. This recommendation stems from the energy requirements of the blower motor during the winter months when renewable energy electrical production is reduced. The Benjamin combination furnace is possibly the most efficient fan-forced

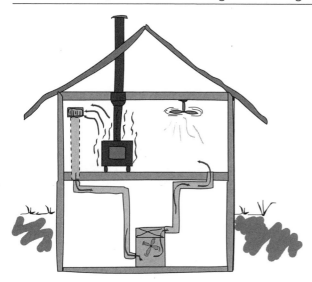

Figure 5.6-16. The air distribution provided by the central furnace fan provides the best method of moving heat away from the wood space heater. A ceiling fan is a good choice for cathedral ceilings or off-grid homes.

model on the market. This unit is similar to the hydronic boiler model shown in Figure 5.6-11.

250 watt motor x 10 hours per day - 2.5 kWh per day consumption

Depending on your location and heating requirements, you may need more or less fan running time per day. This simplified example shows that even a unit having a very small, energy-efficient fan requires a large percentage of your winter electrical power generating capacity. Operating in oil backup mode, the energy required to operate the oil pump and combustion blower will effectively double this energy consumption.

A Word About Outdoor Wood Boilers

The outdoor wood boiler shown in Figure 5.6-10 is becoming a common sight in rural homes and farms in the northeastern United States and Canada. Although these units are popular they are also very inefficient, gobbling up vast quantities of wood while spewing out considerable amounts of atmospheric pollutants because of their poorly designed burning chambers, which cause smoky,

smoldering fires. Many people say that the wood is free, so who cares about efficiency? I personally know several people who own these monsters and I am amazed by the number of hours required to cut, split, haul, and load wood for the cavernous maw of these units. If you add up all of the labor, fuel, and chain saw expenses required to keep these units fed, you will find them anything but cheap to operate.

Many towns and municipalities are outright banning these devices, and rightly so. If you care about the environment, forget about purchasing one of these environmental disasters and consider a solid fuel combustion unit with an EPA certification providing low combustion emissions.

Fan-forced heating is not recommended for off-grid applications.

Wood Fuel – The Renewable Choice

Nature provides us with an endless supply of wood. Dead trees rotting on the forest floor produce the same amount of greenhouse gases and heat as if you had burned the wood in your stove; it just takes a little longer. Sustainable woodlot management requires the thinning and cutting of damaged and dying trees. Even in a small acreage, this "waste wood" will provide sufficient fuel for the largest house.

The species and quality of the wood you burn will have a major impact on the ease of use of your heating system. Freshly cut, wet wood can contain up to 50% moisture by weight. Attempting to burn this wet wood will result in difficult ignition, with reduced heat output and greatly increased pollution. On the other hand, properly dried and seasoned wood will ignite rapidly and provide nearly twice the warmth with less work.

Firewood should be cut and split (at least in half) early in the spring and properly stacked. Piles of wood should be neatly arranged in covered rows, allowing space between each successive row for air to blow through. The summer warmth and breezes will quickly reduce the weight of this

Figure 5.6-17. Propane fireplaces, radiant heaters and freestanding units such as this model require no electricity and can be sized to meet the heating requirements of any house. They also look better than the $5,000 furnace in the basement.

wood by half, saving your back when you bring it inside next winter and at the same time ensuring that the stored energy is not being used to boil off water and sap in unseasoned wood. During the seasoning process, wood will start to crack and split on the ends and turn a grayish color. Picking up two pieces of well-seasoned wood and banging them together will create a clear ringing tone. Doing this with freshly cut wood will result in a dull "thud."

The type of wood you purchase or cut will also make a difference. Softwoods such as white birch and poplar are fine for fall and spring when you are just a bit chilled. However, northern winters require serious heating woods such as maple, oak, elm, and ironwood. These woods have a higher density than softwoods, resulting in higher heat output. Pick up an armful of white birch and an armful of maple. The maple feels twice as heavy as the birch, and guess what? It puts out about twice the energy for the same volume. As wood is sold and trucked by volume, purchasing the hardest

woods will save you money and reduce the number of trips to the woodshed. Just remember to get at least a little mix of the softer woods for those fall and spring days when the sun is providing some of the energy heating mix.

People living in areas where hardwoods are not available or where heating loads are lower will have to opt for softwoods or pellet fuels.

Purchasing Firewood

Your local dealer will offer you a confusing blend of softwood, mixed wood, mixed hardwood, fireplace cords, face cords, and full cords, green or seasoned. It is important to understand what all this means; otherwise your experience with heating in winter may make you think you are back in colonial Pennsylvania.

Firewood is sold in "face cords" or "full cords." The wood is typically delivered pre-cut in 16" (41 cm) logs. When stacked in three tight rows, a full cord of wood is created that should measure 4' deep x 4' high x 8' long (1.2m x 1.2m x 2.4 m). A face cord is, as the name implies, one row or "face" of a full cord, which is equal to one-third of a full cord (16" deep x 4' high x 8' long / 0.4m x 1.2m x 2.4 m). But be careful. If the logs of the face cord are cut into 12" (30 cm) lengths, it will yield only a quarter of a full cord. Likewise, be

Figure 5.6-18 Firewood should be properly stacked and covered and allowed to season for at least six months to drive off moisture. Dry firewood should be a grayish color and have cracks on the ends.

careful when purchasing firewood that is sold "by the pile" or truckload. It is next to impossible to judge volume.

If firewood is sold "green" it means that the wood is not seasoned. If you can get a discount for buying green firewood and have the time to season it, go right ahead. Wood can also be purchased in 4' or 8' lengths. Provided you have the tools, time, and stamina to cut and split your own wood, the savings may be well worth the effort. To quote Thoreau, "Wood heats you twice, once when you cut it and once again when you burn it."

How Much Wood Do You Need?

The best way to figure this out is to try. Our well-insulated home uses only two to three full cords of mainly hardwood (and a bit of softwood) per year. A smaller turn-of-the-century stone home just down the street, which has no insulation to speak of, uses six full cords. If you have an inefficient outdoor boiler, you may use 3 to 4 times the amount of wood used by a high-efficiency indoor stove. It just depends. The best thing to do is to buy more than you think you may require. Covered wood will not rot and should last up to three years, after which it will start to decay. So purchase a bit more and let it sit like money in the bank. You can always use it next year.

Cleanliness and Ashes

There is no doubt that wood heating is messier than heating with natural gas or electricity. Wood must be brought in to the house and ashes removed, creating a path for wood chips and dust. If white shag carpet is your thing, then forget about wood heating. A reasonable plan to ensure cleanliness is to place the wood stove and wood storage area as close to a door as possible or put the heating unit in the basement. Install tile or other easy-to-clean flooring between the stove and door to allow for quick sweep-ups after loading the wood box or emptying ashes.

Most modern stoves are equipped with integral ash buckets, making ash removal less of a chore. I clean out the ash boxes twice a week during the heating season, creating an ash pile the equivalent size of 2 to 3 large garbage cans. The ashes can be spread on icy roads to increase traction, added to garden soils, or simply spread throughout fields.

Ecology and Wood Heating

Many people argue that heating with wood moves the source of pollution from the "clean-burning stove" to the chain saw and splitter. While there is no doubt that forest management and wood harvesting takes energy (plenty of it according to Thoreau) the environmental impact can be managed with the proper attitude and correct tools and supplies.

The best way to reduce wood harvesting energy consumption, whether it's your energy or chain saw fuel, is to reduce wood consumption. A home that is properly constructed or retrofitted

Figure 5.6-19. Most modern stoves are equipped with integral ash buckets, making ash removal less of a chore. I clean out the ash boxes twice a week during the heating season, creating an ash pile the equivalent size of 2 to 3 large garbage cans.

for energy efficiency, along with the most efficient wood heater you can afford, can reduce wood consumption by a factor of 2 or 3 times.

Proper wood lot management is an important area to consider and consultations with your local forestry management association will ensure sustainable harvesting practices.

As for the gasoline-powered tools, consider the following options:

- Purchase chain saws or other gas-powered tools that are equipped with catalytic converters to reduce emissions.
- Small two-cycle engines are notorious for spewing oil and other air pollutants. Manufacturers such as Jonsered (www. jonsered.com) offer several environmentally friendly products which have their own Environmental Product Declaration, meaning that each product will be documented for manufacturing, packaging, recycling, and waste disposal.
- Purchase biodegradable chain saw oil and grease. These products won't damage the ecosystem or your lungs. Chain saw oil is vaporized during the sawing process, creating

Figure 5.6-20. To reduce the environmental impact of harvesting wood, use sustainable wood lot management practices, ethanol-based gasoline in catalytic converter-equipped chain saws, and biodegradable vegetable-based chain oils and greases.

an oil-based aerosol. Biodegradable oil products lessen the health impact of breathing this potential hazard.

- Do not leave engines idling unnecessarily.
- Use ethanol-based gasoline to reduce CO_2 emissions.

You cannot eliminate the energy requirement and the pollution created in harvesting wood, but with a little care and attention to detail these can be dramatically reduced.

Wood Stove Heat Recovery

Generating thermal energy in the summer is akin to stealing candy from a baby. Low heat demand for baths, space heating, and clothes drying (you do use a clothesline, don't you?) means that a solar thermal water heating system should be able to do the job.

Once the calendar flips over to November, the task becomes quite a bit more difficult in most parts of North America and especially in the Northeast United States and in Canada. It is during the period from November to March that most off-grid folks rely on propane for some or all of their thermal energy. Grid-connected people, too, purchase more fuel in cold weather, but those using a wood heating system might wish to consider the concept of wood stove heat recovery.

The idea is simple enough. Store a small percentage of energy to use on the proverbial rainy day and you may not need to purchase any supplementary fossil fuels at all. As anyone who has sat next to a wood stove will attest, even the smallest of units will deliver a tremendous amount of energy into the area where they are located. The trick is to capture some of that intense thermal energy, store it, and then redistribute the heat on an as- and where-needed basis.

The simplest way to capture this energy is with a wood stove that includes a heat recovery system as part of its standard design. The Heartland Oval model cookstove shown in Figure 5.6-12 is one example, although other commercial models offer this optional feature. Alternatively, an indoor or outdoor wood-burning boiler can be used. For those who are a bit more adventurous and wish to consider reducing their carbon footprint to near zero, converting an existing wood stove might also be an option, although care must be taken to ensure that this does not affect home insurance policies or local building code regulations. Consider the following an experimental demonstration of the technology.

Whether you decide to use a commercial boiler or modified stove, the harvested energy must be stored and delivered to the desired thermal loads. This issue will be dealt with following the discussion of the technology behind wood stove heat recovery.

The Vermont Castings Consolidated DutchWest model shown in Figure 5.6-21 is the largest wood stove manufactured by the company. After reviewing the design of this unit, I realized that there might be two means of adding heat recovery to the design without lowering internal firebox and flue temperatures. (A lowering of smoke temperature will increase the possibility of creosote condensing on the stove walls and chimney, which under certain conditions can lead to a stove or chimney fire.)

Externally Mounted Heat Exchanger

With the top cover removed, it was easy to see an area where a heat transfer pipe might be added to the stove (Figure 5.6-26). Once the basic design was decided on, two holes were drilled in the back plate of the stove, allowing the half-inch copper pipe of the heat exchanger to pass through.

The aluminum-finned copper heat transfer pipe was dry fitted to the wood stove, cut, and soldered to make as long a loop as possible. Be sure to remove the plastic fin supports or they will melt shortly after the stove is lit.

Once the heat exchanger was mounted, the plumbing supply pipes were routed to the stove. In my example, the installation was a retrofit job and I was unable to hide the pipes in the floor cavity, so I used commercial pipe hangers to hang the exposed pipes along the wall/ceiling transition

Figure 5.6-21. Whether you decide to use a commercial boiler or modified stove, the energy that is harvested from the unit must be stored and delivered to the desired thermal loads. The Vermont Castings Consolidated DutchWest model shown here is the largest wood stove manufactured by the company and is the model selected for modification for heat recovery testing.

area in the basement and covered them with pipe insulation.

The heat exchanger was connected to the plumbing system and tested for leaks. Once testing was finished, I painted the heat exchanger with flat black, high-temperature paint to aid in heat absorption.

Subsequent testing with only the "hot air" external unit did not provide sufficient energy for my intended application on its own. I did find that it would add some energy to the domestic hot water system, but I decided that a heat exchanger internal to the firebox would be required.

Figure 5.6-22. The top plate of the Consolidated DutchWest model is removable, allowing access to the catalytic converter.

Figure 5.6-23. This view of the wood stove with the top cover removed shows the insulated cover of the catalytic converter at center.

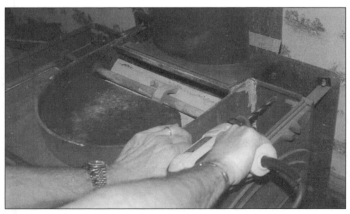

Figure 5.6-24. The back plate of the wood stove was drilled to accept the half-inch pipes of the heat exchanger.

Internally Mounted Heat Exchanger

The internally mounted heat exchanger design uses a non-finned, stainless steel tube mounted directly on the wood stove firebox. This tubing resists high temperatures and is capable of operating at extremely high pressures, well beyond the requirements of this application.

I bent the tubing into the shape shown in Figure 5.6-30. This design allowed the heat exchanger to remain as high in the firebox as possible to ensure that it did not impact wood stove filling and to leave a large clearance area around the catalytic converter. (Cooling of the flue gases around the catalytic converter reduces its efficiency and increases smoke and particulate output.) Two holes were drilled in the rear of the stove and the pipe was inserted. The clearance area between the tubing and the stove wall was filled with stove refractory cement, which is available at all hardware stores.

It is not possible to sweat solder stainless tubing to copper. Figure 5.6-31 shows a compression/transition fitting which compresses against the stainless steel tubing and provides a half-inch National Pipe Thread (NPT) connection to attach the tubing to the copper plumbing pipe.

The two heat exchangers were plumbed in series so that "cooler" return water would flow through the external exchanger first and then through the internal firebox-mounted unit. This provided the highest possible heat collection, as the water flowing in the system had to travel the longest possible distance through the wood stove heating circuits.

Plumbing Design Considerations

The major goal of this project was to determine if it might be possible to stop using propane fuel for our off-grid house. I was well aware that our solar thermal system would cover most of the load during the summer season, but I wanted to use all available energy sources used in our home on a year round basis to make sure the heat recovery project would not fail.

Referring to Figure 5.6-33, you will see a

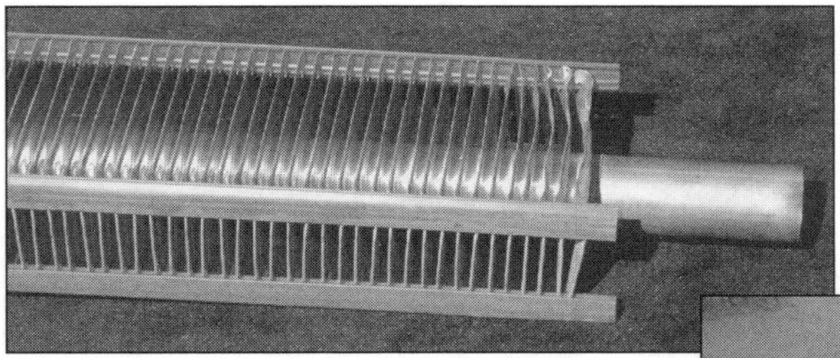

Figure 5.6-25. The heat exchanger is fabricated from the "core" of a standard hydronic heating baseboard heater. These low-cost elements come with aluminum heat transfer fins bonded directly to the copper pipe.

Figure 5.6-26. The heat exchanger was soldered into position, ready for painting and plumbing connection.

Figure 5.6-27. The plumbing pipes were fed through the floor to the heat exchanger ports.

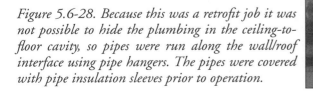

Figure 5.6-28. Because this was a retrofit job it was not possible to hide the plumbing in the ceiling-to-floor cavity, so pipes were run along the wall/roof interface using pipe hangers. The pipes were covered with pipe insulation sleeves prior to operation.

Figure 5.6-30. The internal (firebox mounted) heat exchanger is fabricated from a length of half-inch stainless steel hydraulic tubing. Any automotive supply shop or hydraulic fabrication shop can make any custom size and shape of heat absorber tube.

Figure 5.6-29. Once the system is soldered in position and checked for leaks, the heat exchanger is spray painted with high-temperature flat black "bar-b-que" paint to increase energy absorption.

schematic view of the hydronic heating system used in our home. If there are sufficient solar resources, the solar thermal system will add its own energy through the heat exchanger loop installed in the tank (see Chapter 5.3).

Incidentally, prior to installing the wood heat recovery system, I recorded that the propane system did not run between the months of April and October provided that we had showers and not baths, proving that the solar thermal and PV/ wind diversion energy was capable of providing 100% of our thermal energy needs for this seven-month period.

After installing the wood heat recovery system, I initially turned the propane water heater to the lowest "cut-in" temperature that Lorraine and I considered reasonable for our domestic hot water. Once the system started proving itself, we concluded that there was no need for the propane unit to run at all and it was simply turned off, remaining available to be restarted should the need arise. Having the propane water heater installed increased our hot water storage capacity to a total of 100 gallons (380 liters), 60 gallons in the solar thermal tank and 40 in the propane heater tank.

In most off-grid homes there are times when electrical energy production exceeds demand.

For example, a sunny, windy day almost always produces excess energy. Likewise long stretches of sunny summer days will see the batteries fully charged by ten o'clock in the morning. Once the batteries are full, the voltage regulation equipment will essentially stop feeding the batteries and the excess energy is wasted. To capture this energy, I have installed a bypass circuit that transfers it to a 24V, 1500W electric heating element installed in the first storage tank. Many direct current voltage regulators such as the Outback MX-60 contain diversion load control functions (Figure 10-7).

Grid-connected homes do not have a load

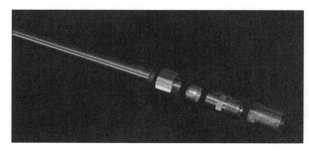

Figure 5.6-31. It is not possible to sweat solder stainless tubing to copper. This picture shows a compression/ transition fitting which compresses against the stainless steel tubing and provides a half-inch National Pipe Thread (NPT) connection to attach the tubing to the copper plumbing pipe.

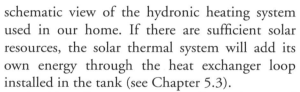

diversion feature since all of the electrical energy they produce can be sold to the electrical utility.

In this off-grid system, cold water from the house pressure system flows into the first storage tank, bottom right. Hot water then flows out the top, traveling to the "cold" inlet on the propane backup water heater already installed in the house.

Hot water from the propane tank continues to flow from the hot water outlet and into the house supply circuits. For the sake of simplicity, I have not shown the water tempering, pressure/temperature relief, or service valves in this overview.

An electrically operated water valve is shown connected to the house hot water supply and is labeled "hot water dump." The electric valve is operated by a device known as an aquastat (Figure 5.6-37), which is an adjustable thermostat installed in the solar thermal water storage tank. Its function is to monitor the storage water temperature and open the hot water dump valve prior to the pressure/temperature relief valve opening. I installed this as a primary fail-safe control to simply dump the very hot excess water outdoors. (Once demand is satisfied, there really is no other place to use the energy.)

During the first year of operating the system this dump valve was activated several times, and I decided to install a third water storage tank between the solar thermal and propane water tanks. (Again, for the sake of simplicity, this additional storage tank is not shown.) A 60-gallon electric hot water tank was used without any electrical connections. With this addition, the system supply and demand has so far remained in balance, ensuring that we have plenty of hot water without having to dump any excess.

Hot water now flows through a one-way check valve that prevents cold supply water from the pressure system from trying to work against the circulation pump feed direction.

The circulation pump is next in the circuit. I have installed a very low-energy-demand Grundfos water circulator pump which has a

Figure 5.6-32. The heat exchanger pipe is shown in the operating firebox. If you look closely, you can see that creosote has condensed on the relatively cool heat exchanger tubing. It will burn off from time to time and has no effect on the operation or performance of the heat exchanger. Also note that the tubing is looped around the intake of the catalytic converter (center), ensuring that flue gases are not cooled prior to entering the device.

power requirement of 25W, about the same as that of two compact fluorescent lamps. This pump is controlled by a thermal switch located in the wood stove flue and is activated any time the woodstove is hot.

Hot water then flows to the hot water control manifold shown in Figure 5.6-34. This circuit may appear a bit complex at first glance but is in fact fairly simple. Hot water from the circulation pump is supplied to the inlet of a three-way motor-operated valve at the point labeled "COM" (common). This valve has two operating positions, "open" or "closed," allowing the water to flow either to the branch circuit at the top (valve open) or to the lower distribution circuit (valve closed).

When the valve is in its normal open or standby position, water flows from the pump to the top branch circuit and continues to the return path circuit at the bottom of Figure 5.6-34 and Figure 5.6-33. From here, hot water enters the first wood stove heat recovery unit, absorbs energy, travels to the second stove, absorbs more energy, and returns to the "cold" inlet of the solar thermal storage tank. This path will continue endlessly provided either one or both of the wood stoves is providing heat.

It may seem rather wasteful having two wood stoves plumbed in series where one stove may be on and the other off and heat is lost as hot water travels through the cool stove. Having said this, the energy is not really lost; it is simply transferred to another location in the house, and from simple observation these losses seem to be minimal.

Another concern is the issue of water being blocked as it travels through the wood stove heat exchangers. A more efficient system would have water pass through a control valve which

directed water to the stove that was operating while bypassing the cool unit. The problem with this rather obvious logic is that if the control valve were to fail and the stove were to light, an explosive situation might occur as a result of steam pressure buildup. Better to lose a bit of efficiency and remain safe!

Whenever energy is required for room heat, operating the clothes dryer, or heating the hot tub, the directional control valve moves from open to closed, allowing water to flow through the distribution branch. A zone valve located on the branch opens, allowing hot water to flow through the selected device. The zone valve used

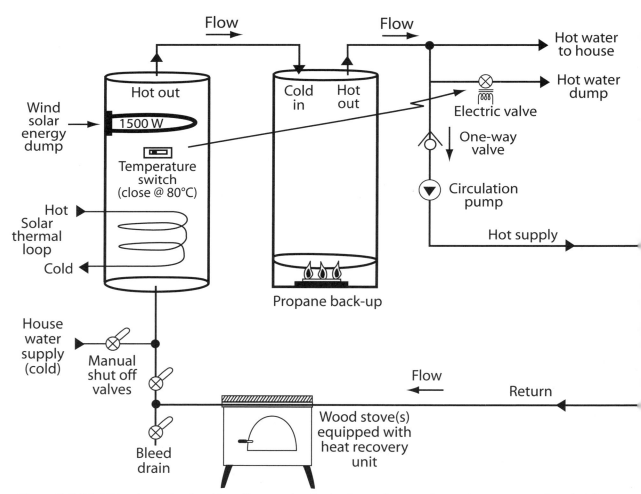

Figure 5.6-33. This schematic plumbing diagram details the thermal-energy-harvesting system used in the Kemp residence. Energy from the solar thermal system and wood stoves as well as excess energy from the PV and wind systems are all collected and distributed throughout the house demand system.

is a Skinner 7000 series general purpose direct-lift two-way valve that provides close to full-flow operation and consumes relatively little energy. These valves are available in several pipe sizes and coil-operating voltages using either alternating or direct current.

- Zone valve one allows hot water to flow through the hydronic baseboard heater located in the basement exercise room area.
- Zone valve two allows water to flow through a small radiator purchased from an automotive supply shop. The radiator is installed at the air intake of the clothes dryer. The dryer is operated in air-only mode.

- Zone valve three causes hot water to flow through a stainless steel heat exchanger that provides heat to the hot tub circulation system.

In each case, the device requesting heat activates the selected zone valve. A room thermostat activates zone valve one, the activation of the clothes dryer motor operates zone valve two, and the circulation pump of the hot tub activates zone valve three. Obviously, this selection of devices can be changed at will depending on your particular requirements.

Water leaving the selected zone then returns to the wood stove heat recovery circuit and back to the solar thermal storage tank, completing the cycle.

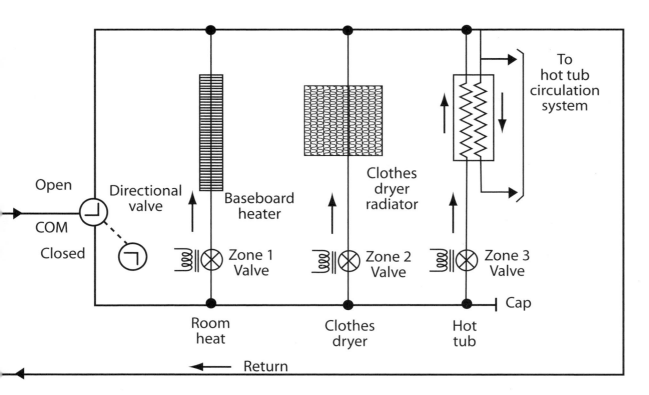

Figure 5.6-34. Hot water from the collection system described above is distributed on demand through this plumbing distribution system. In the author's home, hot water is used to provide energy for exercise room heating, the clothes dryer during the winter months, and the outdoor hot tub.

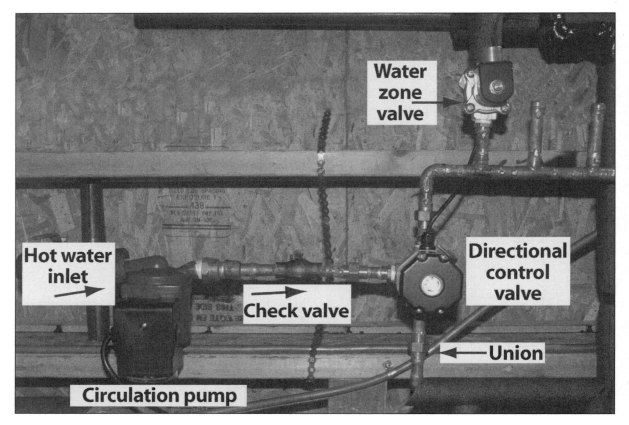

Figure 5.6-35. This pictorial view of the heat distribution system shows the physical layout of the schematic diagram detailed in Figure 5.6-34. Hot water from the propane water heater enters the circulation pump bottom left, passes through the one-way check valve, and is directed into the common point of the motor-operated valve. When there is no heat demand, water passes to the lower outlet of the valve and is fed to the wood stove heat recovery system. When heat is required, it passes out of the top outlet through a zone valve to the fixture demanding heat. (Directional control of the motor-operated valve is denoted with a small "⌐" which indicates flow from the common to top outlet when the valve is "open" and a " ⌐ " symbol indicating flow from the common to lower outlet when the valve is "closed." The direction of the water flow is indicated by the symbol. Note the pipe stub next to zone valve one, which was later connected to zone valve two.

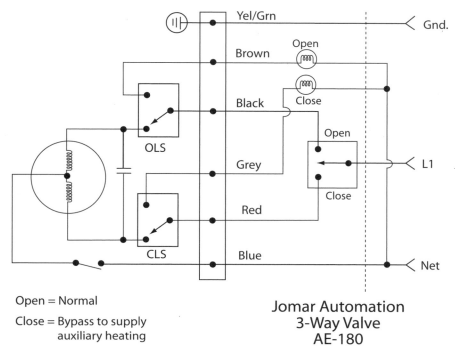

Open = Normal

Close = Bypass to supply
        auxiliary heating

**Jomar Automation
3-Way Valve
AE-180**

Figure 5.6-36. This schematic diagram is for the model AC-180 motor-operated valve manufactured by Jomar Automation. Wiring to the left of the central terminal strip is within the valve. Wiring to the right is by the end client and is shown with a toggle switch to operate the valve. Using a motor-operated ball valve provides full flow (no restrictions) through the valve ports and requires only electrical energy to move it from one outlet to the other, typically within 12 seconds. Applying 120VAC to the black terminal will "open" the valve, while applying the same voltage to the red terminal will "close" the valve. Optional 120V indicator lights may be added at the brown and gray terminals to indicate when the valve has moved to the fully open or fully closed position respectively.

Figure 5.6-37. An adjustable thermostat known as an "aquastat" is installed in the solar thermal storage tank, measuring water temperature and closing a switch whenever the water temperature reaches a desired set point. The aquastat operates an electric dump valve that opens, dumping hot water outside the house. The emergency pressure/temperature relief valves should not be used for this function except as a backup feature.

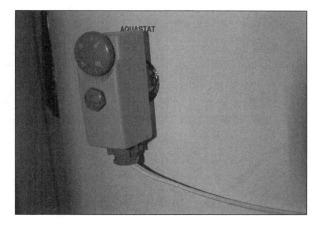

Figure 5.6-38. This photograph shows a typical hot water pressure/temperature relief valve or PTR valve. One PTR valve should be installed in every storage tank and at the exit piping of a heat recovery unit installed in a wood stove. These devices automatically open whenever the water temperature exceeds a preset limit or the working pressure inside the tank is exceeded, preventing a very dangerous steam-powered explosion. Ensure that PTR valves are plumbed to a dry well under the floor of the house or to an outside dry well and that the plumbing is angled towards the plumbing exit to prevent water or condensation buildup from freezing, potentially blocking the safety device. An insect screen is also required on the pipe outlet.

Electrical System Overview

Fortunately, the system does not require complex control electronics to make it operate. In fact, most of the control can be done using common off-the-shelf parts available at most hardware stores or electronic component suppliers such as Digi-Key (www.digikey.com).

Starting from the left side of the schematic the supply voltage is connected to a top and bottom "rail" with a protective earth ground. The circuit draws approximately 0.3 amps at maximum power; therefore a 1-amp slow-blow fuse should be connected in series with the LINE connection before the device is plugged into a household electrical outlet.

Voltage is supplied to one of the stove flue thermostats, which remain in their open position until the stove is operated. When the flue gas temperature reaches the operating set point the switch closes, applying voltage to the circulation pump and causing water to flow through the system as described above.

Simultaneously, voltage is applied to control RELAY1 and fed to the brown "open" connection of the motor-operated valve. The valve then moves to the open/recirculate position, absorbing heat from the wood stoves without feeding any of the secondary appliances connected to the zone valve(s). If an optional 120V indicator light is connected to the motor-operated valve brown

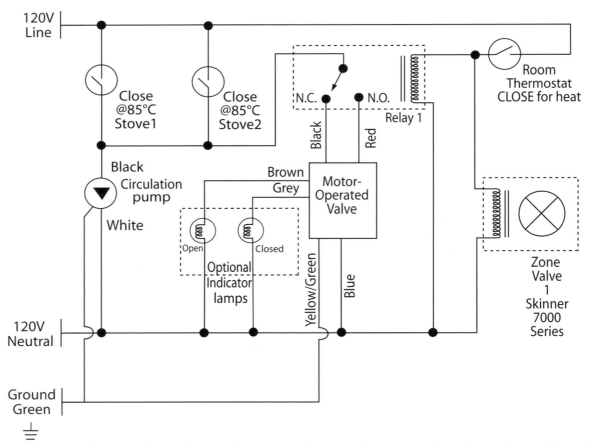

Figure 5.6-39. The electrical control circuit for two wood stoves and one zone valve is shown here. Additional zone valves can be added by duplicating the relay control circuit. Although this design operates on 120VAC it can be converted to work on 12, 24, or 48VDC as well.

wire it will illuminate, indicating an operating mode.

Should a zone request heat, for example by the room thermostat closing, RELAY1 coil energizes, causing voltage to be applied to the red wire of the motor-operated valve. In this case, the valve rotates to the position which allows water to flow to the zone valve supply manifold. Simultaneously, voltage from the thermostat is applied to zone valve 1, allowing hot water to flow through the baseboard heater as described above. If an optional light is connected to the motor-operated valve grey wire, it will illuminate, indicating that one or more zone valves have requested heat. Additional zones may be added by duplicating the connection of the relay and zone valve pair.

Conclusion

Although I have only had one year of operating experience with this system, the results have been quite satisfactory. I shall digress for a moment to explain that Lorraine was in the horse stable during the late winter and didn't hear the propane supply truck pull into the property. She was busy mucking out, delivering a wheelbarrow of manure to the pile, when she noticed the delivery man rolling up his hose to leave. "You didn't use much propane this year," he shouted over to her as he left the yard *without* dropping off an invoice.

I guess you could say that is all the proof we needed that the system works!

The only major concern I have with the design is that when multiple zones require heat at the same time the system is not able to keep up with demand unless both wood stoves are operating. This problem can be countered by either adding a longer heat exchanger pipe in the primary wood stove to collect more heat or investigating means of reducing thermal demand.

It is also possible to add a thermal heat recovery system to the exhaust of your backup generator, although this would require the addition of an antifreeze loop and heat exchanger to prevent freezing of the heat transfer liquid. Further, since we operate our generator so little I am not convinced of the value of this strategy for our home. However, off-grid homeowners who are starting without the benefit of PV panels or a wind turbine and use a diesel generator to charge batteries might find this concept intriguing. As a rule of thumb, for each unit of electrical power generated 1.2 units of heat can be captured.

5.7

Space Cooling Systems

Central air conditioning is by far the largest and least efficient load in the home. It really does not belong in an off-grid home.

The best way to keep your home cool is to stop the heat from getting inside in the first place. This might sound simplistic, but a well-insulated home with shading that blocks the summer sun is well protected against overheating. Open windows at night to create a cross-breeze and cool the house. In short, follow all of the guidelines discussed in Chapter 2, "Energy Efficiency," paying particular attention to:

- light-colored roof and exterior walls
- upgraded insulation in walls and roof
- upgraded radiant barrier insulation in attic rafters
- low-emissivity (low-E) windows that are not required to aid in winter heating
- window shading
- nighttime cooling using fans and natural cross-flow ventilation

Evaporative Coolers

For those lucky enough to have relative humidity levels below 30% in the summer, an evaporative cooling unit may be the ticket. These devices are very common in the southwestern region of the U.S. and use very little electrical energy.

Working in the same manner as perspiration cooling our bodies on a hot day, air blown through a wet pad is cooled as the water evaporates. An electric fan draws outside air to this wet pad, humidifying and at the same time cooling it. The conditioned air is blown into the home causing hot, stale air to be expelled outside. Because cooling only works with dry outside air, these systems do not work well where summertime relative humidity exceeds 70%.

Most evaporative coolers are rooftop installations that use a bottom discharge blower to channel conditioned air into the home. Rooftop-

Figure 5.7-1. For those lucky enough to have relative humidity levels below 30% in the summer, an evaporative cooling unit may be the ticket. These devices are very common in the southwestern region of the U.S. and use very little electrical energy. Several home-sized models are shown as well as portable coolers for camping and other off-grid applications. (Courtesy Southwest Solar)

mounted units are less expensive than ground-mounted configurations. Evaporative coolers can also be installed as add-ons to conventional refrigeration-type central air conditioning or window-mounted units.

Evaporative coolers require large amounts of moving air to cool a home. For desert climates a unit capable of supplying three to four cubic feet per minute of airflow is required for every square foot of floor area. For most other climates this airflow can be reduced to two to three cubic feet per minute.

Refrigerant-Based Air Conditioning

Most A/C systems are sized far larger than needed. This is done in an effort to make sure that "you're getting what you paid for": a very fast, obvious cooling of the indoor air. It also ensures that you don't call the installer back because the A/C isn't working well. On the other side of the coin, a large unit cools the air but does not have sufficient time to reduce indoor humidity levels. A smaller unit running for a longer period will ensure lower indoor humidity and temperature.

All air conditioners sold in North America must have an EnergyGuide label. Use this label as a guide to compare the energy efficiency of different models.

An alternative labeling program is known as the Electrical Efficiency Rating (E.E.R.) and is calculated by dividing the BTUs of "cooling power" by the watts of electrical energy used. An air conditioner with a higher E.E.R. rating is more energy efficient and less expensive to operate.

Don't get caught up in a quest for the lowest-cost air conditioner at the expense of energy efficiency. This is false economy, as any first-cost savings will be absorbed by higher operating charges time and time again. These rules apply to window air conditioners as well as central cooling systems.

Room air conditioners are less expensive to operate than central units if you only need to condition one or two rooms. When selecting a room air conditioner, match the area to be cooled to the rated capacity of the unit as shown in the accompanying table.

Although these units are very large loads on an off-grid system, they may be just the ticket to bring down the humidity and temperature to a reasonable level. One plus for these small units is that they tend to be used only when there is a surplus of energy. This normally occurs on those long, hot summer days that make the photovoltaic panels so happy and energy productive.

Area to Be Cooled (Ft²) / (M²)	Capacity (BTU per hour)
100 to 150 / 9.3 to 14	5,000
150 to 250 / 14 to 23	6,000
250 to 300 / 23 to 28	7,000
300 to 350 / 28 to 33	8,000
350 to 400 / 33 to 37	9,000
400 to 450 / 37 to 42	10,000
450 to 550 / 42 to 51	12,000
550 to 700 / 51 to 65	14,000
700 to 1000 / 65 to 93	18,000

Table 5.7-1 This chart indicates the capacity of the air conditioner required based on the size of the area to be cooled.

Central Air Conditioning Energy Use
kWh per 500 Hours of Operation

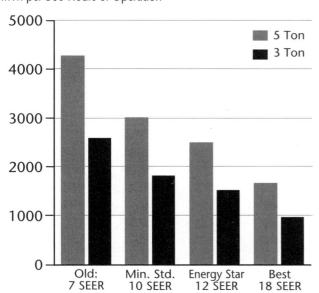

Figure 5.7-2. Purchasing an energy-efficient air conditioning system will reduce operating costs by more than half. Purchase models carrying the ENERGY STAR label or with a higher efficiency level based on EnergyGuide or EnerGuide ratings. (Source: Washington State University Extension Energy Program)

Figure 5.7-3. Now you see it, now you don't. This one-ton (12,000 BTU) air conditioner is required only occasionally. Installing it permanently in a wall-mount design as shown here reduces the amount of encroachment on your living space and allows for easy hiding.

5.8

Earth Energy with Geoexchange

Geoexchange technology is more commonly known by its many nicknames such as geothermal, heat pump, and ground-source heating. No matter what you call it, geoexchange technology offers a virtually endless supply of renewable solar energy that is stored just below the earth's surface since approximately 50% of the sun's energy is absorbed into the ground.

The basic concept is easy to understand. In winter, heat energy is drawn from the earth through a series of pipes called a loop. An antifreeze solution circulates through the piping loop, carrying the earth's natural warmth to a compressor and heat exchanger inside the home, hereinafter called a "heat pump."

The heat pump concentrates the earth's energy and transfers it via duct work or a hydronic heating system for space heating requirements. If domestic hot water is also desired, the heat pump can supply energy directly to any thermal water storage tank as discussed in Chapter 5.3.

Heat energy may also be extracted using open or closed loop groundwater transfer. In this configuration well or lake water is supplied directly to the heat pump and energy is extracted, with the cold discharge water returning to the lake or second well. Open loop systems may require annual coil flushing to remove mineral buildup if the source water is "hard."

In summer, the process is reversed; heat is extracted from the air inside the home and transferred back to the earth or water.

According to the U.S. Environmental Protection Agency and Natural Resources Canada, "They are the most energy-efficient, environmentally clean, and cost-effective space conditioning systems available." Because geoexchange systems burn no fossil fuels onsite, there are no locally produced atmospheric emissions. If the electricity to run the geoexchange unit comes from hydroelectric sources, greenhouse gas and other emissions are reduced to zero.

There are approximately 500,000 installations in the United States and 35,000 in Canada, resulting in annual energy savings of approximately 4 billion kWh of electricity, eliminating 20 trillion fossil-fuel BTUs and slashing greenhouse gas emissions by approximately 3 million tons (2.7 million tonnes).

Air-to-Air Heat Pumps

An air-to-air heat pump operates in the same manner as a geoexchange heat pump with the exception that heating and cooling energy is transferred between indoor and outdoor air rather than between indoor air and the earth or well water. The common central air conditioning unit is an example of a unidirectional heat pump that can provide only space cooling.

Air-to-air heat pumps are not as efficient as their ground-source cousins and are better suited to mild climates if large heating loads are anticipated. When operating in heating mode, energy extraction from the outside air diminishes as the temperature drops. At a temperature close to the freezing point of water, the unit can no longer extract heat from the outside air. When this occurs, a standard resistance-heating element, similar to those used in baseboard heaters and electric furnaces, is activated. Although heat provided by the electrical element is no more efficient than that of any other electric heating source, the overall seasonal heating efficiency of the heat pump *and* resistance heater must be taken into account, and it will always be greater than the heating element operating alone.

Measuring Heat Pump Efficiency and Value

Heat pumps use electrical power to operate a compressor which extracts approximately two-thirds of the home's required heating energy from the ground. Because this energy is free and the heat pump puts more energy into the home than

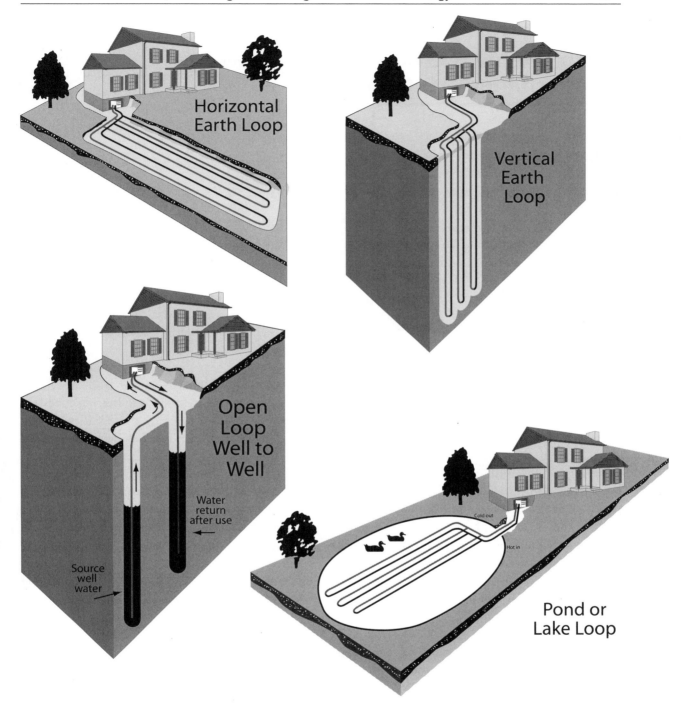

Figure 5.8-1. Geoexchange heat pumps are available in two broad categories known as open loop (bottom) and closed loop (top). Open loop systems extract heat directly from water pumped between two wells or from a lake or stream. Closed loop systems use a large array of pipes buried in the ground through which a circulating liquid returns heat to the compressor unit located inside the home.

Figure 5.8-2. The air-to-air heat pump is a cross between a central air conditioning unit and a geoexchange heat pump. Heating and cooling energy is transferred between indoor and outdoor air rather than between indoor air and the earth or well water. The outdoor component contains the compressor and heat exchange coil while the indoor unit contains the air handler, blower, and second heat exchange coil. Insulated refrigerant lines run between the two components.

it consumes from the electrical grid, heat pumps have an energy efficiency rating of greater than 100%. For every kWh of electricity supplied to the heat pump, between 2.8 and 6.7 kWh of energy are supplied to the home. This efficiency factor is known as the *coefficient of performance* (COP) and is calculated by dividing the heat output (in watts) by the energy input.

Heat pumps are able to operate in either a heating or cooling mode of operation. Therefore, when shopping for a system you must make coefficient of performance comparisons for both conditions. In the sun belt of the United States, cooling mode will be used much more frequently than heating mode. In the northeast and throughout much of Canada the opposite is true. When shopping for a heat pump, purchase the model with the highest overall energy efficiency,

bearing in mind that it is better to have higher efficiency in the most frequently used operating mode should price or model selection criteria permit.

Earth energy systems have a reputation for being very expensive compared with traditional heating systems. However, this argument often lacks factual analysis. A conventional natural gas furnace and central air conditioning unit have approximately the same capital cost as a heat pump, but the cost of the gas lines and chimney must be factored into their installation cost.

The major difference in installation cost involves the ground loop or well water exchange unit as well as excavation or drilling labour. These costs can vary significantly depending on climatic conditions, heating and cooling demand, and which heat extraction loop technology is required.

The larger the heating and cooling load of your home the larger the loop will be. Moist, dense soil is better than light, dry, sandy soil for conducting heat into the system loop. Poor quality soil requires a correspondingly larger ground loop.

Several heat pump manufacturers provide software to aid in the assessment of soil type and loop size. In addition, a thorough understanding of local soil conditions is required.

After installation, heat pump operating and maintenance costs are considerably lower than those of conventionally fueled systems. Approximately two-thirds of the energy produced by a heat pump is free, reducing utility bills by 60% or more. Maintenance costs are lower because there is no combustion of fossil fuels to gum up the unit and all components are mounted indoors, protected from weather and vandalism. In fact, the only regular maintenance required is to clean the air filter, assuming you have a forced-air system.

According to Natural Resources Canada, independent research has shown that geoexchange heat pumps can be expected to perform for fifty years or more provided they are professionally installed.

It is impossible to calculate the capital and operating costs of a heat pump (or conventional heating/cooling system) without performing a heat loss calculation for the home. As discussed in Chapter 2, an energy-efficient home will always be less expensive to operate, and upgrading an inefficient home will be more cost effective than installing a larger heating or cooling unit and having to pay excessive energy bills because of it.

Historically, the majority of general contractors and architects will choose the lowest first-cost mechanical heating/cooling system. The lowest-cost system will usually yield the least comfort, cost most to operate, and often be installed by the lowest-cost contractor. A mechanical contractor who is the least expensive installer usually can't afford to stay current with the proper training to understand how to design and install a well-balanced system and provide

follow-up consultation and thorough warranty service. Given that a geoexchange heat pump will last for many decades, a quality installation is advised.

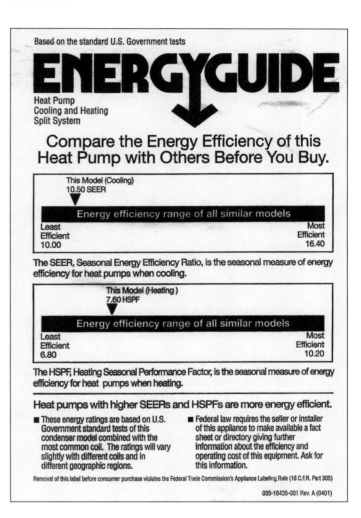

Figure 5.8-3. When shopping for a heat pump system, look for the EnergyGuide label in the United States and the EnerGuide label in Canada. Heat pumps operate in either cooling or heating mode, necessitating an efficiency graph for each.

Figure 5.8-4. During site preparation a closed loop heat extraction field is excavated to a depth below frost line. (Courtesy Major Geothermal)

Figure 5.8-5. A heat extraction field prepared with a backhoe, requiring less expensive excavation and soil removal than the method shown in Figure   5.8-4. (Courtesy Major Geothermal)

Figure 5.8-6. This close-up view of the closed loop heat extraction pipes shows how they are anchored to the earthen walls prior to backfilling. (Courtesy Major Geothermal)

Figure 5.8-7. Urban homes may not have sufficient land area to install excavated heat extraction fields. Where lot sizes do not permit horizontal heat extraction fields, vertical drilled boreholes may be used. Here, workers are shown installing the supply and return pipes into the well. (Courtesy Wayne Smith)

Figure 5.8-8. This picture shows a large heat exchanger being lowered into an artificial pond. Because source water does not flow through the heat exchanger piping, this design eliminates maintenance required because of mineral buildup and fouling associated with older open loop or well-to-well designs. (Courtesy Major Geothermal)

Chapter 6
PHOTOVOLTAIC ELECTRICITY GENERATION

The process of capturing and using solar energy is as old as time. Like most things in the modern world, the simplicity of capturing the sun's energy has been elevated to new technological heights. Unlike solar heating collectors that are used to warm fluids running within the collector, photovoltaic cells magically convert light from the sun directly into electricity. The photovoltaic cell used in renewable-energy systems is definitely the product of rocket scientists, powering communications satellites to ensure everyone on the planet can watch reruns of *Friends*.

The term photovoltaic is derived from the Greek language "photo," meaning light, and "voltaic," voltage which assists the flow of electricity. Friends simply call them "PV" cells for short. Bell Laboratories discovered the PV cell effect in the 1950s. It didn't take the folks at NASA very long to figure out that PV cells would be an ideal means of producing electricity in space. Many missions later, PV cells have improved in performance and have come back down to earth in price. Nowadays, the technology is used in watches, calculators, street signs, and renewable-energy systems for the urban homeowner.

What is Watt?

For the home renewable-energy system, PV products are relatively standardized, allowing even a novice to make accurate comparisons between product lines. There are currently four major product technologies that should be seriously considered for home use: single crystalline, polycrystalline, laminate (roof shingles) and string ribbon cell. Other cell technologies such as "thin film" are an option provided product warranty and manufacturer

Figure 6-1. Photovoltaic cells are the backbone of renewable-energy systems for homeowners. Since prices and supporting technologies have improved, hundreds of thousands of homeowners are now living lightly on the planet. (Courtesy Sharp Solar)

Figure 6-2. PV technology is everywhere you look. Watches, calculators, and even street signs use this ultra-reliable technology to provide electricity wherever the economics makes sense.

Figure 6-3. PV cells are similar to transistors, except they're larger. Individual cells are polished, interconnected, and mounted in a frame, creating a PV module.

financial strength to honor the warranty period are acceptable.

PV Cell Construction

PV cells are transistors or integrated circuits on steroids. Most people have seen the latest microcomputer chip used in PCs. It's a silicon wafer about the size of your thumbnail that holds several million transistors and other electronic parts. PV cells start out the same way as chip circuits, but they are kept in the oven until they're much larger, approximately 4" (10 cm) in diameter. The baked silicon rods are sliced into

thin wafers which are polished and assembled with interconnecting electrical wires.

If we were to take one wafer, expose it to bright sunlight, and connect it to an electrical meter, we would measure 0.6VDC of electrical pressure. A voltage higher than the battery rating is required to "push" electrons and charge the battery. For example, to charge a 12VDC battery, at least 15VDC is required, plus some additional voltage for electrical losses in the system. For this reason, PV cell manufacturers typically connect 36 cells in series to create an additive voltage. (Maybe you should reconsider reading Chapter 1.3 now.) A grouping of PV cells thus arranged and mounted in a frame is known as a module.

36 cells in series x 0.6 VDC per cell
= 21.6 Volts Open Circuit

This voltage appears to be a bit higher than our target voltage of "just over 15V." An interesting phenomenon of PV cells or other electrical generators is a reduction of voltage when the cell is under load, for example when charging a battery. In addition, heat also causes PV cell voltage to drop. When the voltage of a PV cell or series of cells is measured without a connected load, we call this the "open circuit" voltage. Manufacturers often refer to this as "VOC."

When the module is at its maximum-rated power output, the voltage is less than the open circuit rating. This is known as the "voltage at maximum load" and is typically 17VDC for a 12V unit, double for a 24V model.

In order to complete an electric circuit, we must have a source of voltage and current flow.

Figure 6-4. Flexible solar modules are lightweight, easily transported and rugged making them ideal for boats, RVs and trips to the cottage. They can even be fabricated as roofing shingles if you are concerned about "curb appeal." (Courtesy PowerFilm, Inc.)

A PV module will cause current to flow within the cells and out of the supply wires to the connected load. As an example, a Sharp (www.sharpusa. com) model NE165 (shown in Figure 6-3) has a rated current of 4.77 amps (A) at a rated voltage of 34.6VDC under ideal conditions and rated illumination.

34.6VDC x 4.77A current flow
= 165.04 watts of output power

This module is therefore rated at 165W. However, there's one small exception: if it's dark or very cloudy outside, the power output is zero. Obviously light intensity has to play into our equation. The standard light intensity should naturally be that of the sun, as this is the source of energy the panels will use. But where should we measure our sunlight? In Alaska during winter, or perhaps the Bahamas in July? PV manufacturers have taken the liberty of helping us out here. PV lighting intensity has been standardized by industry agreement so that all manufacturers are using the same "solar energy" levels to prepare their data tables. Without this standardization, it would be impossible to determine which PV panel would be the best for your application.

Our sun has very generously decided to output a nice round 1,000W of energy per square meter at noon on a clear day at sea level. In order to test the intensity of PV panels, light sources have been developed that create this same level of light intensity or *flux*. The light is beamed onto the test PV panel, and its electrical ratings are recorded in the product data sheets. As a consumer, this greatly helps determine the value of one PV panel relative to another.

PV Output Rating Caution

Use caution when comparing module power ratings. Manufacturer power ratings are based on ideal sunshine conditions, which rarely occur in the real world. It is wise to derate module power ratings by 20-40% based on local atmospheric conditions. Check with local system owners or a reputable dealer to determine the "Real Power Rating" of your proposed installation.

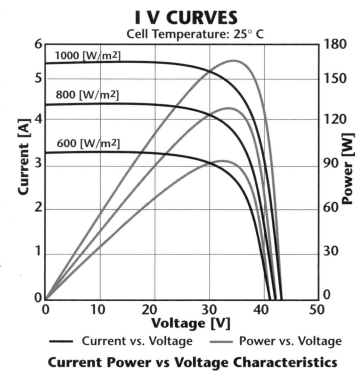

Figure 6-5. The "IV Curve" has nothing to do with a visit to the hospital. Current (I) and voltage (V) of a typical PV cell are plotted against the illumination or "solar brightness." Increased illumination increases module power output.

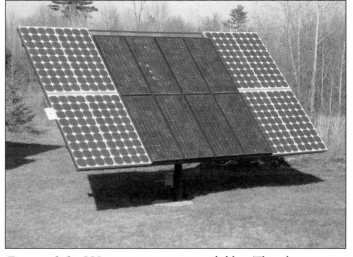

Figure 6-6. PV systems are upgradable. This homeowner installed new single-crystalline modules to increase the capacity of the original polycrystalline units located in the center of this array.

Manufacturer Module Rating x 0.6
= Real Power Rating
…OR…
Real Power Rating ÷ 0.6
= Manufacturer Module Rating

PV module output can also be increased using a technology known as maximum power point tracking. Chapter 10 covers this technology and will discuss how to ensure that your expensive PV panels always operate at their peak efficiency.

PV Module Lifespan

Almost all PV module manufacturers provide a written guarantee for 20 to 25 years or more. The manufacturers are obviously quite sure that their products will stand the test of time. The reason for this certainty is the same one that explains why old transistor radios last so long. The semiconductor technology of the cell wafer results in very little wear and degradation.

The standard warranty term from Siemens states that any module that loses more than 10% power output within 10 years or 20% within 25 years will be repaired or replaced. (This is known as a *Limited Liability Warranty*, more commonly known as the fine print; be sure to read this detail carefully to ensure you understand the warranty terms.) Cell technology and quality of workmanship are very high in the industry, so be sure to purchase cells with the best possible warranty for your money.

Let's start the ball rolling with a question: How much energy do you need?

The cells themselves are quite fragile. To protect them from damage and weather, the cells are bonded to a special tempered-glass surface and sealed using a strong plastic backing material. (Laminate or flexible "roof shingle" systems such as those offered by Uni-Solar (www.uni-solar.com) replace the glass surface with a tough, flexible polymer.) The entire module is inserted into an aluminum, non-corroding housing to form the finished assembly. Once a grouping of modules, called an array, is mounted to a roof or to a fixed or tracking rack, it should stay put forever.

PV Module Maintenance

Let it rain. That pretty well summarizes what you need to do to maintain your PV array. In the wintertime, ice and snow may build up on the glass. Don't smash the ice to remove it, or you run the risk of smashing the glass too. A quick brushing with a squeegee will take off the loose, highly reflective white snow. Once this coating is removed, the sun will quickly warm the panels, melting any ice even at –4°F (-20°C) or lower.

PV Module Installation Checklist

Place your PV module in the sun and collect power—that's it, that's all? Well, not quite. Although PV modules are well designed and last a long time there are several issues that you should consider before choosing where and how to install them:

1. Calculate your electrical-generation requirements.
2. Ensure that the site has clear access to the sun.
3. Decide whether to rack or track.
4. Eliminate tree shading.
5. Consider snow and ice buildup in winter.
6. Ensure that modules do not overheat in hot weather.
7. Decide on the system voltage.
8. Locate the array as close to the batteries as possible.

Step 1 - Calculate Your Energy Requirements

This is square one in your quest for a supply of renewable energy. Let's start the ball rolling with a question: How much energy do you need? If you skipped over the chapter on energy efficiency, it's quite likely that you are consuming significantly more electricity than necessary, resulting in equally high generation system requirements and costs. Remember the rule of thumb: for every energy dollar you put into energy efficiency, you reduce electrical generation capital costs by between three and five dollars. You cannot afford to sidestep the energy efficiency process. Sorry!

Start by checking your electrical bills over the last few months or call your utility to find your average monthly electrical consumption. A figure of 20 to 40 kilowatt-hours per day (kWh) is average and may be considerably higher if you have electric heat, electric hot water, or central air-conditioning.

If your current electrical consumption is a bit on the high side you will quickly find out that you need pockets as deep as Bill Gates in order to generate enough electricity to power your house.

To quickly recap, let's consider the case for compact fluorescent light bulbs one more time. As discussed earlier, a compact fluorescent light bulb uses approximately 4.5 times less energy than a standard incandescent bulb. Assume for a moment that your home requires 600W of lighting in the main floor area. If this light is generated using incandescent bulbs, we need 600W of PV panels to power them. If PV panels are selling for $4.00 per watt, the cost is $2400. With compact fluorescent lamps, however, we need only 130W of bulbs to create the equivalent brightness, for a PV module cost of $520. Efficiency is **ALWAYS** cheaper than additional generation.

Another factor in determining the size of your PV generation system involves any grants or low-interest loans that may be available in your jurisdiction. The California renewables buy-down program offers a grant of up to 50% with a maximum dollar limit. Political realities being what they are, it is always wise to get onto the gravy train while you have the chance.

Lastly, discuss this decision with your dealer. Get advice about system sizes for your roof area, equipment package prices, and other market variables that are impossible to define in this book.

As there is no standard or typical system size, we will assume an electrical-energy-generation objective of 4,000 watt-hours (4 kWh) per day.

Figure 6-7. This grid-interconnected, ultra-high-efficiency home is equipped with a 2.4kW roof-mounted photovoltaic array. It is located in Finland at 63.5 degrees North latitude, proving that photovoltaic panels work in all climates and not just in the Nevada desert. This home uses one-tenth the heat energy and one-quarter the electrical energy of the typical Finnish home. (Courtesy International Energy Agency)

To give you a bit of guidance on this value, Lorraine and I run our household on between 3 and 5 kWh per day, less if we are traveling, more if we have friends over for a late-night party. If you have teenagers who stay up all night, maybe you should stay grid-interconnected. To define "typical" is pretty hard, but here are some values that I can share from other installations:

- **Weekend cottage running a few lights, small T.V., boom-box stereo, and no fridge**: a 12VDC system is fine with two to four PV modules and a small battery bank. This system will generate 200W peak power and 1,000 watt-hours per day energy.

- **Full-time, seasonal cottage with four people, requiring refrigeration and limited water pumping**: a 12V system is fine, possibly using a mix of direct current and alternating current outlets for various loads. If the cottage will "grow" over time, wire for 120VAC now. Use a small inverter for 1,500W peak power and up to 2,000 watt-hours per day of energy. The system will require a PV array of approximately 500W peak output.

- **Full-time residence with four people, large energy-conserver philosophy, no dishwasher or clothes dryer, a small television and stereo, high-efficiency clothes washer, and 120V refrigerator**: wire system for 24V batteries and provide an inverter with 2,500W peak power and up to 3,000 watt-hours per day energy. The system will require a PV array of between 800 and 1,000W peak output with a backup generator for "dark months." Depending on the location, more PV may offset the need for a generator provided winter sun hours are able to provide 100% of the daily load.

- **Full-time residence with all the electrical (high-efficiency) goodies, possibly including a home theater system, dishwasher, gas clothes dryer, central vacuum, washing machine, and electric refrigeration**: system may be wired for either a 24 or 48V battery. Use 24V for total daily energy consumption below 7,000 watt-hours per day and 48V for larger systems. Provide an inverter of between 2,500W minimum and 4,000W peak output or larger, depending on final peak loading. Will require a PV array of 2,000W peak or higher depending on load. For off-grid systems a backup generator will be required.

- **On-grid residence wishing to sell electricity to the utility company:** as many watts of generation as you can afford or are dictated by your site. There is no hard-and-fast rule for on-grid system sizing. Discuss this matter with your local retailer to determine system constraints.

Sunlight and PV Energy Generation

The amount of sunlight we receive varies from day to day, depending on clouds, rain, humidity, and smog, but also as a function of the seasons. In most locations, there will be a surplus of sunlight and resulting electrical energy production in the summer, with the opposite being true in the winter. It may be tempting to calculate the energy production of the PV panels using the worst sunlight hours of the year in an attempt to cover all of your energy requirements. However, this will increase the cost of the system and generate unusable, excess energy in the summer. Let's take a look at how this would work. Assume you live in the Rochester, New York area. Turn to Appendix 5, *Solar Illumination Map for North America (worst months)*. Find the Rochester area and note that the amount of solar illumination in sunlight hours per day for the worst months of the year is 1.6 hours. We can now plug in some numbers to calculate the size of PV array we require. (Watt-hours per day divided by sun-hours per day calculates the theoretical PV panel size. The theoretical size must then be derated as noted below).

*4,000 watt-hours/day ÷ 1.6 sun hours/day
= 2,500W PV panel*

Don't forget to derate the manufacturer ratings (assume a 20% derating factor):

*2,500 real power rating ÷ 0.8 reality factor
= 3,125W manufacturer rating*

A quick search of the web for PV panels reveals nothing even closely approaching that figure. The largest 12V module you can find is from Sharp and is rated 123W with a list price of $685.00. Kyocera has an 80W model on sale for $375.00. This is starting to get complicated and expensive.

PV modules are no different than batteries in the way they can be connected and their capacities increased. Having read Chapter 1.3, you will recall that batteries may be connected in *series* or in *parallel* to increase the voltage or current respectively; the same is true with PV cells. Assuming a PV array with 12V output, modules can be grouped in parallel to increase the current and wattage rating:

3,125W array ÷ 123W PV module = 25 modules

or

3,125W array ÷ 80W PV modules = 39 modules

With so little sunlight in the Rochester area in winter, this PV array will require a lot of modules

4 PV modules connected in parallel

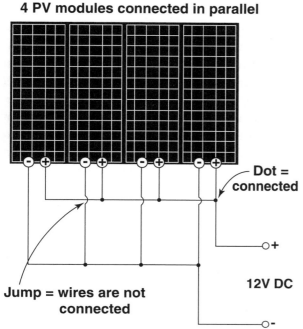

Dot = connected

+

12V DC

Jump = wires are not connected

-

Figure 6-8. PV modules may be connected in parallel to increase wattage in the same way as connecting two small batteries in parallel increases their energy capacity.

to support 100% of the electrical load. Now let's take a look at the cost.

25 – Sharp 123W modules x $685.00
= $17,125.00
39 - Kyocera 80W modules x $375.00
= $14,625.00

Although these figures are **examples only**, they do reveal an interesting phenomenon. You would expect that since each PV array makes the same amount of power the cost would be the same. Welcome to the world of retail. This is no different than purchasing a car; it's best to shop around.

I will digress a moment to explain how to comparison-shop for PV modules. If you are the sort who cannot understand people who drive a few miles to buy tomatoes at another store because they cost ten cents less, just wait a minute. PV module comparison-shopping is easy and will save you more than enough money to pay for this book and possibly put your kids through college.

The energy-to-cost ratio allows for a very quick comparison of all module types. The calculation is quite simple: divide the cost of the module by the rated wattage. Using the PV modules from the example above:

Sharp 123W modules @ $685.00 ÷ 120W
= $5.71 per watt
Kyocera 80W modules @ $375.00 ÷ 80W
= $4.69 per watt

Depending on the warranty and on the installation cost for the additional modules required to make the same amount of power, the Kyocera modules are probably the least expensive based on the "$/Watt Factor." Now back to Rochester.

It takes quite a few modules to make 100% of the energy required to power a home when the winter sun hours are minimal. The opposite is true in the summer. Refer to Appendix 6, *Solar Illumination Map for North America (yearly average)*. You will find that the number of sun hours per day has nearly doubled. The effect is obvious: when the sun hours per day are peaking, the required number of PV modules in your array will be half the number required at the worst time of the year.

Many system owners and dealers use the monthly averages for one complete year to calculate your energy requirements. Using this approach, generation and consumption of electrical energy will never match on a day-to-day basis.

If you are using your PV modules in a summer cabin or cottage, you are in luck. Use the calculations for the best sun hours per day. If you visit the cottage for brief visits in the fall and winter, possibly add another panel or two.

For year-round homeowners, the choices are a bit more complex. You can spend the extra dollars and purchase all of the panels you require or consider a complementary renewable resource such as wind power generation. A third option, which tends to be the most common solution, is also available. Add a backup propane, gasoline, or diesel generator to help boost production in the worst of times. We will review all of these options in greater detail in later chapters.

The correct number of panels to make up your array will depend on more than just simply spending the money to purchase all you need. For starters, the weather seems to have a mind of its own. One year it's sunny all winter, the next you wonder if the sun will ever come out again. Charts like those shown in Appendix 5 and Appendix 6 are averages, which really means that you should be prepared for variability. Even if you purchase all the PV modules you can possibly mount on your roof, it is still a good idea to have a backup generator or alternate source, just in case. A thousand PV panels will not operate a single appliance if it's cloudy and dark!

Speaking of mounting panels on your roof, this is the next concern with the number of panels you select. You have several choices about where to mount your PV array. The roof is one of these places. The more panels you have in the array, the larger and heavier it becomes, placing limitations on some of the best mounting locations. If the panels are mounted on a sun-tracking unit, the limit is 16 to 20 modules per tracker.

What is the correct number of panels to purchase? There is no "correct" answer. If you have deep pockets, by all means purchase all the modules you require to offset the darkest winter days. If you are like most people and live on a tighter budget, start small and expand your system later as you can afford it. When you win the lottery or your great aunt leaves you some valuable stocks and bonds, splurge and purchase the additional twelve modules that you really need. The additional panels can always be added later, like those shown in Figure 6-6. Just remember to plan for the expansion at the start of the installation. This will lower future costs and headaches.

Step 2 – Ensure the Site Has Clear Access to the Sun

PV modules must face the sun to make electricity. As every romantic knows, the sun sets in a brilliant

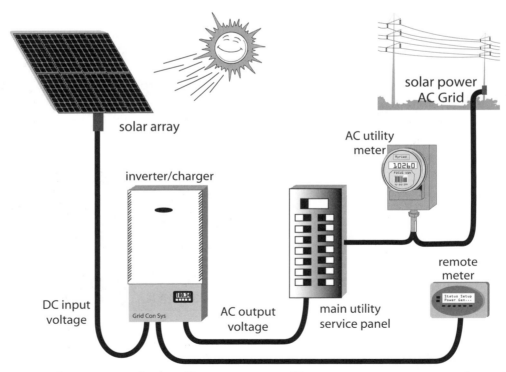

Figure 6-9. A grid-interconnected renewable energy system will have a positive impact on your electrical bill as well as on the environment, regardless of its capacity. Most designs try to balance generation with daily consumption.

display of color in the west. My mother-in-law tells me that it does the same when it rises in the east, although I cannot personally attest to this statement. Where, then, do you face the modules? The quick answer is to point the panels directly at the sun. However, with the sun moving from east to west during the day and progressively higher in the sky as summer approaches, this might be a rather difficult task.

You have two options: mount the modules to a fixed location pointing *solar south* or install a tracking unit that automatically aims the panels directly at the sun. Either way, make sure the location you choose is not impeded by trees, buildings, or other obstructions. Site locations can be difficult to assess due to the seasonal variation in the sun's track, leaves on deciduous trees, and neighboring buildings. To help simplify the assessment of a site, consider purchasing or renting a Solar Pathfinder to eliminate any guesswork. This device works on the well-known principal observed by every schoolchild with access to a magnifying glass. As the sun tracks across the sky, light enters the magnifying and focusing dome of the device (see Figure 6-10). The concentrated

sunlight falls on specially sensitized paper burning a copy of its path. Gaps in the marking indicate obstructions that will cause unacceptable shading of the PV panels.

Step 3 – Decide Whether to Rack or Track

Roof Mounts

Perhaps the simplest mounting system is to attach the panels to a solar-south-facing roof as shown in Figure 6-11. *Solar south* differs from magnetic south as a result of a phenomenon known as magnetic declination. Appendix 4 shows the correction of compass readings based on your geographic location. If you live in Florida, for example, there is no change between magnetic and solar south. If you live in Alberta, the error in the compass-magnetic versus solar south reading is so large that it will affect your system's energy production if you use fixed-mount PV arrays.

Another problem with fixed-mount arrays is their inability to change their angle of view throughout the seasons. The winter sun barely

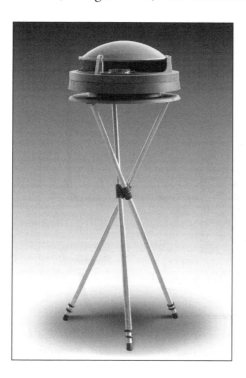

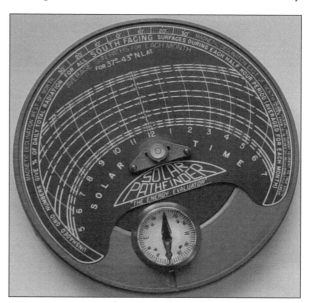

Figure 6-10. The Solar Pathfinder is used to determine magnetic and solar south (using the data in Appendix 4) as well as to record the sun's track across the sky using the specially sensitized paper shown right. (Courtesy Solar Pathfinder)

Figure 6-11. Attaching a PV array to a solar-south-facing roof is one of the least expensive mounting methods. The downside can be getting on the roof to brush off a foot of snow in the middle of winter.

scrapes across the horizon in most of the northern United States and Canada. During the summer, the change in the earth's angle places the sun almost directly overhead. If your roof mount

or other fixed mount cannot be adjusted for summer-to-winter sun angles, a good rule of thumb is to set the array at an angle from the ground equal to your latitude. Mounting racks, which have a summer/winter angle adjustment, should be lowered by fifteen degrees for summer. Likewise, raise the angle towards the vertical by fifteen degrees for winter.

Ground Mounts

If the thought of climbing onto your roof to wipe snow off the PV array has you going woozy, then ground or pole mounting is for you, a simple and inexpensive method of mounting the panels. The unit shown in Figure 6-13 was purchased, but you can also build a unit yourself. Select galvanized steel or aluminum to limit corrosion. When designing your mounting system, make sure to include a hinge assembly to allow for seasonal adjustment. It is possible to make a ground mount using preserved wood or cedar, but keep in mind that the panels will almost certainly outlast the wood.

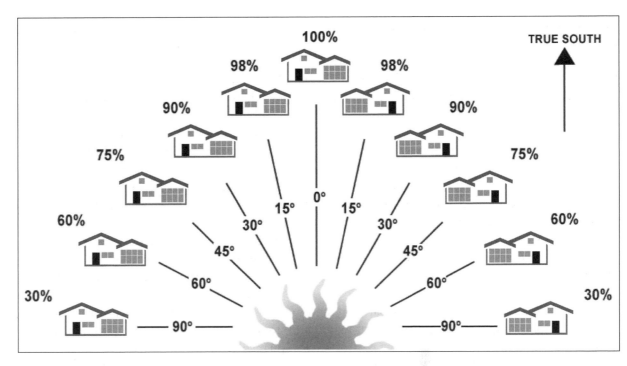

Figure 6-12. PV panels should be mounted facing as close to solar south as possible. PV panels mounted 45° east or west of solar south will reduce potential electrical output by 25%.

A word of caution regarding ground mounting: PV modules are expensive and may have a tendency to walk away. Ensure that you have security bolts (or a good guard dog) to prevent theft. If possible, place the mounting legs in concrete footings to make sure that everything stays in its place. Snow is another problem. If you live in an area where "sweeping" the array requires a snow blower rather than a broom, then ground mounting is not for you. Be careful to ensure that lawnmowers and playing children will not send rocks or other debris flying at the modules. Although it is very likely that no damage will result, there is no sense tempting fate.

Tracking Mounts

The tracking mount shown in Figure 6-14 is the most advanced means of pointing your PV panels at the sun and has the added benefit of being hypnotic to watch as it scans the sky. PV arrays comprising up to twenty panels are mounted on the tracker, forming a billboard-sized unit. (A sixteen-module array measures 16' x 8' or 4.9 m x 2.4 m.) The tracker is designed to move to an easterly location when it gets dark. At first light it follows the sun on its westerly track across the sky. At day's end, the unit returns to the easterly heading to repeat the process. Electronically controlled trackers can be fitted with a manual seasonal-adjustment device or with an automatic version.

Passively controlled trackers are also available. The Track Rack manufactured by Zomeworks (www.zomeworks.com) uses heat from the sun to operate a tracking mechanism. Although this mechanism is simpler than those used in active tracking units, the Track Rack model will go to sleep facing west and may require an hour or two of valuable sunlight before it will start tracking correctly on cold days.

Should you use a tracker? The debate rages on, but the following are considerations that may affect your decision:

- Trackers increase summer PV production by up to 50%. Winter production is improved

Figure 6-13. The commercial ground-mounting unit is ideal for smaller PV arrays and is also a good option where cost is a concern. (Courtesy Zomeworks)

only 10-20% due to the lower, smaller arc of the sun. If your energy consumption is higher in summer than in winter, perhaps due to the operation of a spa, swimming pool, or air conditioner, then tracking will be a benefit.

- The further north you are the less sense it makes to track in the wintertime. This is especially true along the United States/Canada border area.

- If your site has a limited window of sunlight—less than six hours—then tracking will not greatly improve system performance.

- Trackers are not cheap. The cost of the tracker may be used to purchase a fixed rack and more PV panels, which might offset the loss in non-tracking production.

- Trackers add a degree of complexity to the system. The bits and pieces are just one more thing to have to maintain.

Regardless of which type of PV mounting system you decide to use, keep in mind that they take up a fair bit of area and make wonderful kites or sails in high winds. Ensure that proper mounting and foundation work has been undertaken in compliance with the manufacturer's installation instructions. If you want to harness the wind, don't use your PV panels.

Figure 6-14. Tracking units such as these increase summer electrical production by up to 50% but offer little improvement during the winter.

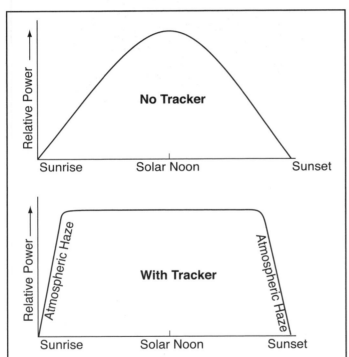

These graphs demonstrate the relative power output of a PV array equipped with and without a tracking unit. The top graph indicates that solar energy reaching a fixed array and its corresponding electrical power output increase slowly until the sun is perpendicular to the array at solar noon. Power output then decreases until sunset. With a tracking unit, solar energy rises very quickly as the tracker is always oriented towards the sun. The effect of dust and atmospheric haze and interference prevents power output from rising to its maximum point immediately. Array power output will remain at its maximum point throughout the duration of the day, until just before sunset.

Step 4 – Eliminate Tree Shading

Although you want to protect the south side of your house from the summer sun, the last thing you want to do is to shade even a very small portion of your PV array. Partial shading will cause a disproportionate reduction in electrical generation in the shaded area. Keep trees well clipped in the sunlight window (between the hours of 9 a.m. to 3 p.m.), keeping in mind both winter and summer sun tracking. If you are unsure about shading, consider evaluating your site using the Solar Pathfinder shown in Figure 6-10.

Step 5 - Consider Snow and Ice Buildup in Winter

If you live in a snowy area, take winter snow and ice buildup into account when considering PV location. Although PV angles at this time of year are almost vertical, snow and ice will stick to the array. If a coating of fresh, white, powdery snow is covering the array, it may take several days for it to fall off without brushing—and you may not want to brush off PV arrays if they're mounted on a second-story roof.

Step 6- Ensure That Modules Don't Overheat in Hot Weather

PV module electrical output fades (just like everyone I know) as the mercury rises. In order to ensure peak operation of the array, modules must not be seated directly on the roof surface. An air gap of 2-3" (5-8 cm) will allow cooling air to circulate under the array, providing maximum power output.

Step 7 – Decide on System Voltage

For off-grid systems, there are three commonly used system voltages: 12, 24, or 48. Although it is possible to select differing voltage and power levels, the industry uses the following guidelines.

System Size	PV/Battery Voltage
Up to 2 kWh per day	12V
From 2 to 7 kWh per day	24V
Over 7 kWh per day	48V

The PV industry has standardized on modules with 12 or 24V nominal outputs. As discussed above and as shown in Figure 6-8 we can interconnect PV modules in parallel to increase current flow and system wattage without changing the voltage of the panels. Figure 6-15 illustrates a mixture of series- and parallel-connected PV panels. In this arrangement there are two rows of four PV panels. Both rows are interconnected in series, increasing the system voltage. If you look carefully and pretend that each PV module is a battery, you can follow the series connection. The positive of one module connects to the negative of the next and so on. When each 12V panel is connected in a series grouping of four, the voltages are additive, resulting in a 48V array.

Once both rows of panels are at the same system voltage, which in this example is 48V, the array may be connected in parallel. Examine the wiring connection at the point indicated in Figure 6-15. The negative lead from each row is connected, as is the positive lead. Together, the two rows of 48V arrays will deliver additional current flow (power) to the load.

Step 8 - Locate the Array as Close to the Batteries as Possible

Have you ever connected 2 or 3 garden hoses together, turned on the tap, and wondered why the sprinkler was dribbling an anemic spray of water even though the pressure at the house was fine? What you have witnessed is resistance to flow. This phenomenon applies not just to garden hoses but to electrical circuits as well. The resistance to flow of electricity works in the same manner. Low-voltage (pressure) circuits require very large wires to carry the necessary current in the cable. It stands to reason that the further the PV array is from the battery or inverter, the greater the loss of electrical energy will be. Increasing the pressure or voltage of the electricity will help, but distance is a bad word when it comes to low-voltage direct current circuits. Keep the wire runs as short as possible and maximize wire size to prevent unnecessary power loss.

8 PV modules connected in series and parallel

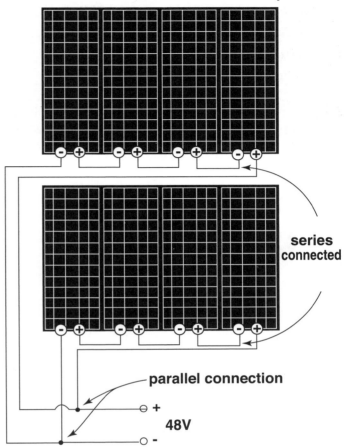

series connected

parallel connection

48V

Figure 6-15. PV panels are just like batteries: they can be interconnected in series to increase voltage and in parallel to increase the current of the array.

As a side note, this is the reason that the North American electrical grid transmits energy at very high voltages, reaching 750,000 volts in some areas. Even at this high voltage electrical losses through the system can be extreme. The province of Ontario has approximately 15,000 miles (24,000 km) of transmission grid. The electrical losses on this system are equal to the energy output of three large nuclear reactors! Distributed generation using off-grid or grid-interconnected photovoltaic systems places the consumption and generation of electricity at the same location, thereby eliminating transmission losses. At a couple of billion dollars apiece for nuclear generating plants, not to mention fuel disposal, decommissioning costs, and safety liability, PV panels mounted on the nation's roofs are a bargain.

Conclusion

PV systems are simple and reliable. Where a home or cottage is just a little too far from the electrical lines, there is no contest. It will make economic and environmental sense to use this technology without further consideration.

With electrical rates rising and governments committed to "greening the grid," there is little question that all of North America will be pushing PV in the same manner as California is now. According to Environment California, recent proposals before the state legislature would have 50% of all new homes running on solar energy within the next ten years. Although this goal is lofty and well ahead of any other jurisdiction in North America, we still have a long way to go before we catch the leaders, Japan and Germany.

Figure 6-16. The Conde Nast building in Manhattan is a magnificent example of what the future will look like for urban dwellers. This building, equipped with advanced energy-efficient designs, fuel cell electricity generation, and photovoltaic panels, pumps green electricity into the New York State grid. Over the coming years, commercial buildings, condominiums, and apartments will achieve an energy goal of almost net zero. (Courtesy Fox & Fowle Architects / Andrew Gordon Photography)

Chapter 7
ELECTRICITY FROM THE WIND

I have been told by a number of people that while women seem to like PV technology, men tend to gravitate towards wind turbines. Perhaps for the men the appeal lies in the big tools, concrete, propellers, and spinning stuff. Whatever the case may be, mankind has been capturing the wind for eons. All school children know about the quaint Dutch windmills of old. What they may not know is that citizens of Germany, Sweden, and Denmark are also amongst the world's foremost users of wind electricity generators.

Years of low fossil-fuel prices and certainty of supply convinced most North Americans that there was nothing to worry about. The Europeans, on the other hand, have been worrying for years. Most European countries rely on imported oil and have high population densities. Conserving electrical energy has become a way of life there.

Fortunately we are waking up on this side of the pond and as the alarm bells are ringing wind turbines are coming of age. Sites such as King Mountain in West Texas are good examples of how North Americans are catching up. Like everything else in Texas, this site is big, with 214 turbines installed. Together they provide an output of 278,000,000W (278MW), sufficient to power 80,000 super-sized Texas homes or 3% of Denmark's electricity requirements. These units are not toys.

It is doubtful that you will require this much power for your renewable-energy-powered home, but this does indicate the scope of what wind power can do.

Figure 7-1 (above). This 10kW wind turbine is typical of larger home-based machines. With a blade diameter of 23 feet (7 m) and the right amount of wind, this unit can provide sufficient power for almost any home. (Courtesy Bergey Windpower)

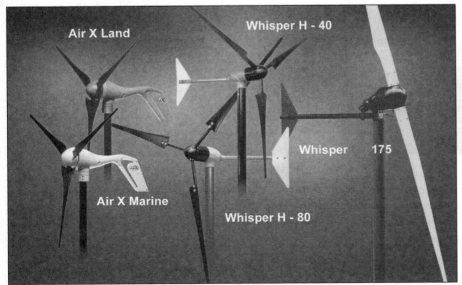

Figure 7-2. Wind turbines are available in a wide variety of sizes designed to suit just about every application. (Courtesy Southwest Wind Power)

Home-Sized Wind Turbines

The market today includes many manufacturers of wind turbine systems. There is also a market for rebuilt and used machines that were installed prior to rural homes being connected to the growing electrical grid in the early 1900s. The Resource Guide in Appendix 3 will provide you with a listing of the major manufacturers and rebuilders of new and refurbished equipment.

When people first see a wind turbine their reaction is to ask, "How big is it?" This can be a misleading question, since size in this case involves more than just the height of the tower or rated electrical capacity.

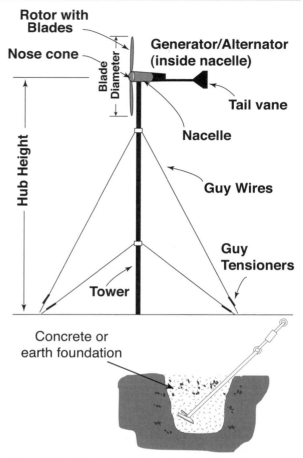

Figure 7-4. All horizontal-axis wind turbines share similar components.

Figure 7-3. Wind turbines are used extensively in this obviously off-grid installation. What better harmony than using the wind to sail the boat and make electricity at the same time? (Courtesy Southwest Wind Power)

Wind turbines require a more detailed assessment than PV modules when it comes to determining what "size" of turbine is required for your application. Let's begin by taking a unit apart in order to understand how they are made and how they operate.

The majority of home wind turbines are horizontal-axis machines similar to the one shown in Figure 7-4. The names of the major components are common to all wind turbines regardless of size. A tower structure supports the

wind turbine and raises it high above the earth's surface. As the tower height is raised, the speed and smoothness of the wind increases: "the higher the tower, the greater the power." Typical tower heights range from 60 to 100 feet (18 to 30 m). The tower may be self-supporting or more commonly and less expensively have a series of guy wire cables to support it. A concrete pad that accepts the downward weight supports the tower and prevents it from skidding sideways. The guy wires provide all of the support and are secured by anchors to earth or concrete foundations. Cable tensioning devices are installed on each guy wire to remove slack and prevent tower motion.

An important feature of the wind turbine is the rotor and blade assembly in the centerline of the turbine, consisting of the blades on the

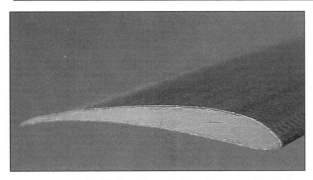

Figure 7-5. The blades of a wind turbine are shaped in a form similar to that of an aircraft wing. Wind rushing past the blades causes aerodynamic lift, imparting a torque or rotational force that powers the generator. (Courtesy Bergey Windpower)

outside and the generator/alternator on the inside. A nose cone is attached to the rotor to give the wind a smooth surface to pass over as it travels through the blades. A nacelle or cover protects the generator and support components from the weather. Finally, a tail vane assembly is used to keep the rotor and blade assembly pointed into the wind. On some models the tail vane also provides a means of slowing the blades down in dangerously high winds by automatically performing a function known as furling. On models such as the Bergey Excel shown in Figure 7-1, a manual furling winch is also provided, allowing the unit to be shut down during severe storm seasons.

In Figure 7-6 a close-up view of a small 1,000W turbine assembly shows the rotor assembly with the blades removed from their sockets. The generator wiring is just visible at the rear of the rotor assembly. If you follow the wires from the generator, you will see them terminate at a black module known as the rectifier bridge, which converts the alternating current from the generator into direct current suitable for connection to the battery bank. The alternating current is supplied as 3 phase (which is similar to most industrial building services) at the top of the rectifier, while the direct current output is at the bottom of the rectifier. Following the positive and negative lead wires from the rectifier, you see that

they connect to a section of the tower assembly. As the turbine spins on the top section of the tower, it would quickly twist the power cables running down to the house. To prevent this, the wiring is fed to a "slip ring" assembly which acts as brushes, conveying the electrical power from the rotating "yaw tube" to the stationary tower head assembly. The tail vane mounts to the rear post, completing the assembly.

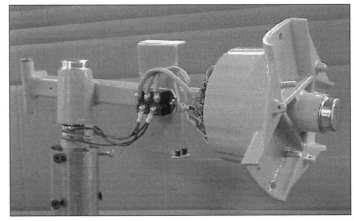

Figure 7-6. This picture details the inside frame, rotor, and generator of a small 1kW wind turbine with the blades and nacelle cover removed. (Courtesy Bergey Windpower)

Wind Turbine Ratings

Now that you have been formally introduced to a typical home-power-sized turbine, we need to clear up some of the details regarding size. Unlike the debate over certain biological issues, size in a wind turbine does count. However, the diameter of the wind turbine blades is the only size that is truly important. You may recall from Chapter 1.2, "What is Energy?" that in order to convert kinetic energy to electrical energy we have to capture as much of the "moving fluid" as possible. The blades catch the kinetic energy of the wind. A unit with a smaller blade area will capture a smaller amount of energy; it's that simple. Because the blades sweep a circular path we have to do a little bit of fancy mathematics to calculate the area they cover. Assume our turbine blades have a diameter of 10 feet (3 m), giving them a radius (½ the diameter) of 5 feet (1.5 m).

π x Radius ^{2}	=	*Area Swept by Blades*
3.14 x 25	=	*78.5 Square Feet (7.3 m²)*

Machine manufacturers like to use power ratings that are not as well standardized as they are for PV modules. This makes comparison-shopping between units difficult. Calculating the rotorswept area of different machines and comparing those values will provide an even keel for calculating power output. Keep in mind that rotor-swept area is in a linear relationship with power. Simply stated, a doubling of the sweep area doubles the power output.

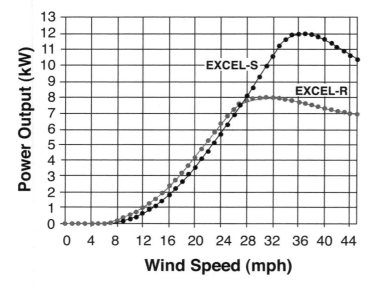

Figure 7-7. An alternate way to compare wind turbine output is to look at the power curve rating. This chart shows the electrical power output for a given wind speed. (Courtesy Bergey Windpower)

If the math is too much trouble, an optional means of comparing machine ratings is to look at the "power curve" graph. The graph shown in Figure 7-7 is for the large home-sized turbine shown in Figure 7-1. This graph details the electrical power output in kilowatts based on a given wind speed in miles per hour. For example, at a wind speed of 24 miles per hour (38.5 kph), this unit outputs 6,000W of electrical energy. An excellent way of cross-checking manufacturers'

figures is to compare both power curves at the same wind speed as well as the rotor-swept area. The ratio between the two should be approximately the same: a machine with twice the rotor-swept area should have a power curve with twice the power output.

The power curve graph also details some other interesting facts concerning all wind turbines. You will notice that at low wind speed, below approximately 8 miles per hour (13 kph) the turbine output is almost zero power. Many dealers will talk about units that start in low-wind conditions or begin generating power earlier than others. This information is not factually supported by the physics of wind energy. The turbine has nothing to do with "starting to kick in earlier." Sure the unit might spin a bit in low-wind periods, but it will not output any energy. Energy is related to wind speed and you need speedy wind to make energy.

This brings up another important fact about wind energy. Wind speed is not linear to power; it is cubic-geometric. Simply stated, if we double the wind speed we do not get double the power. Take a look at the power curve in Figure 7-7. Follow the Excel-S data line at a wind speed of 16 mph (25.5 kph) and we derive an output of approximately 2,200W of power. Now if we double the wind speed to 32 mph (51 kph) we do not get 4,400W as you might expect. Check the curve; our output is now up to a whopping 10,500W!

What happened? This is a classic example of the non-linear relationship between wind speed and energy in the wind; the theoretical relationship is cubic. This means that for a doubling of wind speed, we get an increase in power of eight times (2 cubed = 2 x 2 x 2 = 2^3 = 8).

This is a very important issue for those considering installing a turbine at "the site by the lake that always has a breeze." I hear this kind of statement over and over again, where people seem to think that their site is going to be perfect for wind, only to later find out that things just didn't work out that way. With wind, the only rule of thumb I would dare to even consider mentioning is that if it is *always* too windy to stand around outside, you *might* just have enough wind to consider installing a turbine. I cannot overemphasize how important a proper wind site assessment is to ensure a satisfactory and profitable installation.

Rating Jargon

The power curve also details information about how the unit behaves at various wind speeds. Referring to the Bergey model Excel-S power curve in Figure 7-7, you see a shape to the curve that is similar but not identical to the units produced by many other manufacturers. Relevant points on the curve are:

- **Start-up speed**. This is the wind speed when the rotor and blade assembly first begins to rotate. The start-up speed is irrelevant, as this low rotational speed will provide no usable power output.

- **Cut-in speed**. This is the wind speed when the generator actually begins to produce **usable power**. Cut-in speed is one of the most misunderstood terms in wind power generation. There is no power in slow wind; if the unit starts at 3 mph (5 kph) it will not be generating useful levels of power. So if the turbine's cut-in speed is being touted as exceptionally low and superior to that of every other model, keep your money in your wallet.

- **Rated speed/power point**. The manufacturer's data gives the nominal power rating of the unit at a given wind speed. The Bergey Excel-S is rated 10,000W and that occurs at approximately 31 mph (50 kph) wind speed. In practice, the rated speed/power point is of little value, as the machine will likely spend very little of its life at this wind level. Additionally, each manufacturer rates its turbines at different wind speeds, skewing the data values and making comparison difficult. Remember, the rotor-swept area and blade diameter are the most important factors.

- **Peak or maximum power**. This is the maximum power the unit will generate. The Excel-S is capable of producing 12,000W at 36 mph (58 kph) wind speed.

- **Furling speed**. The furling speed is the wind speed at which the unit enters a "self-protection" state. When wind speeds are too high, the turbine must be able to lower its rotational speed and resulting power output. In most small wind turbines, the tail and nacelle "hinge" so that the blade/rotor section turns out of the wind. In some models the rotor points upwards like a helicopter until wind speed subsides.

Ratings are important, but the most important consideration is to determine at what point on the power/yearly energy curve the turbine will work at *your* site. Don't become overly confused by the extra jargon that will become immediately unimportant the minute you switch the unit on and you find out you do not have enough wind to make the investment worthwhile.

Wind Resources in Your Area

The wind tends to vary greatly over the span of a year and depending on where you live. There is almost no wind in central Florida, but it never stops in Cape Cod. If a small change in wind speed causes a large change in electrical power output then we must be careful to capture the highest wind speeds to get high power output for an extended period of time to generate sufficient energy. Almost everyone I have talked to about installing a wind turbine tells me that they are on a hill or near a lake and have "great wind." But just what do they mean by "great wind"? A 9 mph (14.5 kph) breeze is pretty strong. At this speed it will make holding your umbrella a bit difficult or whip your tie around your face, so it must be a good wind turbine site. Think again. Looking at the power curve for the 10,000W Bergey Excel-S model in Figure 7-7 you will find the power output is nearly zero.

You can't take chances installing a turbine and hoping it will work out. This is the surest way

Tower Height	Average Wind Speed						
	8mph (13 kph)	9mph (14 kph)	10mph (16 kph)	11mph (18 kph)	12mph (19 kph)	13mph (21 kph)	14mph (23 kph)
60 ft (18 m)	330 kWh	480 kWh	670 kWh	870 kWh	1,110 kWh	1,350 kWh	1,610 kWh
80 ft (24 m)	430	620	840	1,100	1,370	1,670	1,960
100 ft (30 m)	490	700	950	1,220	1,510	1,820	2,130
120 ft (37 m)	550	780	1,050	1,340	1,650	1,970	2,280

Table 7-1. This table details the relationship between wind speed, tower height, and monthly energy production in kWh for the Bergey Excel-S turbine. (Courtesy Bergey Windpower)

to electrical and possibly financial bankruptcy if there ever was one. Having someone tell you there is plenty of wind is just not going to work. A wind turbine is a big investment and a careful site study should be conducted before you commit your dollars and time.

There are two typical methods of determining the characteristics of the wind at a given site. The first method relies on wind maps prepared by or on behalf of government; the second entails a site measurement appraisal for a number of months or longer. However, before we discuss wind mapping and surveys we first need to review what is meant by wind assessment.

A wind turbine without wind is like the proverbial fish out of water. Wind makes a wind turbine operate and, within reason, the more wind the better. But what wind are we talking about? Wind requirements vary with the seasons and depend in part on the type of system you are running. Late fall and winter tend to be dark and rather tough on a PV system, so adding a bit of wind energy to boost things up is helpful then. On the other hand, a system which relies on wind as the main or only source of energy requires a steady supply of wind all year round, especially in the fall and winter when more time is spent indoors and energy needs tend to be higher.

Manufacturers publish tables that estimate the monthly or yearly energy production in kilowatt-hours at differing wind speeds when the turbine is installed on a tower of a specific height. An example taken from the Bergey Excel-S is shown in Table 7-1.

As the wind speed increases, so does the amount of energy produced. But why does the height of the tower have anything to do with energy production? Tower height relates indirectly to wind speed. Think of wind as water for a moment. If water were streaming over your wind

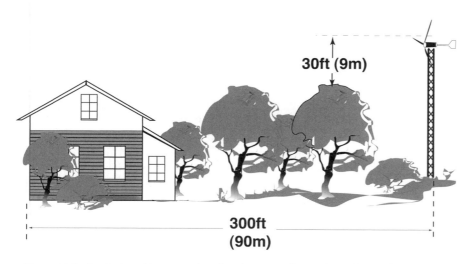

30ft (9m)

300ft (90m)

Figure 7-8. A wind turbine must be placed in smooth air clear of any obstructions. This is one of many reasons that urban wind turbines tend to be difficult to install and operate.

turbine and it were close to the ground, you'd find that trees, rocks, buildings, and other obstructions would cause the water to create rapids and turbulence and generally slow down. As you moved above our "river bed" example, there would be fewer obstructions and the water would start to move much more smoothly. At some point there would be minimal obstructions in the water's path and the flow would become very smooth. At this point we would have what is called *laminar flow*, which not only increases water and wind speed as well as energy production but also prevents "rough air" from stressing the turbine, ensuring longer life.

A general rule of thumb is to place the turbine on a tower of sufficient height that nothing within a 300-foot (90 m) radius is within 30 feet (9 m) of the height of the tower. For example, the house and trees shown in Figure 7-8 must be at least 30 feet shorter than the tower if they are within a 300-foot radius of the tower. Keep in mind that trees grow, so allow for this when determining tower height—or get pruning equipment to maintain this requirement.

The effect of tower height on turbine power cannot be overemphasized. At 8 mph wind speed, the Bergey Excel-S power output increases from 330 kWh to 550 kWh when tower height is increased from 60 feet to 120 feet. In other words, doubling the tower height increases power output by two-thirds.

Before you decide to rush out and purchase a 700-foot (213 m) super-tower, bear in mind that tower cost, foundation work, and wiring expenses will tend to throw cold water on your plans. A tower of 60 to 80 feet is a good start for smaller units under 1,000W. Up to 120 feet is recommended for larger turbines, but work within your budget.

Wind Mapping

Table 7-1 shows energy output at varying wind speeds, which are yearly average wind levels measured at a standard 33-foot (10 m) height using a device known as an anemometer (see Figure 7-

9). One version of this device consists of a series of cups that are arranged so that greater wind speed causes the cups to spin faster. The spinning shaft is connected to a speed-detection device, the output of which is fed to a data logger or recording instrument. The wind speed data is collected for long periods of time and then plotted using a mathematical "Rayleigh distribution" curve. The curve plots recorded wind speed according to frequency of occurrence. The highest point on the curve is the speed at which the wind blows the most often. When this data is recorded over a 12-month period, the result provides average annual wind speed as well as accurate data for particular periods throughout the year.

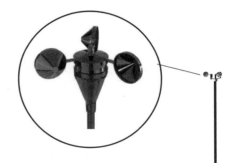

Figure 7-9. The anemometer records wind speeds at given locations around North America. The data is tabulated and developed into maps such as the ones contained in Appendix 8.

At my home, the PV array and tracker unit provide well over 100% of our summer and fall energy requirements, allowing me to turn off the wind turbine during this time period. Allowing the turbine to remain "parked" reduces potential mechanical wear, increasing its life span.

On the other hand, our neck of the woods seems to be completely devoid of sunlight for the entire month of November and the first half of

December. It is therefore critically important to our household to have sufficient wind power at that time to make up energy that the PV system lacks.

Maps are available which detail wind speed during annual, seasonal, and monthly periods. These maps are compiled by and for the government for a variety of purposes such as crop planting, weather forecasting, and wind turbine installations. A typical wind energy map is shown in Figure 7-10, detailing mean annual wind resources in the United States. A full overview map for North America is shown in Appendix 8.

Wind atlases are available online at:

United States: rredc.nrel.gov/wind/pubs/atlas/

Canada: www.windatlas.ca/en/index.php

Wind speed maps are broken down into speed and "power class." The typical speed rating is in meters per second, which can be converted to other units.

Power class numbers that range from one to seven indicate relative energies in the wind based on average annual wind speed for that area. Power class 1 is considered unacceptable for wind turbine installations. Local geographic considerations such as large bodies of water, hilly or mountainous areas, and tree cover can change the average data up or down from the values indicated. Another consideration in assessing the accuracy of the mapping is the distance from the recording site to your location. Wind mapping is a reasonable method of determining relative wind speed, but it is not foolproof.

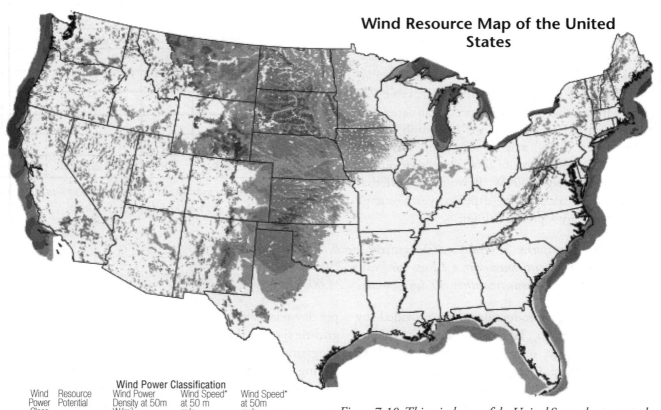

Wind Resource Map of the United States

Wind Power Classification

Wind Power Class	Resource Potential	Wind Power Density at 50m W/m²	Wind Speed* at 50 m m/s	Wind Speed* at 50m mph
3	Fair	300 - 400	6.4- 7.0	14.3 - 15.7
4	Good	400 - 500	7.0 - 7.5	15.7 - 16.8
5	Excellent	500 - 600	7.5 - 8.0	16.8 - 17.9
6	Outstanding	600 - 800	8.0 - 8.8	17.9 - 19.7
7	Superb	800 - 1600	8.8 - 11.1	19.7 - 24.8

*Wind speeds are based on a Weibull k value of 2.0

Figure 7-10. This wind map of the United States shows annual average wind speed. Each geographical area is assigned a class depending on the average wind speed in the area. A minimum power class for off-grid homes is Class 2.(Courtesy http://www. windpoweringamerica.gov/wind_maps.asp)

Site Wind Measurement Study

Perhaps the best way to assess your site is to take your own site survey. This is recommended if a larger turbine may be installed for profit or if the wind mapping data for your location is questionable. Companies such as NRG Systems (www. nrgsystems.com) offer numerous anemometer, data logger, and tower kits. Renewable energy equipment dealers may also provide a rental or "test and analyze" service as part of your turbine package. Wind data analyzers are battery-powered and contain a data logger that will record wind patterns and provide numerical information that will allow you or your dealer to calculate the potential energy contained in the wind.

Once you have determined the appropriate wind speed data for your area, you can apply this information to the turbine's power curve. Using the Bergey Excel-S as our model turbine, let's work through a sample energy calculation.

Assume that your site turns out to be a power class 2, with an average annual wind speed of 6 meters per second. If you consult Table 7-2 you will see that this wind speed translates into 13.4 mph (21.6 kph). Locate 13.4 mph on the wind speed or "x" axis of the graph shown in Figure 7-7, and correlate this to a power output of approximately 1,000W. It would appear that we can now calculate the expected yearly energy output (energy = power x time):

output. Likewise, below-average wind speeds will have less impact on turbine power output.

As a general rule of thumb, using average wind speed to calculate power and energy output of any turbine will significantly underestimate the true power rating of the turbine. Using the Rayleigh distribution curve data that is derived from a site survey will allow more accurate energy calculations based on the frequency of occurrence of wind speeds that are both above and below the statistical average.

Possibly the most interesting element in these calculations is how the estimated power output

Wind Speed meters/second	Kilometers per hour	Miles per hour	Quality of Site for Wind Power	Power Class
4	14.4	8.9	Not Acceptable	1
5	18	11.2	Poor	1
6	21.6	13.4	Moderate	2
7	25.2	15.7	Good	3
7.5	27	16.8	Very Good	4
8	28.8	17.9	Excellent	5
8.5	30.6	19	Excellent	6
9.0	32.4	20.1	Excellent- HI	7

Table 7-2. Wind speed is usually given in meters per second. Use this chart to convert speeds to miles or kilometers per hour.

a) *1,000W output x 24 hr/day x 365 days/year = 8,760,000 watt-hours/year*
b) *8,760,000 watt-hours/year ÷ 12 months/year = 730,000 watt-hours/month*
c) *730,000 watt-hours/month ÷ 30 days/month = 24,000 watt-hours/day*

The energy production data provided by Bergey, shown in Table 7-1, shows a higher monthly energy output than what we have calculated above. Comparing average wind speed with instantaneous power output of the turbine **does not** take into account the effects of wind speeds that are higher or lower than the average. The relationship between wind speed and power output are not linear; wind speeds above the average will contribute significantly to increased power

per day relates to energy consumption needs. For grid-tie systems this is not a problem, as most tariff programs will purchase as much energy as you are willing or able to supply. Off-grid systems tend to consume less than 10 kWh per day and care should be taken to ensure that energy consumption and production are reasonably matched.

The Bergey Excel-S provides more than twice the typical off-grid system energy requirement, even when it is located in a "moderate" wind regime. You should run system-sizing calculations for a smaller, less expensive turbine that will more closely match your energy requirements. You may have to run through this calculation or compare estimated energy output charts a few times for different models, but the exercise will pay dividends by helping you purchase a machine that is neither over- or under-sized for your application.

Figure 7-12. This Air 403™ (now updated to Air X™ and Air Breeze™) is the largest wind turbine that should be considered for roof mounting, although the manufacturer suggests proper tower mounting and having a site of at least one acre in area. If you need a bit of "white noise" to sleep at night, this might be one way to get it! (Courtesy Southwest Wind Power)

Figure 7-11. A rural homeowner has better control of the surrounding environment than his urban cousin and can ensure that the wind is smooth and continuous without the effects of neighboring buildings and trees. (Courtesy Southwest Wind Power)

The Urban Wind Turbine
a.k.a. The Urbine

Wind turbines are springing up like dandelions after a spring rain, dotting the world's rural skyline. The key word here is "rural" given all of the limitations of the urban turbine or *urbine* if you will pardon lexicon.

Now that we know a bit more about the workings of wind turbines, we need to ask an important question: are wind turbines suited to urban applications? This is one of the most hotly debated topics discussed around the office water cooler, perhaps second only to which stock might be the next Google. Without a doubt, wind turbines have been installed in some of the most interesting, unusual, and controversial places on the planet such as the tops of trees or billboards. There is also a segment of society that believes wind turbines can and should be installed in urban or suburban environments.

Locating a Wind Turbine

Another major difference between PV and wind-based systems is their size, height, and space requirement (Figure 7-8). You can a bolt a few PV panels on almost any roof, whether in the city or country. Wind turbines, on the other hand, will attract attention for miles around. If those "miles around" happen to be forests or grazing lands, then only the cattle will pay attention (nosey neighbors notwithstanding). But don't even consider installing a wind turbine until you have all the i's dotted and t's crossed.

Figures 7-13a and 7-13b. Wind turbines and universities seem to go hand in hand. This Lacota™ urbine is located at the University of Toronto, Canada and was installed in approximately 2005. Even the most ardent environmentalist must admit that the turbine does not improve the vista and that PV panels would be more appropriate for this location. (Courtesy True North Power)

Not everyone will share your enthusiasm for wind, especially if your neighbor is located just a whisper away behind the wall of an under-built home located on a 40-foot-wide lot.

Consider the plight of Doug Findlay. Apparently, all Doug wanted to do was put up a small, 33-foot-high (10 m) wind turbine in his backyard. Irate neighbors fought the project and the local planning board rejected his building application. Mr. Findlay decided to appeal to a municipal board which has the power to review and overturn such cases. After spending thousands of dollars, he is still no closer to installing his wind turbine.

The fact that the turbine would not have operated properly owing to the rules discussed earlier apparently did not deter him. Nor did it occur to him that installing benign PV panels on the roof would have caused no issue with neighbors or planning boards and still would have produced the green energy he was longing for.

The effective tower height (actual tower height minus the working "ground" level of roofs and local buildings) must be considered to ensure that the turbine (or urbine) is located in laminar airflow. This issue cannot be overemphasized.

Installing a wind turbine on the roof of a typical suburban home will generate appreciable noise as well as static and dynamic mechanical loads that the house is not designed to withstand.

Responses to wind turbine noise are also extremely varied. Some people (usually the owners) feel that the noise is in harmony with

Figure 7-14. This large wind turbine is located on Lake Ontario in Toronto, Canada and was North America's first real urban turbine. Since many major cities are located along bodies of water, a traditionally good place for wind resource, it is most economical for urban dwellers to leave wind power to the experts and purchase it through a green energy program in the local area.

nature, while others insist that Spitfires flying during wartime were not nearly as audible. You may not mind the noise, but will your neighbor?

People keep trying to develop the perfect *urbine*, with numerous tests being conducted at universities and research centers around the world. Paul Gipe, a leading expert on wind turbine technology, cites the case of a wind turbine installed on the roof of a Northwestern University engineering building. The noise generated by the turbine was surprisingly loud, causing the faculty to terminate its use.

The moral of the *urbine* story is simple. If you are going to consider the installation of an *urbine*, tread carefully and assess the risks before venturing down this road. It is a difficult and uncertain path

to follow and one that should be avoided if other energy technologies are available.

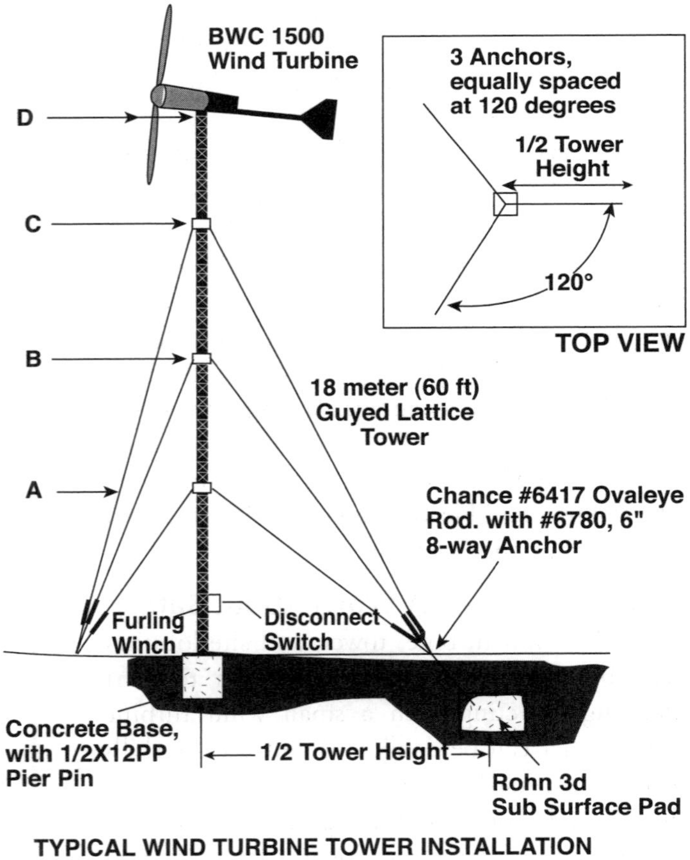

A Proper Location

Assuming that you have passed all of the pre-screening details discussed above, the location of a wind turbine will generally be determined by lot size, cable distance to the main building, and tower construction. The best location, however, is where the winds blow strongest, with consideration for the thickness of your pocketbook. Wind speed and smoothness (or laminar flow) will be better over unobstructed grazing land than in tall forested areas. Tall trees in a woodlot or forest increase the "working" ground level to the tree height, which in effect reduces tower height. Follow the general site requirements in Figure 7-8 to ensure proper placement.

If the tower is freestanding, such as the one shown in Figure 7-1, it can be installed almost anywhere that will accommodate the concrete base. More often than not, the tower will be either a tubular or tripod lattice guyed tower as shown in Figure 7-15. Guyed towers are relatively inexpensive and tend to be the most popular. The guy wires, however, necessitate considerably more room for installation. A good rule of thumb is to allow a minimum clearance radius of one-half of the tower height for guy wire placement.

A power feed cable is required to connect the wind turbine to the battery bank or inverter. An aerial cable may be run between the tower and house, but more typically a direct burial cable or a cable in a PVC plastic conduit is used. The selection is based on preference as well as on obstacles in the cable path. It may be simpler to provide an overhead cable than to blast rock or dig up a paved laneway. Regardless of which method is chosen, there are specific electrical code requirements to ensure fire and shock safety. These items will be discussed in more detail in Chapter 13.

Zoning and building permits may be required depending upon your proximity to urban areas. Wind installations occur almost exclusively in

TYPICAL WIND TURBINE TOWER INSTALLATION

Figure 7-15. Guyed lattice and tubular towers are commonly used for small wind turbine installations. Their main drawback is the room required for guy wire clearance.

rural areas where zoning restrictions are minimal and building permits are relatively easy to obtain. Regardless of whether the installation is rural or near an urban center, however, the best advice is to plan things properly from the start.

A neatly drawn site plan showing property lines and setback distances is a good starting point. Building officials will almost certainly not know anything about wind turbines. This may help or hinder your cause. It may help if building inspectors are curious and willing to assist with something "neat." On the other hand their lack of knowledge may also cause problems. You may be asked whether a strong wind might blow the tower down or if aircraft will become tangled in the unit. It pays to do a bit of homework.

- Provide some pictures of local radio/TV antennas that are of similar or greater height. A wind turbine tower may seem odd, yet most people probably pass a 200-foot microwave tower just down the street without even noticing it.

- Ask your tower manufacturer to provide pre-engineered drawings based on suitable installation guidelines. Your building inspector is used to walls and roof trusses, not towers. Pre-engineered drawings will provide some reassurance that the unit isn't going to crash through your roof during the next windstorm.

- If you have neighbors, be sure to get them onside first. A wind turbine shows a level of stewardship and concern for the environment that should be welcomed. But if George next door thinks it will kill all the birds and interfere with his TV signals and reruns of *Baywatch*, he might mount some opposition and you will have to convince him that these concerns are unfounded. Another common concern is noise. This is generally not a problem provided the neighbors are at least 200 feet (61 meters) away. When it's windy enough to activate the turbine, other sounds like trees blowing in the wind will tend to drown out the sounds of the turbine.

- Check for rights-of-way or legal easements such as underground gas or telephone cables deeded to your property. If you are unsure, check with your lawyer and show her your site plan. No one wants to pay a legal bill, but it is a lot less expensive than having to move your tower over a couple of feet.

- Height restrictions may come into play if you are within a certain distance of an airport or if there are special zoning ordinances. Zoning variations should not be a problem, especially if there are radio towers or buildings in the township that are of similar height. Although aircraft interference is a minor issue, it is a good idea to check with your local FAA office in the United States or NavCan in Canada. Paint and lighting restrictions may be enforced in some areas, but usually only when the tower is over 200 feet (61 m) in the United States and 80 feet (24 m) in Canada. Both agencies will provide a letter indicating whether or not markings are required.

You have every legal and ethical right to install a wind turbine, just as you do if you want to plant a tall tree or erect a TV antenna. Problems with neighbors and building officials occur because of ignorance on their part or lack of planning on yours. You have the advantage of being enthusiastic and better informed. Take the time to transfer this knowledge and educate those around you. The results should lead to an amicable agreement between the concerned parties.

Getting Started: A Word About Safety

Once the wind turbine, tower, and site location have been selected, the next step is to start planning. Erecting even a small wind turbine is a job that can be dangerous and tax the skills of the most mechanically inclined, but don't let this dissuade you. Completing the work yourself rather than hiring a professional is satisfying and will save on installation costs.

There are three common methods of erecting a wind turbine and tower. The first uses a tilt-up arrangement that allows the tower to lie flat on the ground during installation and servicing. When the tower is ready for erection a hand or electric winch pulls the tower to a vertical position. This system is primarily used with smaller wind turbines (<1kW) and shorter tower sections. Small winch-up towers are guyed, while larger "tubular" towers are becoming available which are elegant tapered poles that can be hinged to allow lowering for service.

The second type of installation involves building the tower on the ground and then using a crane to erect the tower in one piece. The advantage of this system is that no one has to leave the ground or work up on the tower. The disadvantage is that servicing requires taking a mortgage on the first-born to pay for the crane. Alternatively, you can climb the tower to perform the necessary work.

The third method requires the use of a "gin pole" and some tower-climbing theatrics that would rival the gymnastics of Cirque du Soleil performers. A gin pole is a boom or pipe that is bolted to the top of an installed tower section and extends upwards to where the next section is to be placed. A ground crew using a pulley mounted to the top of the gin pole raises the next tower section while someone who is strapped to the tower gingerly connects the successive sections. Notwithstanding my personal bias against them, I believe the term "gin pole" reflects the need for such fortification after successfully completing this death-defying act.

"People don't plan to fail, they just fail to plan."

I cannot emphasize enough how difficult working on a tower can be. Climbing a long extension ladder is nothing compared to climbing and working on a tower. A ladder always has some horizontal slant, making this more akin to walking up steep stairs. Climbing a tower with work boots as well as safety climbing and fall protection gear and a bag of tools is great provided you spend your weekends rappelling El Capitan. It can be very tiring and dangerous work. I would stick with the

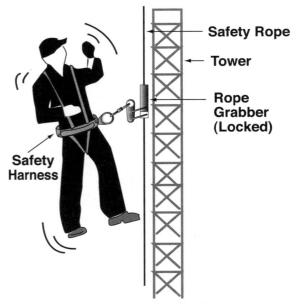

Figure 7-16. Falling from a 100-foot-high wind turbine tower is not something most people would find enjoyable. A safety harness and rope grabber with a short lanyard as shown here make climbing and servicing a much safer task.

tilt tower or mortgage the first-born and hire a crane before resorting to the gin pole system.

At some point it may be necessary to actually climb the tower. Before you get to the top and realize you are too scared to let go and actually do some work, let's discuss climbing gear.

There are too many people who think safety equipment isn't cool or even necessary. The smallest wind turbine tower for home use is in the range of 60 feet (18 m). Falling from that tower would be about the same as falling from a five-storey balcony; it is unlikely that you would survive. Visit your local safety supply store and for a few dollars purchase an approved full-body climbing harness. The Occupational Safety and Health Administration in the United States and the provincial workplace health and safety boards in Canada certify climbing gear suited to this work. Do not purchase a "safety belt" or other non-approved device. One slip off the tower rung and you could be dumped upside down and slide right out of your safety belt.

A safety harness is of no value if it isn't

connected to something, preferably you and the tower. A short section (< 6 ft/2 m) of nylon rope can be attached to the harness and then successively clipped to the tower with a snap hook and unclipped as you climb. While steel snap hooks and approved nylon rope or lanyard will work, there are drawbacks to this system.

The first problem is that constantly clipping and unclipping the hook impedes your progress, and there is a tendency to not use the clip until reaching the tower top. Climbing the small round legs of a tower can be difficult and slippery, and during the descent you may be quite tired after spending time at the top. Safety is just as important on the way down as it is on the way up.

The second problem involves the length of lanyard required between the tower and safety harness. A lanyard that is 6 ft (2 m) long will allow you to drop that same distance before breaking your fall. This may be enough of a drop to cause serious back or muscle injury with a difficult recovery. Also, a lanyard of steel wire will not give, causing an abrupt stop and causing further injury.

The best value for your money and your life is a device from Flexible Lifeline Systems (www.fall-arrest.com). These units comprise a long nylon rope or lifeline attached to the top and bottom of the tower and a "rope grabber" that is attached to the rope and then clipped to the full-body harness. When you climb up the tower, the rope grabber slides along, allowing easy ascent.

Should you fall, your body weight causes a cam inside the rope grabber to pinch the lifeline and break your fall, as shown in Figure 7-16. During a normal descent, the rope grabber must be lifted and slid down the lifeline one section at a time through the descent.

This system will limit your freedom of motion at the top of the tower, as the lifeline is affixed to the tower top section. To provide a bit more freedom, a short lanyard may be attached between the rope grabber and the safety harness.

In addition to protecting yourself from a fall there are a few other points of safety to be considered:

- Always carry tools in a fitted and sealed pouch. You will need both hands free while you're climbing the tower structure.
- Wear leather gloves and non-slip work boots. Tower rungs are very rough and can be slippery from dew, leaves, pollen, and dust.
- When you're climbing the tower, only one hand or foot should be off the tower at a time. For example, release a hand and reach for the next tower rung. Make sure you are gripping the rung before lifting a foot to climb to the next rung.
- Always immobilize the turbine by applying the brake switch and/or furling the machine. Never work on a windy day.
- Wear a hard hat. A light breeze can easily cause the turbine to pitch around or "yaw," striking you in the head.
- Never work on the tower alone. Make sure a buddy is on the ground to assist when needed.
- Make sure that ground persons are not standing in the way of errant nuts and bolts. A bolt dropped from 100 feet (30 m) is not much different from a bullet.

Being cheap or stupid when it comes to safety is a no-win proposition.

Tower Foundations and Anchors

Self-Supporting Towers

The self-supporting tower shown in Figure 7-1 is built on a single pad of concrete reinforced with metal rebar. Alternatively, three holes may be

Figures 7-17 a, b, c. The successive steps in pouring a center pier to hold a guy wire supported tower include: a) forming the pier excavation; b) pouring the concrete; and c) aligning the center pin and smoothing the surface. (Courtesy Bergey Windpower)

drilled and filled with a rebar and concrete mix, providing the necessary pads. It is not possible to cover the design of freestanding tower foundations as there are too many variables to consider, including tower height and weight, wind load, and soil type.

The turbine manufacturer is often able to provide foundation plans that have been pre-approved for most local soil conditions.

Guy Wire Supported Towers

Guy wire supported towers (with or without a ginpole) such as the unit shown in Figure 7-15 are the most common design owing to their lower cost. Guyed towers can be tilt-up or installed with a crane or gin pole arrangement. Smaller turbines, typically less than 8 ft (2.5 m) in diameter, can be mounted on a pipe tower that uses very inexpensive 2.5" OD steel pipe. The turbine manufacturer supplies the guy wire and hardware kit, and you purchase the pipe locally.

Lattice towers such as the ROHN product line manufactured by ROHN Industries (www.rohnnet.com/towers-ssv) are more costly than pipe towers but offer the advantage of fewer guy wires and the option of climbing the tower when servicing is required.

The tower mast sits on a compressive foundation known as a pier. The function of the center pier is to prevent the tower from pressing itself into the ground or slipping sideways. It offers no vertical support to the tower.

The guy wires are connected to anchors that are spaced 120° apart from each other as noted in the top view of Figure 7-15. Pipe towers, being structurally weaker, will generally require spacing of 90°. The anchor is placed a minimum of one-half of the tower height away from the center pier.

The pier is most often built with concrete, although it is possible to drill a hole in exposed bedrock and mortar a tower support pin into the hole. It is not necessary to create forms to hold concrete provided the soil will not cave in when pouring begins. Figure 7-17a shows a simple hole dug into the ground with 2" x 4" lumber used to provide a "finished" look to the surface of the concrete. The hole is lined with rebar that helps prevent the concrete from cracking. Rebar is sold in lengths that can be cut with a hacksaw and held together with wire to form a metal "cube." The rebar cube should be supported so that it is completely immersed in the concrete.

Provided your excavation is approximately 1 cubic yard or less, it is possible to hand mix the concrete, although most prefer to have a ready-mix truck deliver and pour the pier and anchors in one shot as shown in Figure 7-17b.

A metal rebar pin is shown protruding from the center of the pier in Figure 7-17c. The pin, cut just above the surface of the concrete once it has set, accepts the base of the tower. The sole purpose of the pin is to prevent the tower from sliding off the pier.

Anchors

Smaller wind turbines less than 60" (1.5 m) may be anchored using "screw in place" augers. As the wind turbine size increases, the augers must be encased in concrete as shown in Figures 7-18a and b. The Chance company (www.abchance.com) is a major supplier of anchors to the electrical utility market. The company Web site provides a question-and-answer support service which can be used to discuss soil types and matching anchor designs. Screw-in anchors may be used in dense clay, sand, gravel, and hard silts. When in doubt, stick with concrete. A hole dug 4 ft (1.2 m) deep (or below frost line) by 2.5 ft (0.8 m) in diameter must be filled with 2 ft (0.6 m) of concrete to cover the auger helix and provide sufficient holding strength. It is important to ensure that the anchor is fixed below the frost level of the local area or it may be subject to "jacking." This is the tendency of buried objects to be pushed towards the surface during the freeze/thaw cycle. Ensure that the concrete hole is lined with rebar to support the concrete.

The anchor must exit the ground at an angle of approximately 45° and point directly towards

Tower Height ft / (m)	Height to "A"	Height to "B"	Height to "C"	Height to "D"
40 / 12	17 / 5	35 / 11		
60 / 18	27 / 8	55 / 17		
70 / 21	22 / 7	44 / 14	65 / 20	
80 / 24	25 / 8	50 / 16	75 / 23	
90 / 27	28 / 9	57 / 15	85 / 26	
100 / 30	24 / 7	48 / 14	72 / 22	95 / 29
110 / 34	27 / 9	53 / 16	79 / 24	105 / 32
120 / 37	29 / 9	57 / 17	86 / 26	115 / 35

Table 7-3. Tower height determines the number and height of guy wires required.

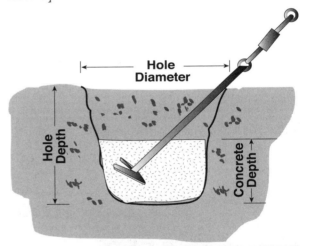

Figures 7-18a and b. A guy support anchor may be either screwed directly into suitable soil or cemented in place if the soil is loose. This anchor hole has a re-bar wire cage for strength once the concrete is poured and is lined with plastic to prevent the dry sand from wicking moisture out of the concrete too quickly.

the center of the tower. This will ensure that the bending force placed on the anchor by the guy wires is equal. If the angle is too great or small, the anchor will have a tendency to bend up or down, reducing its strength over time.

Guy Wires

Guy wires may be supplied as part of the tower kit or they can be purchased in bulk rolls for custom fitting. Use extra-high-strength steel cable for supporting the tower. Refer to the tower manufacturer's data sheets for the diameter of guy cable required for your tower and turbine size. A very conservative estimate is ¼" EHS steel cable for units of up to 3 kW output on a 100 ft (30 m) tower.

The length of guy wire required for a tower will depend on the tower height and the distance from the anchor. Table 7-3 provides the location

Figure 7-19. Guy wires for larger installations are connected to a turnbuckle that allows the cables to be tightened as required. A safety chain or piece of guy wire is threaded through the turnbuckle and preferably key-locked to prevent unintentional loosening.

Figure 7-20. Guy wires can be tightened using a come-along or cable puller. Smaller turbine units with a rotor diameter of less than 60" (1.5 m) do not require turnbuckles. (Courtesy Bergey Windpower)

of the guy wires for levels A, B, C, and D on the installation shown in Figure 7-15.

The guy wire length required to run from the various heights to the anchor can be calculated using a bit of grade 9 trigonometry called the Pythagorean theorem. As my classmates and I were in a deep slumber during this lesson, we should take a moment to review the calculation.

Flip back to Figure 7-15 and you will see that the tower and ground form two sides of a triangle, with the angle at the ground/tower intersection point being 90°. The right angle formed by the two sides and the hypotenuse created by the guy wire form a right angle triangle. Assume that we want to find the length of the guy wire at position "B" for a 60-ft (18 m) tower. Refer to Table 7-3 and you will see that a 60-ft tower requires a guy wire at position "B" of 55 ft (17 m) above grade. The guy radius is one-half the tower height or 30 feet (9 m). With these data we can make the following calculation:

(Tower Guy Height ^{2} + Guy Radius ^{2}) = Guy Wire Length^2

Therefore:

$\sqrt{}$*(Tower Guy Height ^{2} + Guy Radius^2) = Guy Wire Length*

$$= \sqrt{(55^2 + 30^2)}$$
$$= \sqrt{(3,025 + 900)}$$
$$= 62.5 \, ft \, (19 \, m)$$

This length provides the exact distance between the tower and the anchor. An allowance of an additional 5 ft (1.5 m) should be provided to allow for misalignment and affixing the guy to the tower and anchor brackets.

Once the guy wire lengths are determined, carefully unroll the guy cable in a straight line along the ground. Even the slightest kinking will render the guy wire useless, so use caution. And speaking of caution, be sure to wear proper eye protection in case the cable whips around and hits you.

Figure 7-21. The tower and turbine are ready to be lifted by the crane. Note the tower bottom plate center hole and mating pier pin at lower left. (Courtesy Bergey Windpower)

The guy wires may be attached to the tower and anchor using "U" bolts or preformed cable grips. Either method is adequate, although most guy cable suppliers will recommend the cable grips because they can be easily adjusted.

Larger wind turbines require the guy wires to be tightened with turnbuckles, as shown in Figure 7-19. After tightening the cables, thread a safety chain or piece of guy wire cable through the turnbuckle to prevent inadvertent loosening.

Smaller turbine units with a rotor diameter of less than 60" (1.5 m) in diameter do not require turnbuckles. These guy wires may be tightened by pulling them taught using a cable puller or come-along prior to applying the guy wire clamps, as shown in Figure 7-20.

Electrical Supply Leads

Small turbines output direct current at 12, 24, or 48V while large units supply 3-phase alternating current, which is rectified at a remotely located "controller box." In order to transfer this power down the tower, the turbine will have a connection terminal or be supplied with a short section of "flying leads." A direct current machine will be supplied with two wires, one positive and one negative. A machine which outputs 3-phase alternating current will have three wires without any polarity or connection concerns.

The connections between the electrical leads from the wind turbine and the tower wires should be located in a weatherproof junction box mounted near the top of the tower. The tower-to-house lead wires should be weather-resistant flexible armored cable (Teck is one trade name) or else they should be enclosed in conduit. The power lead wires or conduit may be strapped to one of the tower legs using either ultraviolet-protected tie wraps or cable clamps. Wire size and selection will be discussed in Chapter 13.

Figure 7-22. The crane lifting straps are installed at 75% of tower height and wrapped through the rungs to prevent them from sliding up the tower. Note the support cradle used to hold the tower off the ground, allowing turbine installation before the arrival of the crane. (Courtesy Bergey Windpower)

Erection of the Tower with a Crane

Pre-Lift Check List

At this point, the tower should be flat on the ground or slightly raised and supported along its length if the turbine is already installed. We are now ready to prepare for final inspection prior to lifting.

- Guy wires (if used) are attached to the tower and secured. Guys should be run down to the tower. Ensure that cables are in order from lowest to highest and are not twisted.
- Turbine is wired into the junction box at the top of the tower. Connections are made with tower lead wires.
- Tower lead wire is secured to the tower and is sufficiently free to allow tower movement when aligning base plate with pier pin.
- Turbine is securely mounted to tower and bolt torques tested.

- Tower bolts are checked for security and torque.
- Turbine brake and furling are enabled.
- Turbine is tied with rope to prevent it from yawing and hitting the crane or lift cables.
- Check, check, and re-check that all hardware and tools are ready and close at hand. The crane service charges by the hour, even when you have to run to town to pick up a couple of last-minute ten-cent washers.
- Check the weather for a still, clear day and call for the crane.

The Lift

Once the crane is ready, the tower lifting straps are threaded through the tower rungs at 75% of the tower height. The tower is then slowly lifted, making sure that the guy wires, power feed cable, and turbine assembly do not tangle in or hit the crane structure.

If the turbine was not installed at ground level, lift the tower to a comfortable working height and install the unit. (It is preferable to have this step completed prior to the arrival of the crane, as this time is billed at a rate similar to that of a New York lawyer.)

Figure 7-23. The tower and turbine are slowly lifted into position, with care being taken not to tangle the guy wires or cause the turbine body or blades to strike the crane. (Courtesy Bergey Windpower)

Figure 7-24. Once the tower is secured in position, the lifting straps and anti-yawing ropes are removed. This is also a good time to appreciate the view and congratulate yourself on a job well done. (Courtesy Bergey Windpower)

Once the tower is located over the center pier, the guy wires may be secured to their respective anchors as described above and detailed in Figures 7-19 and 7-20. At this point, the leveling of the tower and tightening of the cables is only approximate. Once the tower is secure, you need someone to climb the tower, remove the crane lifting straps, and untie the anti-yawing ropes securing the turbine. Use caution when removing the anti-yawing ropes to ensure that the turbine doesn't yaw and strike you.

The crane may now be packed up and sent on its way—just as soon as you've written a big check.

Hinged Tower Lifting

There are numerous designs of hinged towers on the market today, including guyed or tubular freestanding models. The advantages of a hinged tower are obvious, as it allows all of the installation and service work to be done safely from the ground. In addition, a crane is not required to perform the lift. Hinged towers employ a hand-operated or power winch that acts on a levered section of the tower which is also known as the gin pole.

The following sequence of pictures was taken at the home of Cam and Michelle Mather, who had an environmental engineering class at a local community college assist with the installation work. The steps taken to install this Bergey XL.1 turbine are very similar to those used in the installation of the large guyed lattice tower described above.

Figure 7-25. The tubular tower base hinge is bolted to the center support pad, with successive sections of tower pipe "stacked" one upon the other. The upper section lying on top of the tower is the "gin pole" which will be walked up perpendicular to the tower to spread the force throughout the tower during erection.

Figure 7-26. (Above) At 20-foot intervals, a guy wire support member is slid between overlapping tower sections, providing a secure anchor point for the guy wire and fastening pipe sections together. The upper guy wire bracket (left) is bolted to the top mast low enough to be out of the path of the turbine blades.

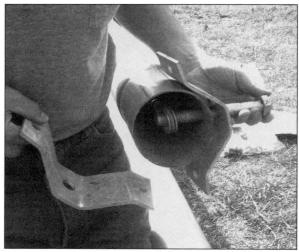

Figure 7-27. The upper gin pole bracket is attached to allow the cable to the winch to be fastened. Notice the bolt going through the gin pole safety wire to prevent the section of the gin pole from separating during the lift.

Figure 7-28. The guy wires are rolled out from their respective spools to their anchor points. The wire must be rolled out carefully to avoid tangling and kinks.

Figure 7-29 The base of the erect tower shows two of the copper grounding wires going to the 10-foot grounding rods prior to being pounded in. In the lower right you can see the wires coming underground from the battery room. They are in a flexible conduit to prevent damage and to allow it to have some play as the tower is raised and lowered.

Figure 7-30 The base of the tower is shown again after the copper grounding wires have been pounded in at the base. They are below grade to prevent being an obstruction to a lawn mower. Note the extra nut left at the top of the bolt which fastens the base to the concrete pad to prevent injury should anyone working in the area accidently trip and fall onto it.

Figure 7-31 The gin pole is walked up into its vertical position prior to the initial lift. It is lifted and walked up until the winch can tighten it the rest of the way.

Figure 7-32 The gin pole requires ropes from the side anchors to keep it in line with the tower during the initial lift. Once the side wires are properly tightened the ropes are no longer required. The initial lift does not have the turbine installed on it. This means less weight for this first pull-up and allows the tower to be straightened and guy wires properly tensioned without the turbine.

Figure 7-33 The tower is ready for the initial lift. The upper bracket is bolted in place. The two heavy-guage wires are secured so they do not drop into the tower during the lift. The gin pole is in its vertical position and is kept in position by the two ropes secured to the side anchors.

Figure 7-34. The foreman watches the lifting process and should be ready to stop the operation if guy wires become snagged or tangled or if the tower becomes unstable. Note the winch wire pulling the tower upright in the center of this photograph. One person is required to monitor each of the side anchors to ensure that not too much force is exerted on the tower or anchor.

Figure 7-35. The view of the initial lift from near the winch.

Figure 7-36 Once the tower is erect the foreman instructs the person on each side anchor to tighten or loosen each set of guywires as required to straighten the tower.

Figure 7-37. The work crew is anxiously watching as the winch operator begins the lift process. An electric drill is "chucked" into the winching gearbox, pulling the guy wire that is routed over the gin pole and connected to the main tower section.

Figure 7-38 The tower is pulled up with this worm gear winch which is bolted onto a bracket attached to the anchor adjacent to the gin pole when it's in down position. The tower requires four fully charged batteries on an 18V cordless drill to pull it up and two to lower it down since gravity does much of the work on the way down. An advantage of gin-pole towers like this is that the turbine can be lowered within 30 minutes for maintenance or if you are in a hurricane-prone area. Note that care was taken to ensure that the wire wound uniformly to prevent jerking when the tower lowered.

Figure 7-39. The turbine with blades ready to be tightened.

Figure 7-40. Using a torque wrench the blades are tightened to 40 ft.lbs. On those days when the wind is howling, you'll be glad you invested in or borrowed a good torque wrench for this part of the job.

Figure 7-41. The turbine is placed face downward to prevent it from yawing or spinning on the tower during the lift operation.

Figure 7-42. Once tensioning and double-checking are complete, the turbine is ready to spin and start supplying electricity to the Mathers' 120-year-old off-grid home.

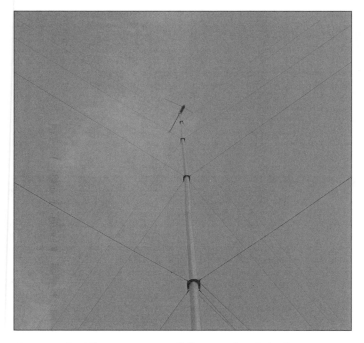

Figure 7-43. The tower is now fully erected and the four guy wires spaced at 90° can be clearly seen in this photograph.

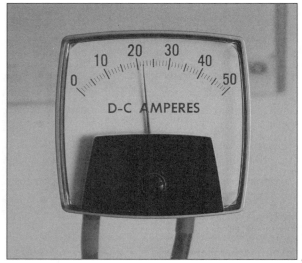

Figure 7-45 Since installing the new wind turbine Cam has cancelled the sports channels on his satellite TV and now spends most nights in his battery room watching his ampmeter track the unit's output. As with fireworks one can't help but "oooo" and "awww" as the needle dances up and down merrily. Unlike solar panels, a wind turbine can mean that the batteries actually have a better charge when you get up in the morning because the turbine has been busy all night. Alright, Cam didn't cancel the satellite TV; he just watches less of it.

Final Assembly

After the tower is raised, or when the crane has left (or stopped billing), set vertical by using a transit sighting scope as detailed in Figure 7-46. A transit may be rented from a local tool company. These units are often seen on sites where construction workers are building roads or checking various angles. A basic unit comprises a tripod stand with an optical sighting scope similar to those used on rifles. The tripod is placed between one and two tower heights away from the center pier and leveled using the integrated water bubble levels and adjustable legs. The vertical crosshair in the scope is then aligned with one tower leg, close to the ground. The scope is then moved vertically up and down, ensuring that the tower leg and crosshair remain in parallel through the entire length.

Guy wires may be adjusted tighter or looser using the turnbuckles to compensate for error. The transit is then moved 90° and checked again from this alternate position. It may be necessary to move back and forth a few times to get the alignment perfect

If you do not have access to a transit, it is possible to use a standard carpenter's level and check for tower alignment at the bottom section. You can then sight up the tower, placing your eye as close to the tower leg or pipe as possible. Bends in the structure will be obvious and can be adjusted by aligning the guy wires as described.

Once the tower is aligned, tighten the guy wires until there is little slack in the cables. A good shortcut for testing cable tension is to strike the cable with a hammer. The cable should sound a guitar-like note and a "wave" will be seen running quickly up and then back down the wire. Don't forget to attach the turnbuckle security chain as shown in Figure 7-19.

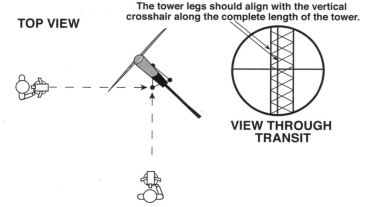

Figure 7-46. Renting a transit is an excellent way to determine if the tower is properly leveled.

Figure 7-47. At least one tower post (more in sandy conditions) should be connected to a grounding rod driven 8 feet (2.5 m) into the soil. (Courtesy Bergey Windpower)

The last step in the installation process is to ground the tower and guy wires. In clay and other moisture-retaining soils, a single 8-ft (2.5 m) electrical grounding rod should be driven into the ground close to or through the tower pier as shown in Figure 7-47. In sandy soils, at least two grounding rods should be installed, each at a different tower leg, to ensure proper lightning protection qualities.

Any guy wires sharing a common anchor should be connected together using a minimum 0-gage grounding wire and galvanized U-shaped bolts as detailed in Figure 7-48. The end of the ground wire may be connected either to the guy anchor or to a separate 8-ft grounding rod.

For both the tower leg and guy wire grounding, be sure to use as short a run of wire as possible and smooth, even bends. Lightning hates sharp corners and may completely miss the grounding rod if severe wire bends are used.

Figure 7-48. Each guy wire should be grounded by attaching a U-shaped steel clamp to the guy anchor and grounding wire. Use smooth bends and as short a run as possible.

Vertical Axis Wind Turbines

Someone is always looking for the better mousetrap, and an alternative to the standard horizontal axis wind turbine (HAWT) is no exception. Possibly the best-known variation on the HAWT is the Darrieus or "egg-beater" turbine designed and patented by Georges Jean Marie Darrieus in 1931.

Although the theoretical design of the Darrieus and other vertical axis models are comparable to horizontal axis machines, there are a number of issues that prevent vertical turbines from reaching their potential.

The Darrieus turbine is designed with the lower section of the rotor located too close to the ground, thus reducing potential power generation. As with any wind turbine, locating the unit close to the ground is akin to installing PV panels in the shade of a tree.

Excessive centrifugal force loading on the blades, inability to self-start, and resonant frequency self-destruction are further reasons these units are not commercially viable. However, that does not mean that people have given up trying.

The GALE™ vertical axis wind turbine is offered by Tangarie Alternative Power (www.tangarie.com) and is available in three models with rated power outputs of 1, 5, and 10 kW. The firm even offers to dress your turbine in the Stars and Stripes to make it more aesthetically pleasing.

Being the ever-inquisitive engineering type, I had to do a bit of research on this newcomer and see for myself how well these units perform. A quick review of the product manuals reveals that, "at 90 mph wind speed [145 kph – a Category 1 hurricane level] the [1 kW turbine] could produce 0.6 kW [while the] 5 and 10 models could produce 5.0 and 10.0 kW respectively." Indeed, I would expect a child's propeller beanie to produce as much energy at such high wind speeds!

To make matters worse, installation drawings and pictures seem to indicate that it is acceptable

Figure 7-38. The Darrieus wind turbine is the best known of the vertical axis wind turbine (VAWT) family. Although many countries have tried to develop VAWTs, there are very few models operating or financially viable. (Photo source: http://en.wikipedia.org/wiki/File:Eoliennes_Gaspesie.jpg)

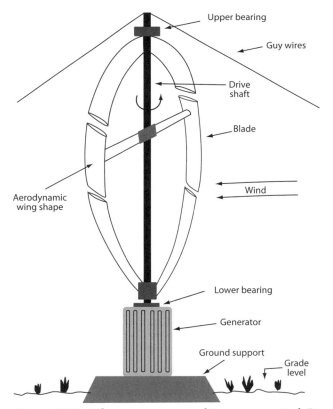

Figure 7-39. The components making up a typical Darrieus VAWT are shown here. The turbine rotor is able to generate power and spin as a result of the energy imparted on the aerodynamic "wing" shape of the blade.

to place the turbine on a light pole standard or projecting just above the roof of a house. As we learned earlier, the "rough" surface of the surrounding area will create considerable amounts of turbulence and reduce the kinetic energy of the wind. The old adage that "the higher the tower, the greater the power" applies to all turbines, including VAWTs.

Let's take a closer look at the power curves.

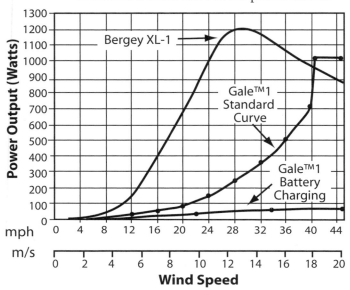

Figure 7-41. This graph details the currently published (August 2009) power curves for a GALE™ 1 VAWT and the Bergey XL1 HAWT. Each unit is intended for small home or cottage off-grid power applications and is rated at 1,000W capacity.

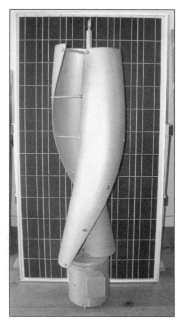

Figure 7-40. The GALE™ VAWT manufactured by Tangarie Alternative Power, LLC is available in three models with rated power outputs of 1, 5, and 10 kW.

The power curve for the Bergey XL1 shows a typical exponential increase in power according to wind speed. This model has an output of 150W at 12 mph, while doubling the wind speed increases the power output of the turbine to its rated capacity of 1,000W.

The GALE™ 1 power curve[1] is highly attenuated, meaning that useful power output does not occur until the turbine is exposed to very high wind speeds. The difference between the two machines installed in the real world is very telling. A review of the Wind Energy Resource Atlas of the United States shown in Figure 7-10 indicates that the highest average annual wind speed (Power Class 7) occurs in small pockets of the country, and even at these locations the GALE™ 1 will generate less than 100W of power. Using the same wind speed data, the Bergey XL1 will generate 700W of power, a sevenfold increase in output.

If you are going to consider using any alternative to the tried-and-true HAWT, make sure you understand the machine's limitations as well as the warranty and the reputation of the manufacturer for its product support. A healthy dose of caveat emptor is strongly advised.

When considering the purchase of any turbine, HAWT or VAWT, check to see if the United States National Renewable Energy Laboratory (NREL) has certified the desired model under the small wind turbine certification program. Although this program was just started in 2008, the NREL expects to continue testing wind turbines to standardized conditions, giving potential buyers of these units confidence relating to the design and quality. To follow the certification work of the NREL visit the website at http://www.nrel.gov/wind/small_wind.html.

Endnotes

[1] Data derived from the interactive graph at http://www.tangarie.com/i/products/power_your_home.html

Recycling a Classic Jake
A Homeowner's Story by Josée Guénette

We purchased our Jake in 2003, three years into the construction of our owner-built log home and a full eighteen months before we would be ready to use it. At that time, I was not really aware that the Jacobs wind turbine, also known as "The Jake," was considered the gold standard in the industry; but my husband was, and he jumped at the opportunity to buy one second-hand, even if it needed a bit of work!

A family in Lanark, Ontario had been relying on this Jacobs turbine to supply their household electricity for many years, but an unexpected grid extension, corresponding with a much-needed maintenance cycle, helped them decide to sell not only the turbine they were using but also a second one they had purchased and not yet commissioned. We bought the larger of the two, a 2.4 kW turbine which dated back to 1958. The current owner had purchased it from a farmer in Saskatchewan. Our deal was Cash & Carry—we paid for the turbine where it stood, including everything up to and including the wires into the battery bank. Moving it was our responsibility, and the next challenge!

The dismantling and re-commissioning of the Jake were feats of assiduous project management, and our success was due largely to planning and family involvement. There were conference calls, tool lists, battle plans in triplicate, personalized tool kits, troop deployment diagrams, and tight schedules. Oh, and a nice picnic lunch, to keep everyone in good spirits.

It took about six hours to get our Jake out of the sky, disassemble the tower, pull up the 4/0 wiring and send it all to our home in La Pêche, Québec. The turbine did need some work: the insulation on the windings was worn, the pickup rings had fused, the original sitka spruce blades needed some paint. During the repairs to the turbine, we reassembled the tower on the ground. It took a lot longer to reassemble it than to take it apart!

But now, triumphantly refurbished and repainted, our Jacobs turbine stands 90 feet in the air and is largely responsible for the electricity in our home. It recharges a NiCad battery bank which was salvaged from a decommissioned military installation. Considering that we started this adventure by decreasing our electricity use where practical, our system is perhaps a good example of Reduce, Reuse and Recycle!

DJ MacIntyre of Le Boisé Alternatives (www.leboise.com) nearing completion of the restoration of a 2.4 kW, 1958-era Jacobs wind turbine for homeowner Josée Guénette and her husband.

DJ is feeling pleased now that the Jake is once again proudly flying at its new home in La Pêche, Quebec, Canada.

The "Jake" is just about ready to fly, looking pretty sharp in a new coat of paint.

Chapter 8
MICRO HYDRO ELECTRICITY PRODUCTION

If you are lucky enough to have a stream, river, or waterfall on your property it is well worth considering micro hydro electricity production. While PV is undeniably the simplest renewable energy source to install and maintain, micro hydro systems (units under 1 kW capacity) can be the least expensive. The density of water is considerably higher than that of wind. This increase in density allows a hydro turbine to be many times smaller than a wind turbine of similar output.

Smaller turbines and fairly simple installation help keep the cost to a reasonable level. In addition, hydro sites are generally able to supply electricity on a 24-hour-a-day basis. A smaller amount of power supplied over a longer period equates to a larger amount of energy. For example, a typical PV array is rated at 1,000W of output power. Over a five-hour period of sunlight, this equates to 5 kWh of energy. A small 200W hydro turbine operating 24 hours per day produces approximately the same amount of energy. On the downside, suitable hydro power sites are not nearly as common as either PV or wind sites.

Understanding the Technology

As we discussed in Chapter 1.2, "What is Energy?," hydro electricity is produced from the kinetic energy in moving water under the force of gravity. The pressure or head and the flow determine the potential energy in water. The vertical distance that water must fall under the influence of gravity determines the head pressure. Head is measured either in feet or in units of pressure such as pounds per square inch (psi) or kilopascals (kPa). It is interesting to note that water standing in a vertical tube will exert a known force at the bottom of the tube at the ratio of 1 psi (6.9 kPa) per 2.31 vertical feet (27.7 inches/0.7 m). This allows head to be easily converted back and forth between vertical height of water and pressure, which, when multiplied times water flow rates, will provide the predicted power output for a given river, stream, or waterfall.

Flow is the quantity of water flowing past a given point in a given period of time and is expressed as a volume. Typical units are gallons of water per minute (gpm) or liters per minute (lpm).

When the flow and head are integrated mathematically the result is power, expressed in horsepower or watts. Once the units of water power are converted to watts of electricity by the turbine's generator, we are back to common electrical units of measure. Because head and flow are interrelated, a high head site with a low flow produces the same amount of power as a site with a low head and a high flow. For example, a site with 100 feet of head and 2 gallons per minute of flow has the same potential energy as a site with 2 meters of head and 100 gallons per minute of flow. This is not semantics. Most sites are located in an either/or situation. A

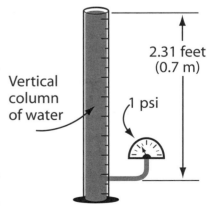

Figure 8-1. A vertical column of water 2.31 feet high (0.7 m) exerts a pressure of 1 psi (6.9 kPa).

Figure 8-2. This turgo wheel turbine and generator are suited for high head and low flow applications. This unit is installed in the Rocky Mountains of British Columbia and provides an output of 500W. (Courtesy Energy Systems and Design)

Figure 8-3. A low head turbine such as this propeller-based model requires high water flow. With 3 feet (1 m) of head, this unit produces 200W of power. (Courtesy Energy Systems and Design)

high head site is generally located in a very hilly or mountainous region, where the water might flow through several hundred to several thousand feet (100 m to 1000 m) of plumbing line. Fortunately a high head site requires little flow, so the pipe and turbine in this instance are fairly small.

The opposite is true of low head sites. Commonly known as "run of river" designs, low head sites can be sized to operate from 2 to 10 feet (0.6 m to 3 m) of head. What we gain in lower head we pay for in more water flow and a larger turbine; nobody rides for free.

Water turbines come in two broad categories and dozens of subcategories that we do not need to worry about. The main consideration with respect to the turbine selection is whether there is a large flow of water and little vertical drop or vice versa.

All turbines have a number of components in common, as do the civil works of the hydro site. Figure 8-4 shows the workings of a low head, high flow hydro turbine manufactured by Energy Systems & Design of Canada (www.microhydropower.com). As the low head turbine must have a large flow, the intake is formed using the large gray plastic "flume." Water enters the turbine heart and passes through a set of guide vanes. These vanes are stationary and are used to force the water to hit the runner blades at the correct angle. In this low flow turbine, the runner blades are shaped like a large boat propeller. As the

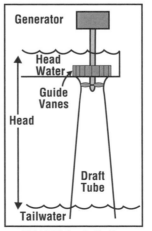

Figure 8-4. A low head turbine is physically larger than its high head cousin but retains many of the same components. (Courtesy Energy Systems and Design)

Figure 8-5. High head turbines tend to be quite small owing to the high water pressure and resulting low flow required to operate the tiny runner blade. (Courtesy Energy Systems and Design)

water falls through the vertical head, the runner blade captures the kinetic energy and causes the generator to rotate, producing electricity. The draft tube forms the exit path for the water. The tapered draft tube acts as a siphon, helping to draw water down the tube and away from the runner blades, increasing turbine efficiency. To increase the head and resulting output power of the turbine, a series of draft tube extensions can be added.

High head turbines such as the Turgo wheel unit shown in Figure 8.2 are considerably smaller owing to the low flow of water required to turn the runner blade. Increased head and resulting water pressure make up for what is lost in flow. For example, a turbine installed with 300 feet (91 m) of head will have a water pressure at the runner nozzles of 130 psi (900 kPa).

300 feet head ÷ 2.31 feet per psi
= 130 psi nozzle pressure

Figure 8-5 shows a water intake pipe plumbed to a series of two shutoff valves prior to entering the turbine. These valves may be used for turning off the turbine during maintenance or when insufficient water flows to operate the system during dry periods. After leaving the shutoff

valves, water is directed through a nozzle prior to striking the Turgo-style runner blade. Nozzle openings can be changed to modify the water flow and resulting output power of the turbine.

The high-pressure water jet strikes the runner blade, imparting a turning effort or torque to the generator shaft, producing electricity. The water then exits the turbine heart by simply dropping away from the open-bottom casting to return to the tailrace downstream.

The civil works comprise the dam, intake, penstock, and tailrace sections of the hydro site. Figure 8-6 details the major components required for a high head site. A dam structure provides a means of inserting an intake pipe and strainer midway between the riverbed and river surface. If water levels are sufficiently deep it may be possible to eliminate the dam and simply install the intake in a submerged strainer or crib.

The dam's outlet pipe is fed to a stop valve that is used to drain the penstock system during servicing or in the event of a broken pipe. Penstock pipes are often buried to prevent mechanical damage and chewing by animals. The penstock is also provided with an air valve at the top and a pressure relief and drain valve at the bottom. A second turbine stop valve and pressure gauge are also provided.

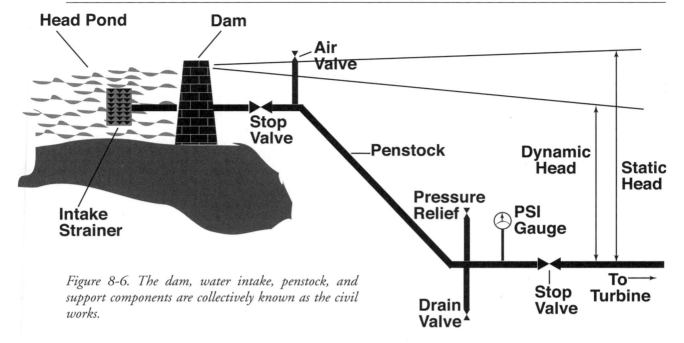

Figure 8-6. The dam, water intake, penstock, and support components are collectively known as the civil works.

When the turbine stop valve is closed, water will stop flowing and the penstock will remain filled with water to the height noted as the "static" or "not moving" head level. As discussed earlier, the static head can be directly translated into a pressure by dividing head in feet by 2.31 to produce psi. The pressure gauge will record this static pressure when the turbine stop valve is closed.

As the turbine stop valve is opened, water will begin to flow down the penstock to the turbine. As this occurs, the pressure on the psi gauge will drop as a result of pressure loss due to friction in the penstock pipe. The resulting pressure drop will lower turbine output accordingly. To circumvent this problem, a correspondingly large penstock pipe may be used. Alternatively, lower friction polyethylene SDR pressure-rated pipe may be chosen over PVC material. Pipe friction losses are noted in Appendices 9a and 9b.

If you have ever been in a home where suddenly turning off a water tap causes the whole house to shake, you have experienced the effects of water hammer, which occurs when moving water is suddenly stopped. Moving water contains kinetic energy (the energy of motion) that must be removed when the movement stops. If the turbine stop valve is suddenly turned off, the entire weight of water in the penstock has to find someplace to let off this energy. The effect is not much different from a freight train slamming into a solid wall: something has to give. Usually it's the penstock that gives. If the water hammer is severe enough, the penstock pipe will absorb the energy and possibly split or explode. To help stop damage to the penstock a pressure relief valve is located near the bottom. Should pressure build due to water hammer, the pressure relief valve is sized to open before the pipe bursts.

An air valve is provided at the topmost section of the penstock to eliminate any air trapped within.

Low head or run of the river sites are generally simpler but still require a dam, intake, and penstock pipe feeding a flume (water delivery box) as shown in Figure 8-3.

What Is a Good Site?

In order to determine if you have a suitable hydro site, it is necessary to perform an evaluation of the available resources. In assessing a particular site,

the variables to consider are:
- water head level
- flow at the turbine intake
- penstock design issues

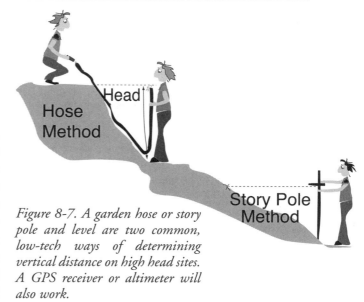

Water Head Level

Head may be measured using several techniques. One method is to hold one end of a garden hose at the same level as the proposed intake from the stream. A first person ensures that the hose remains filled with water. The other end of the hose is routed downhill, causing water to flow out. If a second person slowly lifts the downstream end of the hose, water will stop flowing once the upstream and downstream ends are at equal heights. Measure the distance from the ground to the point where the low-side hose stopped leaking water. This is the head measurement for the first section. Repeat the process with the first person now holding the upstream end of the hose at ground level where the second person was standing. The process is repeated until the final head measurement is completed at the desired intake location for the turbine.

Figure 8-7. A garden hose or story pole and level are two common, low-tech ways of determining vertical distance on high head sites. A GPS receiver or altimeter will also work.

A variation on the hose method is to use a measuring stick or "story pole" and a carpenter's level, working uphill. The story pole is placed at the desired turbine intake location. A carpenter's level is placed against the pole at any convenient height. By sighting along the level, a spot will be noted ('x' in Figure 8-7) which denotes a point of equal elevation. The story pole is moved to this spot and the process repeated until the stream intake location is reached.

A third and much simpler method is to use a global positioning system receiver or accurate altimeter if one is available.

Flow at the Turbine Intake

The easiest method to measure small flows is to channel the water into a pipe and, using a temporary dam of rocks, plywood, or whatever, fill a container to a known volume. By measuring the time it takes to fill the container you will be able to calculate the flow rate.

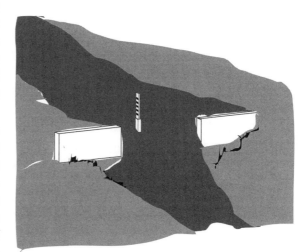

1 to 4 feet between depth ruler and weir

Water height above weir

Stake holding ruler

Figure 8-8. Measuring higher flow levels requires the use of a temporary dam or weir installed in the brook, stream, or river.

For higher flow rates the weir method is more versatile. This technique uses a rectangular opening cut in a board which is set into the stream much like a dam, as detailed in Figure 8-8. The water is channeled into the weir and the depth is measured from the top of a stake that is level with the edge of the weir and positioned several feet upstream.

The resulting measurement can then be applied to the data contained in Table 8-1 that converts the weir measurement to US gallons per minute. Multiply gallons per minute by 3.8 to determine liters per minute.

TABLE 8-1 - WEIR MEASUREMENT TABLE

Table shows water flow in gallons/minute (gpm) that will flow over a weir one inch wide and from 1/8 to 10-7/8 inches deep.

Inches		1/8	1/4	3/8	1/2	5/8	3/4	7/8
0	0.0	0.1	0.4	0.7	1.0	1.4	1.9	2.4
1	3.0	3.5	4.1	4.8	5.5	6.1	6.9	7.6
2	8.5	9.2	10.1	10.9	11.8	12.7	13.6	14.6
3	15.5	16.5	17.5	18.6	19.5	20.6	21.7	22.8
4	23.9	25.1	26.2	27.4	28.5	29.7	31.0	32.2
5	33.4	34.7	36.0	37.3	38.5	39.9	41.2	42.6
6	43.9	45.3	46.8	48.2	49.5	51.0	52.4	53.9
7	55.4	56.8	58.3	59.9	61.4	63.0	64.6	66.0
8	67.7	69.3	70.8	72.5	74.1	75.8	77.4	79.1
9	80.8	82.4	84.2	85.9	87.6	89.3	91.0	92.8
10	94.5	96.3	98.1	99.9	101.7	103.6	105.4	107.3

Example of how to use weir table:
Suppose depth of water above stake is 9 3/8 inches. Find 9 in the left-hand column and 3/8 in the top row. The value where they intersect is 85.9 gpm. That's only for a 1-inch weir, however. You multiply this value by the width of your weir in inches to obtain water flow.

Measuring the flow at different times of the year helps you to estimate your minimum and maximum usable water flows. If the water source is seasonably limited, you will have to depend on a hybrid system or fossil-fuel generator to make up the loss of hydro energy.

You must also keep in mind that a reasonable amount of water should be left in the natural stream path to support life in that section of the stream. An intake screen will ensure that fish and other animals are not driven into the turbine's intake system. When you have removed the energy, the water will be returned to the stream with no loss in total water flow. Such an arrangement not only provides you with renewable energy, it is also environmentally responsible.

Zero Head Sites with Good Flow

After carefully explaining above that head and flow combine to make power, I am going to modify that statement slightly to allow for zero head turbines, which rely exclusively on the flow or run of the river for the extraction of energy. One of the few manufacturers of these products is Boost Energy Systems Ltd. (http://www.ampair.com/ampair/waterpower.asp), who produce the model UW100 turbine shown in Figure 8-9.

Zero head units were originally designed to be pulled behind slower-moving sailboats as a means of charging batteries in these seriously off-grid systems. It stands to reason that if the turbine stops moving and the water doesn't, the same power will be generated.

Installing a unit in a culvert or fast-moving stream will start the outboard motor type of propeller spinning. Throw a ping-pong ball into the stream. If you can keep pace with the ball at a walk the site is too slow. A brisk walk to partial jog indicates a power output of 6 amps at 12 volts potential (1,700 watt-hours per day). If you need to increase your speed to a jog, the power output will be 9 amps at 12 volts (2,500 watt-hours per day).

These units are ideal for locations where the stream will not freeze or for summer cottage installations.

Penstock Design Issues

All hydro systems require a pipeline. Even systems that operate directly from the dam require at least a short plumbing run or flume. It is important to use the correct type and size of pipe to minimize restrictions in the flow to the nozzle(s) or turbine runner wheel. More power can be obtained from the same flow with a larger pipe which has lower losses. Therefore, pipe size must be optimized based on economics. As head decreases, efficiency of the system decreases; it is important to keep the head losses low.

The pipe flow charts shown in Appendix 9 show us that 2" (50 mm) diameter polyethylene pipe has a head loss of 1.77 feet (0.5 m) of head per 100 feet (30 m) of pipe at a flow rate of 30 gpm (113 lpm). This is 17.7 feet (5 m) of loss for 1000 feet (300 m) of pipe. Using 2" (50 mm) PVC results in a loss of 1.17 feet (0.36 m) of head per 100 feet of pipe or 11.7 feet (3.6 m) per 1000 feet.

Polyethylene comes in continuous coils because it is flexible (and more freeze resistant than steel). PVC comes in shorter lengths and has to be glued together or purchased with gaskets (for larger sizes).

Additional Pipeline Construction Details

At the inlet of the pipe, a filter should be installed. This can be a screened box with the pipe entering one side or a section of pipe drilled full of holes and wrapped with screen. If the holes are sufficiently small, the screen may not be necessary. Make sure that the filter openings are smaller than the smallest nozzle used on high head machines. This prevents debris from blocking the nozzles and restricting power output.

The intake must be above the stream bed so as not to suck in silt and deep enough so as not to suck in air. The intake structure should be placed to one side of the main flow of the stream so that the force of the flowing water and its debris bypasses the intake. This also ensures that fish

Figure 8-9. Zero head turbines such as the UW100 Aquair features a special low-speed generator that can be mounted in fast-moving streams or pulled along side a boat. (Courtesy Boost Energy Systems Ltd.)

will not become entrapped in the intake system. Routinely remove any leaves or other debris from the intake.

If the whole pipeline doesn't run continuously downhill, at least the first section should so that the water can begin flowing. A bypass valve may be necessary. This should be installed at a low point in the pipe.

For pipelines running over dams, the downstream side may be filled by hand. Once filled, the stop valve at the turbine can be opened to start the flow using a siphon action. If full pressure is not developed, a hand-powered vacuum pump can be used to remove air trapped at the high point.

At the turbine end of the pipeline a bypass valve may be necessary to allow water to run through the pipe without affecting the turbine in order to purge the line of air or increase the flow to prevent freezing.

A stop valve should be installed upstream of the nozzle and a pressure gauge should be installed upstream of the stop valve so that both the static head (no water flowing) and the dynamic head (water flowing) can be read.

The stop valve on a pipeline should always be closed slowly to prevent water hammer (the column of water in the pipe coming to an abrupt stop). Water hammer can easily destroy your pipeline and for this reason you may wish to install a pressure relief valve just upstream of the stop valve. Water hammer can also occur if debris clogs the nozzle.

Electrical Energy Transmission Distance

Let us assume we have a 12V battery system and the transmission distance from the turbine to the battery bank is 200 feet (60 m) for the round trip. Because of electrical resistance losses in the transmission cable we will require a slightly higher voltage at the generator to charge the batteries. If our turbine is providing 300W of energy, this will translate into a current of approximately 20A at 15VDC.

No electrical conductor is perfect, and a

Wire Gauge	Diameter (Inches)	Ohms per 1000 Feet	Ohms per Kilometer
0000	0.460	0.05	0.16
000	0.410	0.06	0.20
00	0.364	0.08	0.26
0	0.324	0.10	0.32
2	0.258	0.16	0.52
4	0.204	0.25	0.83
6	0.162	0.40	1.32
8	0.128	0.64	2.10
10	0.102	1.02	3.34
12	0.081	1.62	5.31
14	0.064	2.58	8.43
16	0.051	4.10	13.39
18	0.040	6.52	21.33

Table 8-2. Copper wire resistance loss chart measured in ohms per 1000 feet or 1 kilometer.

decision must be made about how much energy should be lost in the wiring. Systems with lower losses require larger, more expensive cable than those with slightly higher losses. An alternative is to raise the battery voltage to 24V or even 48V. Remember that a higher voltage will result in lower current.

A good starting point is to work with a 10% loss. This yields a 30W loss out of the original 300W production assumption. The formula for conductor resistance loss when wire resistance is not known is:

watts of loss ÷ current ^{2}
= *wire resistance in ohms*
= *30 watts loss ÷ 20^2 amps*
= *30 watts loss ÷ 400*
= *0.075 ohms of wire resistance*

This is the calculated wire resistance that will produce a 10% electrical transmission loss. The wire resistance data in Table 8-2 shows losses per 1000 feet of cable, while our sample installation requires only 200 feet round trip:

1000 feet ÷ 200 feet x 0.075 ohms
= 0.375 ohms per 1000 feet.

Table 8-2 indicates that 6-gauge wire has a resistance of 0.40 ohms per 1000 feet:

200 feet ÷ 1000 feet x 0.40 ohms
= 0.08 ohms

Increasing the wire size further reduces the losses in the same manner albeit at a higher cost.

We must also take into consideration that the power loss within a cable is shared between the current and the voltage. Voltage drop in the wire is equal to:

current flow in amps x resistance of wire
= *voltage drop in wire*
= *20 amps x 0.08 ohms*
= *1.6 volts drop*

Therefore if the battery voltage is 13.4 the hydro generator will be operating at 15.0 volts. Keep in mind that it is always the batteries that determine the system voltage. That is, all voltages in the system rise and fall according to the battery's state of charge, as discussed in Chapter 9, "Battery Selection and Design."

Site Selection Summary

After having determined the head, flow, penstock, and electrical transmission issues, the question still remains: will this site be suitable for a micro hydro installation? To make this decision, it is now necessary to refer to the manufacturer's data sheets to determine how well the site will work with the available turbine selection. Hydro turbines, like wind turbines, come with a power curve data sheet, but instead of relating wind speed to power, hydraulic units relate flow and head to generator output power.

The power curve for this model of turbine is based on the assumption that a minimum of 2 feet (0.6 m) must be provided before the unit develops usable power of 100W output (or 2.4 kWh of energy per day). Even at this level of power output, the unit will require 500 gpm (1,900 lpm) of water flow. On the other hand, a high head site will require much less water flow but correspondingly higher head pressure levels.

It will take a bit of juggling to determine which turbine will work best for your site based on the manufacturer's data. As with any complex technology, a dealer who has experience installing these units will be invaluable. Additional turbine-specific installation data is available from the

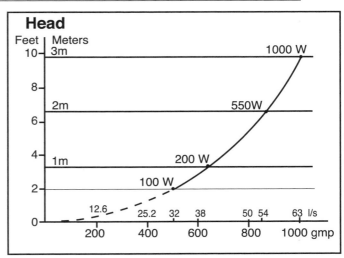

LH1000 Output (Watts Continuous)

Table 8-3. Output of a low head turbine rated 1 kW at various flow and head levels. (Energy Systems and Design)

various manufacturers listed in the resource guide in Appendix 3 .

Connecting the Micro Hydro Parts

Micro hydro systems are a little bit more difficult to understand, design, and install than PV or wind-based systems. Offsetting the complexity issue is the fact that hydro systems can generate more power and energy than can other forms of renewable energy. During the early part of the industrial revolution water power was used to

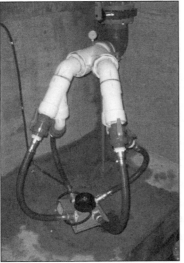

Figure 8-10a and b. Because water has a higher energy density than any other renewable energy source, it requires very small turbines to make a considerable amount of energy. This four-jet turbine fits in the palm of your hand and with sufficient water resources can power any battery/inverter-based off-grid home. (Courtesy Energy Alternatives Ltd.)

operate mills directly, and within a few years those same waterwheels and turbines began generating electricity.

Water has the advantage of being much denser and having higher energy content than wind. Because of this, a hydro turbine of the same capacity as one powered by wind will be many times smaller. The turbine shown in Figures 8-10a and b fits in the palm of your hand and will provide power 24 hours per day for battery charging and inverter systems.

With sufficient water resources, it is possible to power any home with a direct-to-alternating current turbine. Although these systems are still considered micro hydro, they operate on exactly the same principle as mega hydro systems such as Niagara Falls or Hoover Dam. Water falling through a vertical drop or "head" strikes the turbine blades causing a shaft connected to an alternating current electrical generator to spin. The power from the generator is fed via a control system known as a load-shedding governor to the house electrical panel.

With a given amount of water flow through the turbine and a given amount of electrical load connected to the generator, the turbine speed

Figure 8-12. A load shedding governor is designed to monitor the generated power and rapidly remove or apply some auxiliary electrical load to compensate for the varying electrical frequency resulting from changing household loads. Using this technique, the load shedding governor can maintain the fixed electrical load required by the turbine and provide a constant frequency of electrical power to the home. (Courtesy Canyon Turbines Inc.)

and resulting power frequency will remain fixed. However, electrical loads are never fixed. When a refrigerator switches on, a load is placed on the turbine, causing it to slow down and lower the frequency of the generated power. With a direct-to-alternating system, electrical frequency must be carefully controlled and maintained at 60 cycles per second or 60 Hz, as this is the frequency at which most appliances must operate. The slowing of the turbine and generator is therefore not acceptable.

A load-shedding governor is designed to monitor the generated power and rapidly remove or apply some auxiliary electrical load to compensate for the varying electrical frequency resulting from changing household loads. The auxiliary load is typically an electrical water heater which is either mounted in the domestic water system of the

Figure 8-11. With sufficient water resources, it is possible to power any home with a direct-to-alternating current turbine. Although these systems are still considered micro hydro, they operate on exactly the same principle as mega-hydro systems such as Niagara Falls or Hoover Dam. (Courtesy Canyon Turbines Inc.)

home or installed in the flowing river to dissipate excess electrical energy. Using this technique, the load-shedding governor can maintain the fixed electrical load required by the turbine and provide a constant frequency of electrical power to the home.

The water intake is typically located in a small creek or stream at a point with a considerably higher elevation than the hydro turbine. Higher vertical head yields more energy while using less water than a low head site. Although this requires longer penstock runs, it reduces the size of the pipe and turbine.

The intake is designed to extract only a small portion of the river flow, particularly if there is

Figure 8-13. The water intake is typically located in a small creek or stream at a point with a considerably higher elevation than the hydro turbine. A self-cleaning strainer prevents debris, fish, or other materials from entering the system and causing failures due to mechanical damage or blockages. This intake filter provides the water power to operate a 12V, 30A micro hydro system that powers a provincial park ranger station on the coast of British Columbia.

marine life. A self-cleaning strainer prevents debris, fish, or other materials from entering the system and causing failures due to mechanical damage or blockages.

Water is then directed down the penstock pipe, building pressure under the influence of gravity. At the powerhouse, the high-pressure water strikes the turbine blades, causing the drive shaft to spin, operating the electrical generator.

Direct current generators connect to a battery bank and inverter system in exactly the same way as a wind turbine would be interconnected to an off-grid home.

Alternating current systems power the house with 120/240VAC just as in a regular grid-connected home.

Figure 8-14. Where there are no roads, it can take extraordinary measures to deliver the penstock. Here a helicopter delivers 50 ft (15 m) sections of DR17 poly pipe 1,500 ft (457 m) up a mountain.

Figure 8-15. Moving long sections of steel or poly pipe uphill can be demanding work. The job is made easier with a special winch attached to a gasoline-powered chain saw.

Figure 8-17. High-pressure water strikes the turbine blades, causing the drive shaft to spin, operating the electrical generator. This three-phase induction motor is used as a generator and step-down transformer to provide battery charging for an inverter-based system.

Figure 8-16. This view shows the installed penstock running through a mountain forest. As the penstock runs down the mountain, water pressure increases, necessitating the conversion from poly to steel pipe as in this 250 psi (1723 kPa) section.

Figure 8-18. This beautifully designed power house contains the turbine and generator of a mid-sized hydro generating station. In this view, the penstock can be seen entering the building on the right. Water leaves the turbine and exits the power house through the rectangular opening in the concrete foundation.

Tantalus View Retreat - A Mountain Homestead

Janice and Warren Brubacher have enjoyed and nurtured their homestead in the Coast Mountains of British Columbia for over ten years. Unfortunately, the first few years required the use of a small gas generator to power their home, which was not in keeping with their desire for peace and quiet and their wish to live in harmony with nature.

The Brubachers decided to see if the small stream on the property had sufficient water flow to power their home. They contacted Peter Talbot of Home Power Systems (www.homepower.ca) to help with this assessment. Now, several years after the successful installation, the family is living in a beautiful, pristine location with their home powered by the fresh mountain rain.

The family now runs two home-based businesses. Coastal Cedar Creations is Warren's wood artistry business, where he teaches the art of log-home building and produces cedar carvings and one-of-a-kind log furniture and structures. Janice, who is an avid gardener and fabulous cook, operates the guest house inn known as Tantalus View Retreat. Both businesses are on the Web at www.tantalusviewretreat.com or at retreat@telus.net.

Figure 8-19 Janice and Warren Brubacher welcome people to partake in the views, clean mountain air, and country cooking offered in their micro hydro-powered resort.

Figure 8-21. Warren demonstrates the water intake for the micro hydro system. The intake is located 86 ft (26 m) above the turbine.

Figure 8-20. Relaxing in Warren's beautiful handmade cedar furniture and enjoying the view of the vast Coast Mountains has to remind people that we are part of nature and not master of it.

Figure 8-22. The penstock runs through the forest to the intake strainer and sand settling box. The penstock is insulated to prevent freezing where the pipe crosses a creek.

Figure 8-24. An Energy Systems and Design turbine generates 24V at 80A (approximately 2 kilowatts) of power or 48 kWh per day during periods of maximum water flow. This is enough energy to power almost any home.

Figure 8-23. There is almost no sound emanating from the turbine housing, even if you're standing right beside it.

Figure 8-25. The discharge or tail race drops into a creek after a run through 550 ft (167 m) of 4" (100 mm) plastic poly pipe.

Figure 8-26. Janice is seen preparing a morning feast in the kitchen.

Figure 8-28. The living room area has a television, VCR, and stereo system, common fare for any home.

Figure 8-27. The kitchen of the Brubacher home is typical, using a combination of electric and gas appliances. Propane is used to heat the hot tub, as well as for the clothes dryer and cookstove.

Figure 8-29. Warren checks out the hot tub, which is heated with propane but uses the hydro system to operate the pump and pressure jets.

Figure 8-30. The organic gardens provide Janice with the raw materials for her cooking creativity.

Figure 8-31. The couple's home blends with nature at every turn, from the clean micro hydro power to Warren's wonderful woodworking.

Chapter 9
BATTERY SELECTION

Why Use Batteries?

Why use a battery bank in the first place? This is a reasonable question with a simple answer. If you want to go off grid now, there is no other way to store electricity as easily and economically, *period*. While it is possible to store electricity by other means, the technology and cost benefits are completely uneconomic.

What About Fuel Cells?

Automobile manufacturers are starting to use fuel cells, so why not install some of these? Fuel cells are wonders of technology and work well in specific applications, but home use isn't one of them. Besides, fuel cells don't store energy; they simply convert it from one form to another.

Electrical energy from photovoltaic panels and wind turbines combines with water in a device known as an electrolyzer. The electrolyzer generates hydrogen and byproduct gas oxygen. The hydrogen is compressed and stored in a pressurized cylinder until it is required by the fuel cell, where it recombines with atmospheric oxygen, generating electrical power and waste heat. The poor overall efficiency and the expense of this process mean that fuel cells cannot match the economics of today's industrial deep-cycle battery.

Industry fuel cell manufacturer Ballard in conjunction with Coleman produced a small home-sized unit to act as a battery-backup system for computer and office equipment. Unfortunately, safety and supply issues with hydrogen gas as well as capital costs have killed that project for now. If we revisit this in another ten to twenty years the story *might* be a little bit different. However, although fuel cells sound exciting the battery industry is also developing advanced methods of energy storage with lower pricing. The jury is out on which technology will prevail in this particular application.

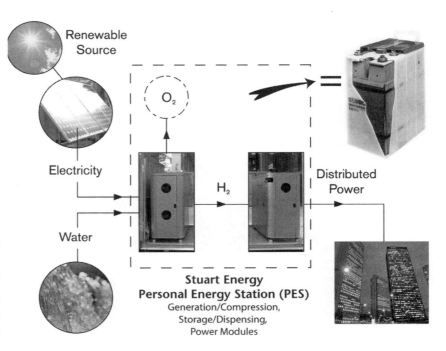

Figure 9-1. Electrical energy from photovoltaic panels combines with water in an electrolyzer device which generates oxygen and hydrogen. The hydrogen must be compressed and stored until it is required. Feeding the hydrogen gas to a fuel cell generates electrical power and waste heat. The poor overall efficiency and the expense of this process mean that fuel cells do not yet match the economics of the industrial deep-cycle battery.

Battery Selection

There are thousands of batteries available, and the more you look the more selection and complexity you will find. Car, truck, boat, golf cart, and telephone batteries are a few examples that come to mind. Then, of course, there are the "special" kinds such as NiCad, nickel metal hydride, and lithium ion, to name a few. What is the correct choice for our renewable energy system?

First of all, let's eliminate a few of the legends and quick fixes that abound:

- Used batteries of almost any size and type are not the answer. It doesn't matter how good a deal you got, the batteries are cheap because there is a problem with them. All batteries die after a given life span or if incorrectly serviced. Sure, they might work for a year, maybe even two or three, but the lower charge capacity of older cells makes this a sour deal.
- Quality battery costs are high. A golf cart battery of the same capacity as a high-quality "deep-cycle" battery may cost less, but the reduction in life span is not worth the hassle except for very small electrical loads or short grid interruptions. Changing batteries is

Figure 9-2. These models from Rolls Battery Engineering (www. surrette.com) are typical of long-life renewable-energy-system batteries. The two batteries in the foreground have the same voltage rating (6 VDC). The model on the right has twice the capacity of the model on the left. (Courtesy Surrette Battery Company)

backbreaking and dangerous work. You only want to do this once or twice in your lifetime, so stick with the best.

- Fancy batteries such as NiCad or lithium ion work well in cell phones and camcorders, but have you ever bought one of these babies? To get a lithium ion battery big enough to operate your renewable-energy system would probably cost as much as your entire house. Be thankful you can have one in your cell phone and leave it at that.

The only type of battery to consider in a full-time renewable-energy system is the deep-cycle lead-acid industrial battery manufactured with liquid or gelled "maintenance-free" electrolyte. These batteries have been around a long time and have been engineered and re-engineered so that they offer the best value for your money. Good-quality liquid electrolyte industrial batteries should last between fifteen and twenty years with a reasonable amount of care. In addition, the batteries are recyclable, and many companies offer trade-in allowances for their worn-out models, which also eliminates an environmental waste issue. The same is not true for NiCad or other exotic blends, which are considered hazardous waste.

How Batteries Work

The typical deep-cycle electrical storage battery, like a car battery, uses a lead-acid composition. A single-cell battery uses a plastic case to hold a grouping of lead plates of slightly different composition. The plates are suspended in the case, which is filled with a weak solution of sulfuric acid, called *electrolyte*. The electrolyte may also be manufactured in a gelled form, which prevents spillage. (Batteries of this type are often sold as "maintenance free.") The lead plates are then connected to positive and negative terminals in exactly the same manner as the AA and C cells described in Chapter 1.3. A single-cell with one negative and one positive plate has a nominal rating of 2V.

Connecting an electrical load to the battery

causes sulfur molecules from the electrolyte to bond with the lead plates, releasing electrons. The electrons then flow from the negative terminal through the conductors to the load and back to the positive terminal. This action continues until all of the sulfur molecules are bonded to the lead plate. When this occurs, it is said that the cell is discharged or dead.

As the cell is discharged of electrical energy, the acid continues to weaken. Using a device called a hydrometer (see Figure 9-4), we can directly measure the strength or specific gravity of the battery electrolyte. A fully charged battery may have a specific gravity of 1.265 (or 1.265 times the density of pure water). As the battery discharges, the specific gravity continues to drop until the flow of electrons becomes insufficient to operate our loads.

When a regular AA or C cell is discharged, the process is irreversible, meaning the cell cannot be recharged. Discharging a deep-cycle battery bank is reversible, allowing us to put electrons back into the battery, thus recharging it. Forcing electrons into the battery causes a reversal of the chemical discharge process described above. When a photovoltaic panel is placed in direct sunlight, it generates a voltage. When the voltage or pressure at the PV panel is higher than that of the battery, electrons are forced to flow from the panel into the cell plate. Electrons combine with the sulfur compounds stored on the plate, in turn forcing these compounds back into the electrolyte. This action raises the specific gravity of the sulfuric acid and recharges the battery for future use. Although there are many types of battery chemistry, the charge/discharge concept is similar for all types.

Depth of Discharge

A deep-cycle battery got its name because it is able to withstand severe cycling or draining of the battery. A car battery can only withstand a couple of "Oops, I left my lights on. Can you give me a boost?" mistakes before it is destroyed, whereas a deep-cycle battery may be subjected to much higher levels of cycling.

Figure 9-5 graphs the relationship between the life of a battery in charge/discharge/charge cycles and the amount of energy that is taken from the cell. For example, a battery that is repeatedly discharged completely (100% depth of discharge) will only last 300 cycles. On the other hand, if the same battery is cycled to only 25% depth of

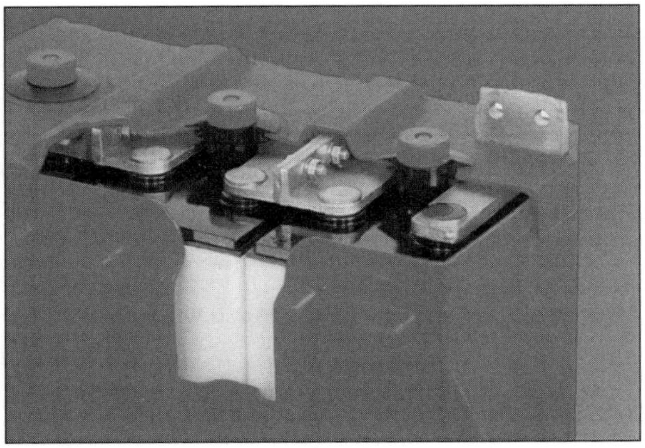

Figure 9-3. All lead-acid batteries comprise lead plates suspended in a weak solution of sulfuric acid. The size of the plate and the acid capacity directly affect the amount of electricity that can be stored. Each cell of the battery can be interconnected with others, increasing capacity and voltage. (Courtesy Surrette Battery Company)

Depth of Discharge %	Specific Gravity @ 75° F (25° C)	Cell Voltage
0	1.265	2.100
10	1.250	2.090
20	1.235	2.075
30	1.220	2.060
40	1.205	2.045
50	1.190	2.030
60	1.175	2.015
70	1.160	2.000
80	1.145	1.985
90	1.140	1.825
100	1.130	1.750

Table 9-1. Using the hydrometer shown In Figure 10-4, it is possible to accurately determine the state of charge, specific gravity, and voltage of each cell in a lead-acid battery bank.

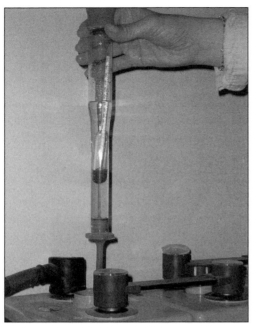

Figure 9-4. The hydrometer measures the density or specific gravity of fluids such as the electrolyte in this cell. The higher the reading, the more electrons are stored in the cell, indicating a higher state of charge.

discharge, the battery will last 1,500 cycles. It stands to reason that a bigger battery bank will provide longer life, albeit at a higher cost for the added capacity. A good level of depth of discharge to shoot for is a maximum of 50%, with typical levels of between 20% and 30%.

A grid-interconnected battery system spends the majority of its life filled to capacity waiting for the next utility failure. Provided power outages are not an everyday occurrence, frequent cycling should not be a major concern. On the other hand, depth of discharge considerations may become a major issue where grid interruptions are frequent. When a lengthy power outage occurs, significant battery-capacity depletion can result from inexperienced operation. Novice operators should consider battery-metering equipment as discussed later in this chapter.

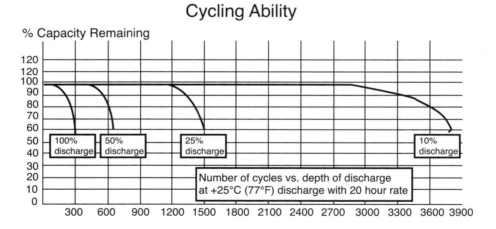

TYPICAL CYCLING PERFORMANCE*	
CAPACITY WITHDRAWN	CYCLES
100%	300
50%	650
25%	1,500
10%	3,800

* Dependent upon proper charging and ambient temperatures

Figure 9-5. This graph relates the life expectancy in charge/discharge/charge cycles to the depth of discharge level. Grid-interconnected systems do not generally cycle deeply or frequently unless the electrical utility is particularly unreliable in your area.

Capacity vs. Operating Temperature

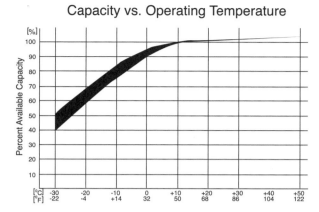

Figure 9-8. Batteries are similar to buckets. Three buckets paralleled together have the same capacity as a large barrel.

Figure 9-6. This graph shows the resulting reduction in battery capacity as the ambient temperature is lowered. A thermometer should be used to correct the specific gravity reading of very cold or overly hot electrolyte.

Operating Temperature

A battery is typically rated at a standard temperature of 75°F (25°C). As the temperature drops, the capacity of the battery decreases as a result of the lower "activity" of the molecules making up the electrolyte. The graph in Figure 9-6 shows the relationship between temperature and battery capacity. For example, a battery rated at 1,000 amp-hours (Ah) at room temperature would have its capacity reduced to 70% at -4°F (-20°C), resulting in a maximum capacity of 700 Ah. If this battery is to be stored outside where winter temperatures may reach this level, the

reduction in capacity must be taken into account. Installing batteries directly on a cold, uninsulated cement floor will also cause a reduction in capacity for the same reason. Be sure to use a wooden skid or other frame to allow room-temperature air to circulate around the battery, maintaining an even temperature.

Freezing is another concern at low operating temperatures. A fully charged battery has an electrolyte-specific gravity of approximately 1.25 or higher. At this level of electrolyte acid strength, a battery will not freeze. As the battery becomes progressively discharged, the specific gravity gradually falls until the electrolyte resembles water at a reading of 1.00.

Figure 9-7. For applications where the battery bank must be stored outdoors in extremely cold locations, consider the more expensive yet longer life NiCad style. (Courtesy Saft Battery Company)

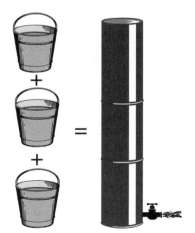

Figure 9-9. Batteries wired in series increase the voltage or pressure. In the same manner, a tall tank will shoot water further than a short tank due to the higher pressure.

For cold-weather applications you may consider a Nickel Metal Cadmium or NiCad battery. Although more expensive than standard lead-acid batteries, the NiCad cell is relatively unaffected by extremes in temperature, particularly where deep-discharge cycles and cold weather threaten to destroy lead-acid varieties.

Battery Sizing

A single cell does not have enough voltage or capacity to perform useful work. As discussed in Chapter 1.2, single cells may be wired in series to increase voltage and/or in parallel to increase capacity.

Batteries are just big storage buckets. Pour in some electrons and the "buckets" will fill with electricity. If cells are wired in parallel, the capacity is increased. If the cells are wired in series, the voltage or pressure rises.

Batteries may be purchased in several voltages. Individual cells have a nominal rating of 2V for lead-acid and 1.2V for NiCad brands. The lead-acid battery shown in the background

of Figure 9-2 has two cells, each of which can be identified by the servicing cap on the top. Using interconnecting plates inside the battery case, the cells are wired in series, providing 2 + 2 volts = 4V total capacity. The batteries in the foreground have three cells, providing an output of 6 V. Using the same approach, a car battery contains six cells, providing a nominal 12V rating. Typical battery voltages for grid-interconnected renewable energy systems are 24V and 48V.

It stands to reason that the more energy consumed, the larger the storage facility required to hold all the electricity. While the capacity of buckets is given in gallons or liters, the capacity of batteries is rated using watt-hours or amp-hours, units of electrical energy. We discussed earlier that a battery's voltage tends to be a bit elastic, with the voltage rising and falling as a function of the battery's state of charge (see Table 9-1). Additionally, when the battery is charged from a source of higher voltage such as a PV array in full sun, the voltage increases further. It is not uncommon for a battery bank with a nominal 24V rating to have a voltage reading of 30V when undergoing equalization charging.

This fluctuation of battery voltage makes it very difficult to calculate energy ratings. You will recall that energy is the voltage (pressure) multiplied by the current (flow) of electrons in a circuit multiplied by the amount of time the current is flowing. The problem with this calculation when it is applied to a battery is deciding which voltage to use.

Energy (watt-hours) = Voltage x Current x Time

Battery manufacturers are a smart bunch. To eliminate any confusion with ratings, they have simply dropped the voltage from the energy calculation, leaving us with a rating calculated by multiplying current x time.

Battery Capacity (amp-hours) = Current x Time

The assumption is that there is no point in trying to shoot a moving target. If we really want to look at our energy storage in more familiar watt-hour terms, we can simply multiply the amp-hour

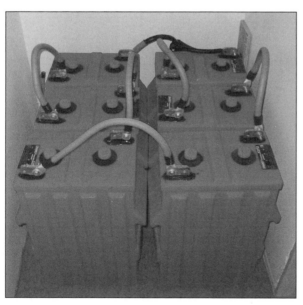

Figure 9-10. This compact battery bank takes up a small amount of room in the utility closet. With a rating of 24 volts at 1,300 amp-hours of capacity, it gives the owners of this home enough energy to ride through several days of grid interruption.

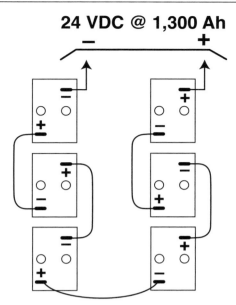

24 VDC @ 1,300 Ah

Figure 9-11. The batteries in Figure 9-10 are shown here in a schematic. Each of the six batteries is wired in series, resulting in increased voltage.

rating by the nominal battery-bank voltage.

Let's look at how this relates to the storage level our battery must have in order to run the essential loads of our home when the grid goes south. If you haven't already done so, it will be necessary to complete the *Energy Sizing Worksheet* in Appendix 7. This form will determine the amount of electrical energy you require in watt-hours per day to operate these essential loads.

For our example, we will assume an energy consumption of 4,000 watt-hours (4 kWh) per day for a 24-hour blackout. This is a reasonable amount of energy for grid-interconnected system users who wish to ride through power failures. In this case, it will be necessary to provide a separate wiring circuit for essential loads such as house lights, a refrigerator, and a few plugs to receive power during a grid interruption. Large or inefficient loads such as electric heating or air conditioning would remain off during the power failure. A detailed discussion of this system configuration follows in Chapter 13.

Remember that we have to take into account the usable amount of energy stored in the battery

bank, not the gross or rated amount. A maximum depth of discharge is typically 50%, with less being recommended.

4 kWh capacity required x battery derating factor of 2
= 8 kWh capacity

Some manufacturers will rate the capacity of their batteries in watt-hours or kilowatt-hours; however, as we discussed earlier, most choose to use amp-hours. To convert watt-hours (kWh) to amp-hours (Ah), divide the desired battery rating by the nominal battery voltage required for your system, typically 24 or 48V (assume 24V).

8 kWh capacity (8,000 Wh) ÷ 24V = 333.33 Ah capacity
≈ 333 Ah capacity

A quick scan of the resource guide data in Appendix 3 indicates that there are no batteries available with exactly 333 amp-hours of capacity. Remember, however, that batteries can be connected in series to increase voltage and in parallel to increase capacity. Two sets of 150 Ah batteries wired in parallel provide 300 Ah of capacity. This is a bit below the calculated rating, but it's close enough. We could also consider 2 sets of 200 Ah batteries, offering 400 Ah. A bit of juggling may be required, but either way we're in the ballpark.

As discussed earlier, under-sizing the battery can lead to over-discharging and shortening the battery's life. If you have to estimate and select sizes, it is always in your best interests to round up. Battery capacity is like money in your bank account: it never hurts to have extra on hand.

Each of the two rows of eight batteries shown in Figure 9-10 is shown in the schematic in Figure 9-11. The columns comprise 3 batteries. Each battery has two cells, as you can see from the two servicing caps on their cases. The nominal voltage of each battery is 4V (2 cells x 2 volts/cell). The batteries are wired in series by connecting successive "-" to "+" terminals, making the voltages additive. Thus, six batteries wired in series multiplied by 4V per battery provide a total bank of 24V.

Each of the batteries shown in Figure 9-10 has a capacity of 1,300 Ah. Wiring the batteries

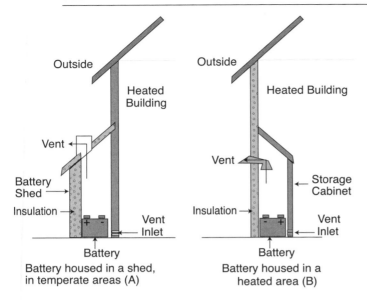

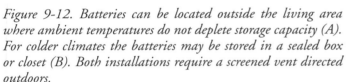

Battery housed in a shed, in temperate areas (A)

Battery housed in a heated area (B)

Figure 9-12. Batteries can be located outside the living area where ambient temperatures do not deplete storage capacity (A). For colder climates the batteries may be stored in a sealed box or closet (B). Both installations require a screened vent directed outdoors.

Figure 9-13. This battery room is power vented. Installing small computer fans in the wall allows the room to be pressurized, forcing hydrogen outside.

in series does not increase the capacity. In order to increase capacity, the batteries must be connected in parallel so that both banks are forcing electrons into the circuit at the same time. (To help visualize this concept, picture a car stuck in a snowbank. One person pushing the car may not provide enough energy to get the vehicle out, while a second person giving a hand may easily free it.) The parallel connection of multiple strings of batteries will be discussed further in Chapter 13, "Putting it All Together".

Hydrogen Gas Production

Connecting a battery to a source of voltage will push electrons into the cells, recharging them to a full state of charge. As a liquid-electrolyte battery approaches the fully charged state it will be unable to absorb additional electrons, causing the electrolyte to start bubbling and emit hydrogen gas. "Maintenance-free" gelled-cell batteries are designed to minimize the possibility of hydrogen buildup.

Everyone knows the story of the Hindenburg dirigible: a few sparks and the hydrogen gas

exploded into a ball of fire. Preventing this in a renewable energy system is actually fairly easy.

The battery bank should be accessible to allow for servicing and periodic inspection. However, safety is of primary importance. Storing batteries in a locked closet, shed, or cabinet is a wise

Figure 9-14. HydroCaps replace the standard vent caps on each battery cell. A special catalyzing material converts hydrogen and oxygen byproducts back into water.

Figure 9-15. In temperate climates batteries may be stored in suitable outdoor cabinets, eliminating any concerns regarding hydrogen production. (PSR cabinet courtesy Outback Power Systems Inc.)

decision. The storage closet may be vented with a simple pipe that can be made with a dryer vent or plumbing pipe (2"/50 mm) in diameter with screening over the exterior opening. Make sure the pipe slopes slightly upwards from the battery room to the outside as hydrogen is lighter than air and will rise on its way outdoors.

For rooms with long or serpentine vent shafts, a power-venting fan can be installed as shown in Figure 9-13. Pressurizing fans are operated by a voltage-controlled switch contained within the inverter; the voltage indicates the state of charge and hence hydrogen production, activating the fans.

An alternate method of controlling hydrogen gassing is to install a set of reformer caps (Hydrocap Corporation) such as those shown in Figure 9-14. These caps contain a catalyst that converts the hydrogen and oxygen by-products of battery charging into water that simply drips back into the cell. The net effect is eliminated hydrogen gassing and lower battery water usage. Note, however, that hydrogen reformer caps control but do not completely eliminate hydrogen gassing, making some supplementary ventilation necessary.

Safe Installation of Batteries

Large deep-cycle batteries are heavy and awkward. Some models weigh up to 300 lbs (136 kg) each. It is important that any frames, shelves, and mounting hardware be of sufficient strength to hold them securely. During installation, make sure you have adequate manpower to lift and place the batteries without undue straining. Tipping a battery and spilling liquid electrolyte (acid) all over you is very dangerous. Follow these rules to ensure a proper and safe installation (See Figure 9-19):

• Batteries must be installed in a well-insulated and sealed room, closet, or cabinet, preferably locked. There is an enormous amount of energy stored in the batteries. Children (or curious adults) should not be allowed near them.

• Remove all jewelry, watches, and metal or conductive articles. A spanner wrench or socket set accidentally placed across two

Figure 9-16. This compact battery bank takes up a small amount of room in a utility closet or other out-of-the-way area in the home. (Courtesy Outback Power Systems Inc.)

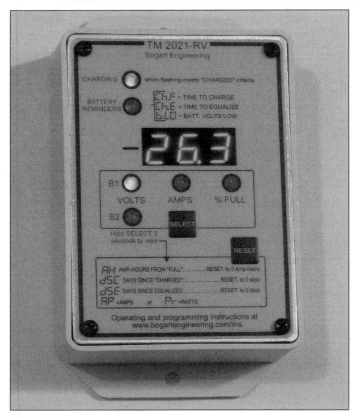

Figure 9-17. A multifunction meter such as this Trimetric model (www.bogartengineering.com) takes most of the guesswork out of battery status monitoring. (Courtesy Bogart Engineering)

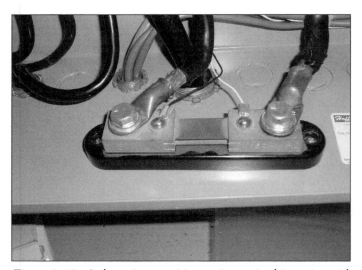

Figure 9-18. A shunt is a precision resistor wired in series with the negative lead of the battery. A small current in proportion to the battery charge or load current is "shunted" to a multifunction meter such as the one shown in Figure 9-17.

battery terminals will immediately weld in place and turn red-hot. This can cause battery damage, possible explosion, and severe burns.

- Hand tools should be wrapped in electrical tape or be sufficiently insulated so as not to come in contact with live battery terminals.

- All benches, shelves, or support structures must be strong enough to carry the massive weight of the battery bank. The batteries shown in Figure 9-10 weigh 900 lbs (400 kg), so use extreme caution when handling.

- Wear eye-splash protection and rubber gloves when working with batteries; splashed electrolyte can cause blindness. Also, wear old clothes or coveralls, since electrolyte just loves to eat your $80 jeans.

- Keep a 5 lb (2.3 kg) can of baking soda on hand for any small electrolyte spills. Immediately dusting the spilled electrolyte with baking soda will cause aggressive fizzing, neutralizing the acid and turning it into water. Continue adding soda until the fizzing stops.

Battery and Energy Metering

Electricity is invisible, making it difficult to determine how much energy is actually stored in your battery bank. While it is possible to use nothing more than a hydrometer to measure the stored energy level, this requires a fair bit of tedious fiddling and working with corrosive acid. A better approach is to install a multifunction meter such as the model shown in Figure 9-17.

These meters operate much like an automobile gas gauge, indicating how full or empty the tank is. Most meters provide dozens of features, the most important being battery voltage measurement, energy capacity in amp-hours and a measurement of the current that is flowing into (charge) or out of (discharge) the battery. These meters can be mounted a long distance from the battery bank, enabling you to read capacity from a convenient location in the house rather than running to the garage or basement.

Meters perform their work through a connection with a device known as a "shunt" (see Figure 9-18), a high-capacity resistor that either bypasses a small amount of the current flowing into or out of the battery or shunts it to the meter.. As the circuit is DC-rated, the direction of current flow is easily recorded. The amount of energy shunted is a calibrated ratio of the total amount of energy flowing in the battery circuit. Increasing the current flow into or out of the battery circuit causes a calibrated current to flow to the meter.

By constantly monitoring the flow into and out of the battery, the meter keeps track of the battery energy level in amp-hours and percentage full. Or it almost keeps track. While there is no doubt that energy meters are very useful and accurate devices, they are not perfect. Batteries are not perfect either. The process of converting electricity into chemical storage within the electrolyte is not 100% efficient. During the charge and discharge cycling of the battery, some of the electrical energy we put in is lost due to conversion inefficiencies. Worse yet, this inefficiency is not the same for adjacent cells, nor does the inefficiency level remain constant over the life of the cell. As a result, the energy meter will require recalibration periodically when the battery is at a known fully charged state (after measurement with a hydrometer).

Electrolyte specific gravity and battery voltage are directly related:

Battery voltage = Electrolyte specific gravity + 0.84

For this equation and the hydrometer reading to be accurate, however, the batteries should be under light or no load, with the electrolyte near room temperature. It is also important to perform this measurement approximately two hours after charging the batteries to ensure that gas bubbles in the electrolyte do not lower the specific gravity reading.

The only way of absolutely knowing the energy level and general "health" of a battery is to use a hydrometer and measure and record the specific gravity frequently. A thermometer should be used to correct the specific gravity reading of very cold or overly hot electrolyte according to the manufacturer's ratings. Correlate the corrected specific gravity to the data in Table

Figure 9-19. Battery installations should be neat and tidy and performed with the necessary tools on hand to measure electrolyte, specific gravity, and temperature. A clipboard and written record of each cell's "health" will ensure long battery life.

9-1 to determine the actual state of charge.

Batteries are charged to a known level a few times a year by performing an equalization charge using grid or renewable energy. During this process the batteries will reach their known fully charged state and the meter may be reset, calibrating the battery level with the meter reading for the next couple of months. Equalization charging will be covered in Chapter 17, "Living with Renewable Energy".

In addition to the "main" features described above, metering may also contain some or all of the following "secondary" indicators:

- number of days since batteries were fully charged
- total number of amp-hours charged/ discharged since installation
- time to recharge battery
- voltage too low for proper operation
- battery charging
- battery fully charged

Summary

In this chapter we have discussed battery selection and safe installation. Chapter 10, "DC Voltage Regulation," discusses the equipment necessary to ensure proper charging and regulation of the battery bank. Future chapters deal with the specifics of interconnecting cells, safe wiring practices, and maintenance. After all, when the grid goes down, your home should be a shining beacon for all to see.

Chapter 10
DC VOLTAGE REGULATION

A charge controller is an important element of a renewable energy electrical system equipped with batteries. Its main function is to act as a voltage regulating device, preventing the over- and with some models under-discharging of battery banks.

The simplest, low-power series controller units are installed between the renewable energy source and the battery bank. These models allow full power from the source to flow into the battery bank until a pre-determined charge setpoint voltage is reached. Upon reaching the charge setpoint, the battery is disconnected from the energy source using a mechanical relay or semiconductor transistor "switch" and will only be reconnected when the battery bank is discharged sufficiently to require recharging.

More advanced models replace the single-step charge control with more advanced pulse width modulation techniques, allowing very accurate 3-stage battery charging algorithms.

Large PV systems require more advanced battery charging control and other optional features that include automatic battery multi-step and equalization programs as well as maximum power point tracking functions.

Most wind and hydro turbines require that an electrical load be connected to them at all times; otherwise they may spin at excessively high speeds, causing damage or destruction. To prevent this from

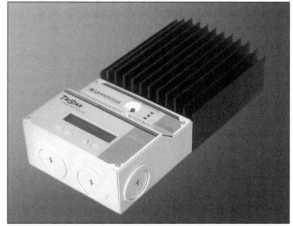

Figure 10-1. The charge controller ensures that your battery bank is properly charged and operating within given parameters. (Courtesy Morningstar Corporation)

occurring, the turbine is connected directly to the battery bank and the charge controller is configured as a diversion regulator, shunting or diverting excess energy from the turbine to a secondary load, such as a water or space heater, which will consume the excess energy production (Figure 10-7).

Grid-interconnected systems not equipped with batteries do not require any form of charge controller, as all of the available electrical energy is pumped directly into the grid. However, most grid-connected inverters contain an "energy boosting algorithm" known as maximum power point tracking, a feature that is explained below.

Series Controller

As the name implies, the series controller is wired in series between the PV array and the battery bank as shown in Figure 10-3. In this arrangement, the voltage output from the PV array is fed through the controller prior to being supplied to the battery. The charge controller monitors the battery voltage and, provided it is below a fully charged condition, power is allowed to flow into the battery. When the battery voltage level is sufficiently high to indicate a fully charged state, the charge controller will taper charging current or completely disconnect the PV array from the battery, slowing or stopping

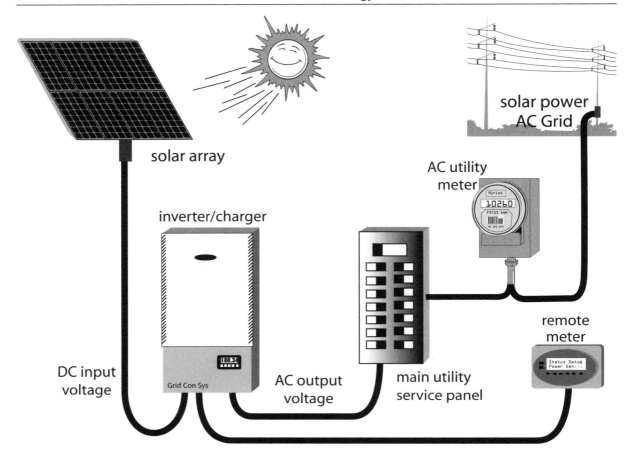

Figure 10-2. The majority of urban, grid-interconnected renewable energy systems are not equipped with battery backup and therefore do not require any DC voltage regulation.

the charging cycle. As the battery is subjected to household electrical loading, battery voltage will drop as energy is removed from the cells. At the recharge setpoint, the controller reconnects the PV array, restarting the charging process. Think of the series controller as an automatic light switch that turns the flow of current to the batteries on and off depending on whether the batteries require additional energy.

During the night, the charge controller performs no function other than to electrically isolate the battery bank from the sleeping PV array. This prevents energy from flowing *backwards* from the batteries into the array, which can occur at night when the battery voltage is at a higher

level than the voltage of the PV array. The large surface area of the PV array may absorb a small amount of energy from the battery and dissipate it as heat. Over the course of a long winter night, this energy loss is measurable and can contribute enough inefficiency to become a concern.

For this reason, most series charge controllers contain a "PV array nighttime reverse-current protection" feature, which disconnects the PV array from the battery to eliminate the possibility of electricity backflow.

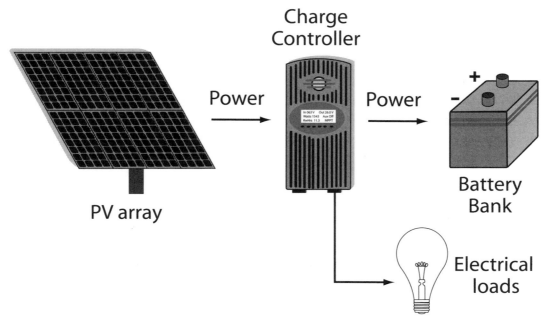

Figure 10-3. The series-connected charge controller is able to connect and disconnect the PV array from the battery, limiting charging current and battery voltage to a safe level. Nothing will prolong battery life like following a proper charging and maintenance regime.

Maximum Power Point Tracking

A word is required about *maximum power point trackers* or MPPT controllers. This technology, which is relatively new to the renewable energy market, is now available in charge controllers and grid-tie inverters manufactured by several companies. All MPPT charge controllers provide the series charge control functions described above. Where these devices surpass standard controllers is in their ability to significantly improve the overall efficiency and power output of a renewable energy source, in essence producing more energy dollars from the same amount of equipment.

The MPPT function is a little bit difficult to understand but well worth a few moments to review. Like any electrical device, the power output of a PV module is the product of the voltage and the current. This can be expressed by multiplying the rated current by the rated voltage under load, resulting in the rated power of the module.

A typical PV panel power curve output is shown in graphical form in Figure 10-5. The

Figure 10-4. This charge controller is rated for 60A of continuous load and is complemented with many features including metering and maximum power point tracking (MPPT). (Courtesy Outback Power Systems)

output voltage of a given module is shown on the "x" axis and the output current on the "y" axis. In Chapter 6 we discussed the fact that a PV module generates a very high open circuit voltage when not connected to a load (i.e. when no current is flowing). This point on the graph is position VOC (Voltage Open Circuit), where zero current is flowing and no work (battery charging) is being done.

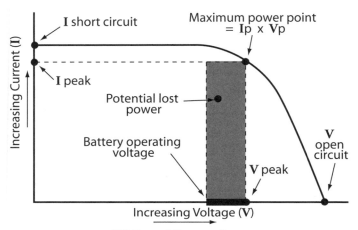

PV Panel Power Curve

Figure 10-5. The maximum power point (MPP) for a PV module or array of modules is the point on their power curve at which current and voltage outputs provide the maximum amount of power (power = voltage x current).

When a load is connected to the panel, current flow increases while voltage begins to slowly drop, tracing along the path of the power curve.

During a short-circuit condition - $I_{short\ circuit}$ (directly connecting the "+" and "-" terminals) will result in high current levels but the voltage will approach zero, thereby producing no useful power, and again no work can be done.

The most efficient point on the curve occurs when the product of the voltage pushing and the current flowing produces the highest value. This occurs at one point, which is known as the maximum power point (MPP). Unfortunately, our battery voltage is unlikely to be located at this point. As we have learned, battery voltage is quite elastic, so there is a high probability that

the PV module output will only transit the MPP occasionally. The difference between the battery operating voltage and the MPP multiplied by the current constitutes lost power and efficiency.

The MPPT controller "sweeps" the power curve of the PV panel or other renewable source output to determine the maximum power point. The MPP is evaluated in relation to battery or, in the case of grid-connected systems, inverter load and determines where the MPP is for the level of

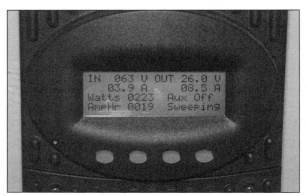

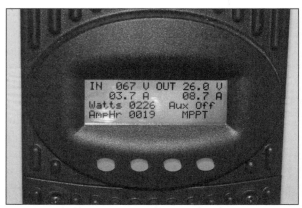

Figures 10-6a and b. The maximum power point tracking (MPPT) controller "sweeps" the power curve of the PV panel to determine how to produce the maximum amount of energy from the available solar resource. The MPPT will periodically re-sweep the power curve to ensure that the system tracks the maximum power point. This photograph shows the display of an Outback MX60 (Figure 10-4) during the sweeping and MPP tracking period. Notice that the array input voltage is more than double the battery voltage. This voltage configuration is typical for MPP tracking regulators.

electrical power being produced at that moment. The MPPT will periodically re-sweep the power curve to ensure that the system tracks the MPP over time.

Testing shows that MPPT systems increase the wattage of a PV array by an average of 15% over one year. For a PV array pushing 800W into the utility grid or battery bank, this will provide an average increase of 120W, which is the equivalent of adding one free PV panel to your array.

For PV-only systems, consider an MPPT controller as a supplement to an active tracking mount to increase daily energy production.

Diversion Charge Controller

The diversion controller method (see Figure 10-7) is used primarily for larger off-grid systems or grid-interconnected designs that incorporate wind or micro hydro turbine generators. In this configuration, the renewable energy source is connected to the battery bank. In this example,

all of the energy produced by the turbine flows into the battery, providing a load connection at all times. This is an important distinction between the two designs of charge controller. Where a PV array can be connected and disconnected at will, many wind and hydro turbines must have an electrical load connected at all times. Removal of the load during operation can cause the turbine to accelerate to a speed where damage can occur.

Compare this to how a car engine operates: driving while holding "the pedal to the metal" results in speeding tickets but no engine damage; however, doing the same with the car in neutral will destroy the engine, as there is no load to limit its rotational speed (rpm).

To ensure a constant load for the turbine without overcharging the battery, the controller will "shunt" or divert excess energy to the diversion load. When the battery reaches its fully charged state, the output voltage of the turbine will rise as a result. The controller senses this increase

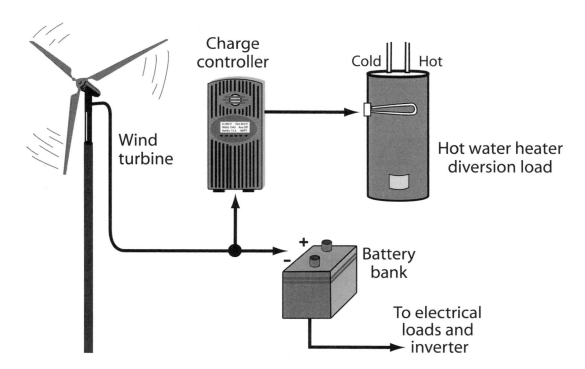

Figure 10-7. The diversion controller strategy shunts or diverts excess energy to an auxiliary air- or water-heating load according to battery voltage or state of charge.

in voltage and shunts the excess energy to the diversion load.

The diversion load in Figure 10-7 is an electric water heater, although an air-heating or other dump load can be used. With this system, the diversion controller is able to divert all or part of the turbine's energy depending on the battery bank voltage and state of charge while maintaining a proper load level for the renewable source.

Voltage Regulator Selection

Regardless of which voltage regulator design is selected, it is important to match its electrical rating to your system. The regulator rating is based on the maximum current that is allowed to flow into the batteries (series regulator) or into the diversion load (diversion regulator). It will be necessary to add the peak charging/diversion currents expected in your system. Once this value is determined a derating or safety factor of 25% is normally added. If your charge or diversion loads require a higher capacity rating than the regulator can supply, multiple controllers may be used to divide the electrical load as shown in Figure 10-9. For example, a 20-panel PV array can be wired as two smaller 10-panel arrays, each connected to its own voltage regulator.

peak charging or diversion current x 25% safety factor = regulator rating in amps

For further technical details refer to Chapter 13, "Putting it All Together Safely."

Charging Strategy (all controller configurations)

As a battery's state of charge increases, its voltage also increases. Likewise, when the battery state of charge falls, its voltage will also drop. The renewable energy source is always designed to produce power at a high voltage relative to the battery in order to provide sufficient "pressure" to cause electrons to flow into the battery. We learned in Chapter 6, "Photovoltaic Electricity Generation," that a typical PV panel will produce 17 volts for a 12V nominal battery system. The 5V difference in pressure allows electrons to flow

from the higher voltage source to the lower one, providing the charging current to the battery bank.

As the battery bank "fills," its voltage continues to rise and upon nearing the fully charged state the battery will start out-gassing hydrogen and oxygen as well as producing heat. This condition is known as over-charging, and if it continues over prolonged, uncontrolled periods of time, damage to the battery will occur.

When the battery voltage is below a level that indicates it is not fully charged, all of the power from the renewable source(s) is allowed to pour in. As the battery voltage rises, the series charge controller rapidly cycles on and off (several hundred times per second), pulsing the generated power into the battery. In a similar manner the diversion controller rapidly cycles power to the diversion load. The net effect is to direct energy in a controlled manner, maintaining the battery voltage within preset limits.

To visualize this concept, go into a darkened room and flip on an incandescent lamp. The room will be lit at maximum brightness. If you rapidly flip the switch on and off, the bulb will alternate between states of full and zero brightness. The effect on your eyes (notwithstanding the flicker) is that the room is seen in a dimmer light. If the light switch could be controlled so that the switch spent more time on than off, the brightness would increase. Likewise, if the switch spent more time off than on, the brightness would decrease. Engineers term this rapid on-and-off switching *pulse width modulation* or PWM.

To understand how the PWM concept is incorporated into a charge control system, let's take a look at a typical 24-hour day in the life of a battery bank during a grid blackout or a typical off-grid day. In this example, we will assume that PV arrays are used. Figure 10-8 shows the effect of battery voltage over one complete night/day/night cycle. The graph shows time on the horizontal or "x" axis, while battery voltage is shown on the vertical or "y" axis.

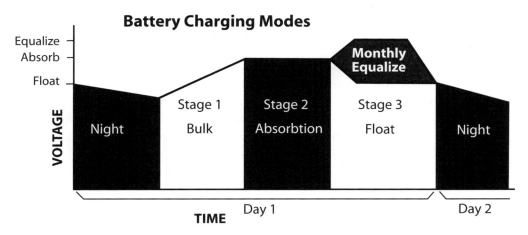

Figure 10-8. A charge controller regulates the amount of energy flowing into the batteries, ensuring the proper charge cycle. Quality charge controllers offer several charging modes rather than a single "bulk-charging" mode.

Nighttime

Starting at the left side of the graph, the battery voltage is shown at its lowest state at night. During the night lights are used, the fridge operates, and other loads draw energy from the battery bank. As we learned in Chapter 9, "Battery Selection and Design," the battery voltage will decrease as energy is removed. During this period, the charge controller also prevents electrical backflow (discussed earlier) and waits patiently for the sun to rise.

Sunrise – Bulk Charge

As the sun rises and shines on the PV array, electrical energy will start flowing and begin charging the battery. The battery voltage will start to rise according to the battery state of charge.

Until the battery state of charge reaches approximately 80%, all of the PV array power is applied directly to the battery, as shown in Stage 1 of Figure 10-8. This charging condition is called "bulk charge" mode. It terminates when the battery voltage rises to approximately 14.6V for a nominal 12V battery bank (consult manufacturers' data sheets for exact charger settings). For 24V or 48V battery banks, multiply the described settings by two or four respectively.

Absorption or Tapering Charge

When the battery reaches an approximately 80%-full state of charge, some of the energy applied to the battery is wasted through "boiling" and out-gassing. Boiling is a slang term used to describe the breakdown of the water in the battery electrolyte into its component elements, hydrogen and oxygen. This occurs when there is more energy applied to the battery than it can absorb.

Hydrogen is a gas that is lighter than air and very explosive when mixed with oxygen and an accidental spark, open flame, or cigarette. For this reason, batteries should be stored in locked and ventilated cabinets. It is possible to reduce hydrogen production while providing the optimum charging current to the battery by introducing PWM absorption current, as shown in Stage 2 of Figure 10-8.

During this stage, charging current is automatically lowered in an attempt to maintain the battery voltage at the bulk setpoint level of 14.6V. Absorption charging continues for a defined period of time, which is typically one to two hours.

Float Stage

Once the battery is fully charged, the charge controller will reduce the applied voltage to 13.4V and begin float mode, as shown in Stage 3 of Figure 10-8. When the battery reaches this stage, very little energy is flowing into it and nearly 100% of the renewable energy produced will be either wasted or shunted to the diversion load. Float mode will remain in effect until the battery voltage dips below a preset amount (typically 80% of full state of charge) or until the renewable energy source stops producing power at night. The process is then repeated for the next cycle.

The excess energy produced during the float-charging stage, which can be a considerable amount, is either wasted when using a series regulator or diverted and possibly used if a diversion regulator is installed. As this wasted energy is non-polluting, there is no concern from an environmental standpoint. However, the amount of energy can be quite considerable and may be used for other useful applications, often offsetting fossil fuels.

Remember that all of this wasting and diverting occurs only during times when the battery bank is full and there is nowhere else for the energy to go.

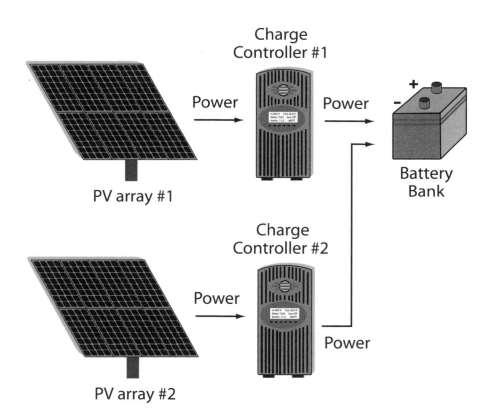

Figure 10-9. If the power rating of your PV exceeds the rating of the charge controller, connect multiple units in parallel as shown. The PV array does not have to be physically split into two groups but can simply be wired to divide the power output so that it is within the maximum safe operating range as recommended by the controller manufacturer.

Equalization Mode

CAUTION!

Sealed or "maintenance-free" batteries do not require equalization charging. Placing them in such a condition can result in fire or cause the battery to explode.

To visualize the equalization process, think of batteries as buckets and the water that fills them as electricity. One bucket can be considered the equivalent of one battery cell.

Over the course of a few dozen charge cycles in a few weeks, energy is taken out of and replaced into the battery several times. This is just as if I asked you to take between two and five cups (500 to 1200 ml) of water out of each "cell" or bucket, representing an average day's electrical load consumption.

Assume that our PV array produces enough energy to replace three cups (750 ml) of "energy" in the cells daily, leaving us with either a small deficit or a gain in volume.

This process is repeated with varying amounts of water being removed (consumption) and three cups of water added (production) daily over the course of a month or two. We will assume that this process continues until such time as the mathematics indicates that the batteries are back to a full state of charge. In our example, the buckets should also be full of water.

In reality, the water in each of the buckets will not be at exactly the same level. A certain amount of spillage and uneven amounts of water removal will leave the buckets with slightly varying amounts of water in them. This is the exact scenario that plays out in your battery bank. Over time there is a gradual change in the state of charge in the cells, which can be determined by comparing the specific gravity of one cell with that of adjoining cells.

If the difference is allowed to continue for an extended period, the cell with a lower than average state of charge will have less available energy stored in the battery bank and can fail prematurely if the condition is not corrected. This is why it is so important to take periodic cell-to-cell hydrometer readings to determine if remediation is required.

To correct this situation, a periodic (typically once per month) controlled over-charge is conducted. Known as an equalization charge, this process is illustrated in Figure 10-8. Chapter 17, "Living with Renewable Energy," discusses how to determine when equalization is required.

Equalization charging is normally conducted early in the morning on a sunny or windy day, allowing the renewable energy source to do the work. Equalization can also be conducted with the energy from a fossil-fueled generator, but why pay for electricity when you can make it yourself? The charge controller is set to equalization mode and the normal bulk charge (Stage 1) and absorption charge (Stage 2) cycles are completed. Upon entering equalization mode the controller raises the battery voltage to a very high level of 15.5V. Equalization mode is maintained for approximately two to three hours, after which battery voltage is reduced and the charge controller automatically enters float mode (Stage 3).

If we go back to our bucket example, the effect of equalization is similar to using a garden hose to add water to the buckets and deliberately over-filling each one of them. When you stop adding water (equalization completed) the buckets are topped up right to the rim.

Where does the "extra" electricity go during the equalization mode? (It does not pour on the floor as in the water bucket example.) During this charging stage, the excess energy applied to the batteries will cause violent bubbling of the electrolyte, producing large amounts of hydrogen and oxygen gas as well as heat and water vapor. It will be necessary to monitor electrolyte levels and battery temperature during this charging stage to ensure that battery parameters remain within manufacturer ratings.

CAUTION!

If your batteries are equipped with hydrogen reformer caps, ensure that they are removed during equalization mode or they will be destroyed.

Diversion Loads

Wind and hydro turbines normally require that an electrical load be connected at all times. Energy that is generated by these sources must go somewhere when the batteries are full. This is where the diversion load comes into play.

A diversion load is any load that is large enough to accept the full power of a renewable source. Although the diversion load may simply waste the excess energy applied to it, a better approach is to put this juice to work. After all, you did pay for the wind turbine, so why waste the "excess" energy other people have to pay for?

Suppose you have a 2kW wind turbine that operates at maximum output for five hours while you and your family are on vacation.

2kW output x 5 hours of operation
= 10,000 watt-hour (10kWh) production

Assume for a moment that the house loads are very low (no fridge or lights on). What can you do with this energy?

This free energy can be used in many ways to help offset energy consumed by other household systems. A common solution is to use the excess energy for space and water heating. If all of your waste energy is produced during the swimming season, consider dumping the energy into a hot tub or swimming pool.

The most common diversion loads are air and water heaters such as those shown in Figures 10-10 and 10-11. For PV-based systems, the majority of excess energy is produced during the summer months when air heating is not required. Wind systems tend to provide maximum power during the late winter and spring periods. Consider the choice carefully. No one wants an air heater on when the mercury is in the 90s. My personal

Figure 10-10. Using large amounts of diverted electrical energy is easy if heat enters the equation. An air-heating element such as the model shown here (available from www.realgoods.com) will absorb up to 1kW of excess power and provide some home heating to boot.

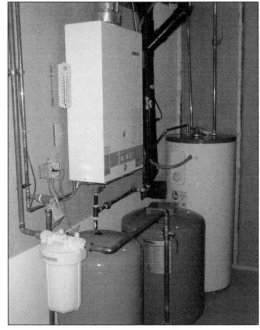

Figure 10-11. This home uses a high-efficiency, in-line gas water heater. The conventional water storage heater in the background is the diversion load. It absorbs waste energy by preheating the cold water fed to the gas water heater.

preference is a water-heating load such as the one shown in Figure 10-11, which will remain equally useful in January as in July.

In this system, a standard electric storage water heater is purchased and the 240V heating elements are removed. New elements are installed which have the same voltage rating as your battery bank or your wind or hydro turbine. The elements are then wired to the diversion-type charge controller or dump-load relay shown in Figure 10-14. The water heater is plumbed so that cold water flows into the electric heater, with heated water flowing out to the "standard" gas or electric model. A schematic view of the system is shown in Figure 10-12.

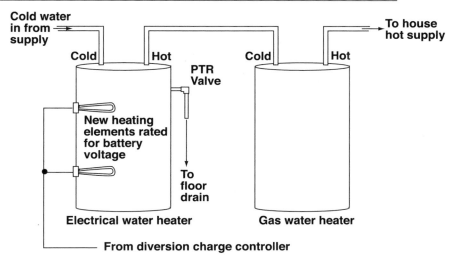

Figure 10-12. A diversion water-heating load is connected in series with a standard gas water heater so that preheated water from the solar dump load will offset energy used by the conventional model. The elements of the water heater are replaced with heating elements that have the same nominal voltage as the battery bank. They are connected to the diversion charge controller.

During normal operation, the cold water supply enters the electric water heater. If there is no excess energy, the cold water entering the tank will absorb some room heat, capturing a small amount of supplementary energy before heading to the regular water heater. As this incoming cold water is below the desired temperature, the regular heater will supply the energy necessary to meet demand.

During periods of excess energy production, the diversion charge controller supplies the low-voltage electric water heater. This energy may heat the water to the setpoint temperature or beyond. Feeding pre-heated water into the regular water heater reduces or eliminates the need for any

further heating, reducing your purchased energy requirements and saving you dollars.

As a further symbiotic design exercise, consider installing a solar thermal system on the diversion tank and reduce energy costs further still.

It is possible for the temperature of the water in the diversion electric heater to rise well above the setpoint temperature, particularly during travel or vacation periods which often occur in the energy-rich summer months. Safety (as well as building codes) dictates that a Pressure and Temperature Relief (PTR) valve be installed on the tank. The outlet pipe should be run to a floor drain or other suitable exhaust to allow very hot water to be safely drained away in the event that the water

Figure 10-13. Alternate voltage water heating elements are available from numerous sources, including Real Goods at www.realgoods.com.

tank temperature rises to an unsafe level.

For the same reason, your plumber should install a buffering valve on the hot water outlet of the final water heater tank. These valves ensure that the hot water supply delivered to the household plumbing fixtures is within a comfortable and safe temperature range. Buffering valves operate by automatically admitting sufficient cold water into the hot water system to ensure that no one is scalded the next time the shower is used.

Extra Features

Charge controllers have come a long way since the days of a simple little box with four wires and a "charged" light attached. With the sophistication of advanced microcomputing technology and software development, charge controllers offer loads of features that are designed to help keep your

batteries in top-notch condition. The specification tables in the appendices allow model-to-model comparison, detailing functions such as:

- user-selectable battery-charging configuration specifications;
- low-voltage alarm and battery disconnection;
- automatic equalization charging or reminder function;
- auxiliary generator control;
- battery and renewable-source energy metering;
- battery temperature compensation adjustment, used when battery electrolyte is expected to vary beyond "room temperature."

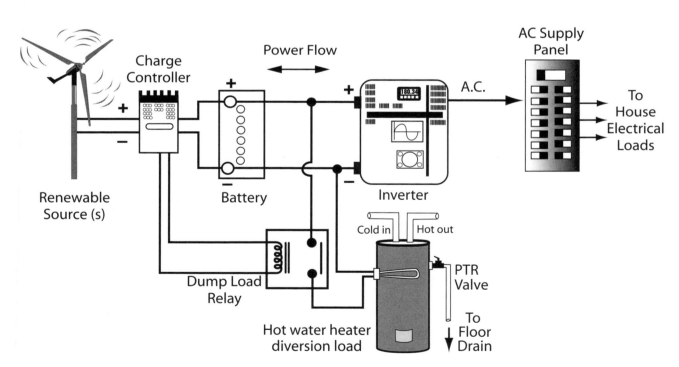

Figure 10-14. Once the battery is fully charged, the charge controller will reduce the applied voltage and begin float mode. When the charge stage is reached, very little energy is flowing into the battery and nearly 100% of the renewable energy produced can be shunted to the diversion load. In order to facilitate this function, a dump-load relay must be added that is rated to switch the water heater element load current on and off. The charge controller will automatically activate the relay based on programmed parameters related to battery state of charge.

Chapter 11
DC TO AC CONVERSION USING INVERTERS

In earlier chapters, we dealt with renewable energy sources that produce electrical energy in direct current (DC) from a PV array. Most small wind and hydro turbines generate alternating current (AC) but convert it back to DC for storage in battery banks.

This chapter explains how to convert DC power back to AC power that can be used with standard home electrical appliances and can also be sold to the electrical utility in grid-interconnected systems.

In the "old days," (think the Summer of Love era) if you wanted to build a renewable energy system you could only do it using 12V direct current appliances similar to those used in many recreational vehicles. Grid-interconnection of renewable energy was as far away as the idea of the Internet. Times change. The Internet is here, and so are very high-quality, low-cost inverters for both on- and off-grid applications.

The basic inverter is a device that takes low-voltage DC power (from a battery bank or directly from another energy source such as a PV panel or wind turbine) and converts it to alternating current at the same low-voltage level. It then "steps up" the voltage to match domestically supplied power from your utility, typically 120V and/or 240V.

Inverters intended for batteryless grid interconnection are similar in design, although they require the DC input voltage to be much higher. This configuration reduces both their size and their cost.

In practice, many inverters offer a host of additional features and functions:

- battery charging capability
- transferability of house power between a generator and the inverter
- low- and high-voltage alarms and disconnection (LVD function) of the battery bank
- energy savings—sleep mode turns inverter "off" yet allows it to come back on at a flick of the first house light switch
- automatic start and stop of a backup generator
- maximum power point tracking for grid-interconnected systems
- full safety protection for both homeowners and utility workers

An inverter installed for battery-backed, grid-interconnected operation is shown in schematic form in Figure 11-2. Energy from the renewable source is stored in the battery and directed to the inverter's internal components. A power supply and controller determine the sequence of events required to make

Figure 11-1. The renewable energy world would be lost without utility-grade sine wave inverters and controls such as this configuration from Outback. (Courtesy Outback Power Systems)

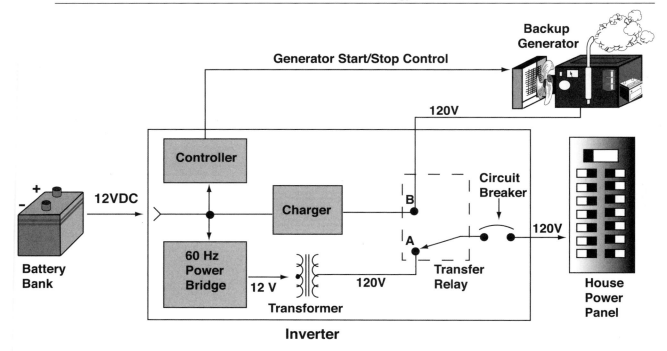

Figure 11-2. The inverter is a marvel of complex technology and is surprisingly easy to operate. A basic inverter for off-grid installation is shown in schematic form.

the unit function as desired. When the inverter is activated, the controller starts a high-powered oscillator or power bridge that generates AC power at the voltage of the renewable source or battery bank, typically 12V, 24V, or 48V.

The low-voltage AC power is set at a frequency of 60 cycles per second or 60 hertz for the North American market. Inverters in Europe and Asia operate at 50 hertz. 60 hertz means that the polarity of the voltage is switched back and forth 60 times per second. Imagine a simple C-cell battery inserted into a flashlight first one way and then the other 60 times per second, as shown in Figure 11-3. If we plotted the polarity of the battery each time it was inserted into and taken out of the battery clips, as shown in "A," we would obtain the plot shown in "B."

The advantage of AC power over direct current is that the voltage can easily be stepped up or down using a transformer.

A transformer inside the inverter receives the low-voltage AC power and steps it up to the utility-standard 120V or 240V. In the inverter shown in

Figure 11-2, the transformer is designed to accept a nominal 12VAC input from the power bridge and convert it to a 120V output. This power is then fed through a protective circuit breaker and into the power panel, which distributes it throughout the house.

If the system is grid-interconnected, the inverter automatically synchronizes its waveform to that of the utility and the electricity is automatically "exported" to the grid via an electrical energy meter.

Inverter AC Waveforms

The electrical utility generates its AC power using mechanical generators connected to steam or water turbines. When a rotary generator creates an AC cycle, the waveform is sinusoidal, as shown in panel A of Figure 11-4. Compare this waveform with that of the simple inverter shown in Figure 11-3 and again in panel B of Figure 11-4. Each waveform has been magnified to show what is known as one cycle, which is one-sixtieth of a second long. Obviously, sixty of these cycles occur in one second.

The main difference between the two waveforms is that the utility-generated power shown in panel A of Figure 11-4 has the voltage ramping up slowly until it reaches a peak level. It then falls back down slowly as the polarity reverses. In contrast, the simple inverter shown in panel B of Figure 11-4 snaps quickly between points of opposite polarity, thereby acquiring the name "square wave." Only the least expensive models that plug into an automotive cigarette lighter and old-style off-grid inverters operate using the simple square wave.

Improvements in technology led to the "modified square wave" (also known as a "modified sine wave," but that's really pushing it), as shown in panel C of Figure 11-4. These inverters still retain a significant share of the market, but they can be used only in off-grid and cottage applications, as the generated waveform is incompatible with the electrical utility.

Inverters using a square or modified square wave can operate 98% of all modern electrical appliances with no problem. However, the fast-switching edges of the waveform can produce a buzzing or humming noise in some items such as cheap stereos, ceiling fans, and record players. One primary advantage of this waveform is the ease of producing it. Inverters utilizing this pattern are robust, electrically efficient, and relatively inexpensive. They are not, however, compatible with the electrical utility system.

A fairly recent development in inverter technology is the sine wave model that outputs a waveform similar to that shown in panel D of Figure 11-4. This digitally synthesized waveform shape is created using a technology similar to that used to record digital sound on a compact disc MP3 file, and we all know how good those sound. In fact, today's inverter technology offers negligible waveform distortion, with frequency and voltage tracking far better than that of even the most expensive utility power station. In the last few years the number of models and manufacturers has exploded, increasing the range of sizes and features while lowering prices.

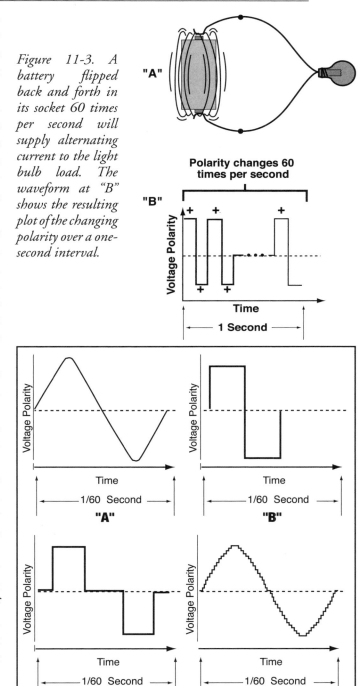

Figure 11-3. A battery flipped back and forth in its socket 60 times per second will supply alternating current to the light bulb load. The waveform at "B" shows the resulting plot of the changing polarity over a one-second interval.

Figure 11-4. A variety of AC waveforms are shown in this diagram. 11-4A is a true sinusoidal waveform as supplied by the electrical utility. 11-4B details a square-wave signal while 11-4C details a modified square-wave signal. 11-4D shows a digitally produced sine wave voltage signal.

Many people argue that although square wave inverters are cheaper than their sine wave cousins they should not be used in the off-grid home because sine wave inverters are better. Two of the homes described in Chapter 4 utilize these inverters without any reported problems. I have used both and have mixed feelings about throwing my two cents' worth into the fray.

Over the last 16 years, I have deliberately kept both types of inverter operating at our house and tested every electrical load that crosses my path. Here is a quick summary of my findings:

- Audio/video equipment has not presented any problems with either inverter with the exception of record players. When plugged into a square wave inverter, the 60-Hz buzz is completely annoying. Aside from this, I have tested all manner of amps, receivers, CD, DVD, tape, satellite, projection, plasma and LCD TVs and found no difference whatsoever.
- Ceiling fans buzz excessively on square wave inverters and work well on sine wave models.
- Our 120V, ¾ horsepower submersible Jacuzzi well pump worked flawlessly when connected to a Xantrex 2500W square wave inverter. When connected to an Outback 3500W sine wave inverter the pump would not start about 30% of the time, tripping the thermal protection device. I was able to correct this by changing the starting capacitors in the pump, but if this problem was not noticed and corrected the pump would fail prematurely. (This problem would be corrected using a 240V pump; however, that would require dual-inverters or a voltage step-up transformer.)
- Lorraine points out that hair dryers and straighteners with an adjustable ionic control burn out quickly when connected to square wave inverters.
- Our Berghoff brand induction cooktop unit did not operate with a square wave inverter.
- Several models of plug-in digital alarm clocks ran too quickly when connected to an Outback 3500W sine wave inverter.

If you have the money, by all means purchase the upgraded sine wave model, but if cost is a concern and you can live with the limitations noted above, the "modified square wave" workhorses still have a lot to offer.

Grid-Interconnection Operation (grid-dependent or batteryless mode)

A sine wave inverter that generates AC power of utility quality should be able to supply power to the grid in a manner similar to that of a  hydroelectric turbine. Ten years ago this was considered heresy by the electrical utilities (and still is by some). Today, however, the demand for cleaner renewable energy (including political head-knocking by environmental and industry advocates) has weakened the monopoly previously enjoyed by utilities. Of course, some areas of North America are slower than others at getting the point, but the change is already occurring and will inevitably continue. Take California, for example: after continuous problems with rolling blackouts, sky-high energy costs, and never-ending smog, the politicians got the message. Now anyone can install a bunch of PV panels on the roof, connect them all through a grid-interconnected inverter, and watch the electrical meter spin backwards. If that isn't enough, the state will even provide an incentive payment to help make it happen.

The grid-connected system shown in Figure 11-5 is a simple, effective means of allowing renewable sources to supply energy to the utility. In this arrangement, electrical energy flows from the grid, through the utility meter, and into the power panel and electrical loads of the home. This is the normal *import* or purchasing mode that most homes are familiar with. During importation of power, the meter records the power usage on one electronic counter.

When the sun starts to shine on the PV array or the wind or water turbine starts spinning, the inverter synchronizes its own sine wave to that of the utility. In order to *export* power, the inverter creates a waveform that has a higher voltage than that of the utility source, causing current to be pushed onto the grid. During power exportation the utility meter records the generated energy on a second electronic counter. At the end of the billing period, the meter reader will record the power imported and exported to your house. With newer so-called "smart meters," the data can be sent from the meter to the central office using cellular phone networks, wireless broadband radio, or other communication means.

For safety reasons, the inverter's internal electronics will perform a series of electrical system tests to ensure that the connection with the utility is within tolerance. Should a disturbance (lightning, brownout, or blackout) occur, the inverter shuts down and waits until several minutes after grid power has returned to normal before reapplying power.

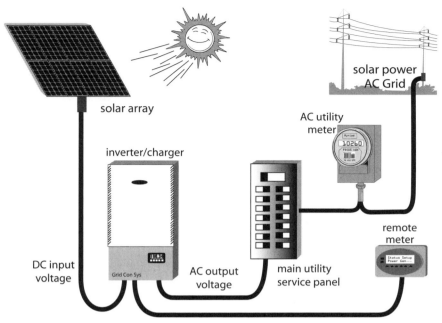

Figure 11-5. *A typical grid-interconnected inverter is shown being fed by a PV array. This simple connection allows the electrical meter to spin backwards when energy is being exported (sold) to the grid. In cloudy times, you import (purchase) energy from the grid in the usual manner.*

Grid-Interconnection with Battery Backup (grid-interactive mode)

Figure 11-7 illustrates the essential components of a grid-interconnected system with battery backup, also known as a grid-interactive mode of operation. Using this arrangement it is possible to provide battery storage to supply power during periods of grid blackout. The major difference between grid-dependent and grid-interactive modes is the requirement for battery storage and a secondary electrical supply panel to feed essential loads within the house. (For those of you considering a grid-interconnected system with battery backup, the section later in this chapter on off-grid inverter selection will apply to you.)

When the utility supply is present, electricity is imported and exported in the same manner as in

Figure 11-6. *The micro-inverter provides all of the features required to convert direct current to alternating current acceptable for grid-interconnection. The micro-inverter concept provides high reliability and simple electrical connection. (Courtesy Enphase Energy)*

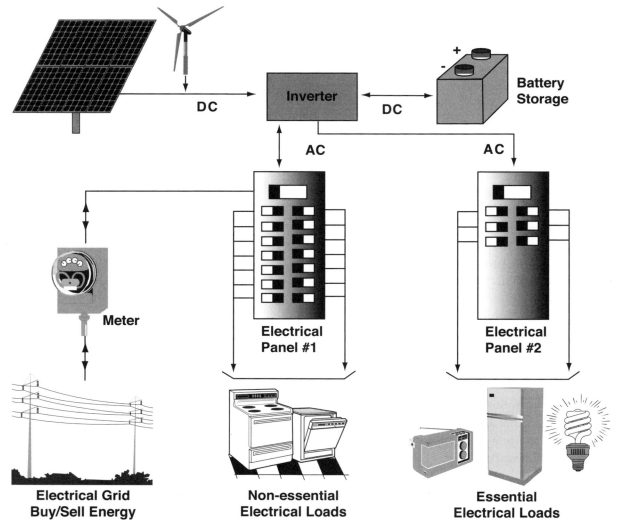

Figure 11-7. A grid-interactive system configuration provides electrical power to essential loads during grid blackout periods. This is accomplished by storing energy in a battery bank and wiring the home with two power distribution panels, one for essential loads and the other for non-essential loads during periods of grid failure.

a grid-dependent system. During times of grid interruption, you will recall that the inverter in a grid-dependent system will disconnect itself and enter a standby state waiting for the grid connection to return to normal. With a grid-interactive configuration, the inverter will disconnect from the electrical utility and reconnect itself to an "essential electrical load supply panel." Using energy stored in a battery bank as well as from any renewable sources, the inverter will generate sufficient power to operate

these connected loads. It is recommended that all the essential electrical loads be of the highest efficiency and lowest wattage, thereby prolonging battery life and corresponding electrical generation during a blackout.

Batteryless Grid-Tie Inverters

Grid-dependent inverters differ from grid-interactive units in several ways. Because there are no low-voltage batteries to contend with, the inverter does not require a step-up transformer, battery charging system, transfer switch, or generator control. This allows the inverter to be made in a variety of sizes, including a micro-inverter model that connects one PV module to the grid at a time, as shown in Figure 11-6.

Because the inverters must comply with utility interconnection requirements, specific safety and regulatory standards are required that do not apply to off-grid systems, including ground-fault protection, anti-islanding, and frequency control. All in all, the capabilities of the small, grid-connected inverter have increased dramatically in the past decade owing to the explosion of sales around the world. Selecting which model to purchase can become a bit daunting, but with the assistance of a reputable dealer and licenced electrician/installer, you will find most models reliable and feature-packed, offering:

- ground-fault protection
- maximum power point tracking
- direct-to-computer metering
- weatherproof enclosures

Take the following steps when considering the ratings factors for batteryless inverters:

- Ensure that the inverter is certified to UL1741 or CSA C22.2 No. 107.1-01 as well as IEEE 1547.
- Match the "watts input DC" from the PV array to the rating of the inverter. This value is supplied by the manufacturer or can be calculated by using the Standard Test Condition (STC) PV rating in watts multiplied by inverter efficiency factor.
- Match the PV array operating voltage window to the inverter. Most grid-tie inverters operate at very high DC voltages of up to 600 VDC. The PV array will require multiple panels wired in series to achieve this voltage, bearing in mind that the real-world output voltage will rise and fall depending on solar illumination. It is very important to ensure that the input voltage does not exceed the inverter operating window in any circumstances. (Remember the adage that it is tough to put the smoke back into the capacitors after the unit is blown.)

Off-Grid Inverters (including grid-interconnected systems with battery backup)

Grid-interconnected systems with battery backup use exactly the same inverter technology as off-grid systems, using sine wave inverters described below.

Off-grid system inverters can range from simple cottage-use 12VDC to 120VAC square wave models to units that work in exactly the same manner as the grid-interactive with battery backup design described above—without the electrical utility connection of course. The major consideration for off-grid applications is choosing a unit that will meet your current and future energy requirements. (If you're planning to have kids in a year or two, get the next size up.) The list of options and features to consider are outlined below.

Inverter Ratings

Important considerations when purchasing an inverter are the electrical ratings of the unit and whether or not you should purchase a "modified square wave" or "sine wave" model. Following is a list of items to look for when shopping for any inverter:

Sine Wave Versus Modified Square Wave Inverter Models

For off-grid or emergency backup power systems that are not grid-interconnected you can choose between either sine wave or modified square wave inverter models. Modified square wave inverters tend to be less expensive than their sine wave counterparts and will operate  almost all home appliances. Your decision will be based on budget as well as on the features you find desirable.

Output Voltage

North American homes are wired to the grid so that loads may be connected at 120V or 240V. Normal wall plugs are rated at 120V, while heavy electrical loads such as electric stoves and furnaces, clothes dryers, and central air conditioning units operate at 240V.

The voltage rating of the majority of inverters is set to 120V, allowing the operation of most household appliances within the capacity rating of the renewable energy system. Increasing the size of the renewable energy source beyond the rating capacity of the inverter necessitates the addition of a second inverter. As a general rule of thumb, systems with a generating capacity of greater than 4kW will require two inverters. When a second inverter is added, the inverters are "stacked," making the voltages additive at 240V.

In most cases the only 240V essential loads that can be reliably connected are domestic water pumps. The vast majority of 240V loads are appliances that draw enormous amounts of electrical energy, like central air conditioning, cook stoves, clothes dryers, swimming pools, and hot tubs. Battery capacity is limited and just because you have many kilowatts of inverter capacity does not necessarily mean the battery can support massive loads.

Inverter capacity refers to the amount of power that the unit can supply continuously.

If a 120V inverter is all that is called for in your application and you require a 240V supply to operate a water pump or other device, an auxiliary step-up transformer may be installed at considerably less cost than a second inverter.

Some larger homes and most businesses require three-phase electrical power, necessitating a minimum of three inverters wired to provide this type of supply. Very few off-grid applications are configured in this manner as the cost of the entire system would be enormous. It is a given that the home or business using 3-phase power is most likely a very large energy consumer to begin with.

Inverter Continuous Capacity

Inverter capacity refers to the amount of power that the unit can supply continuously. The inverter rating should be at least 25% higher than the maximum power delivered to all of the connected appliances that will operate *simultaneously*. Remember that multiple inverters may be wired in series and parallel configurations to increase the capacity as required.

You will recall that power, expressed in watts, is the voltage multiplied by the current. When operating a home, you are likely to use more than one electrical load at a time. In order to determine the inverter continuous capacity required, it is necessary to total all of the electrical loads that are likely to be turned on at any given time. For example, assume that a washing machine (500W), television and stereo (400W), refrigerator (400W), and a bunch of CF lamps (100W) are all on at the same time. The total power requirement of these loads is:

$$500 + 400 + 400 + 100W$$
$$= total\ continuous\ power$$
$$= 1,400W$$

Don't forget to consider intermittent loads that will operate at the same time. In the above example, it is reasonable to assume that a water pump (1,100W) and gas-heated clothes dryer (250W) must be considered. Add a safety factor to this maximum power requirement by purchasing an inverter that is at least 25% larger than the estimated value. Over time, these electrical loads have a habit of creeping upwards.

$$1,400 + 1,100 + 250 + 25\% = 3,440W$$

Remember to calculate the essential load value when filling out the *Electrical Energy Consumption Worksheet* in Appendix 7. When you are shopping for an inverter, you will find that they tend to be sized in "building block" sizes. There are numerous small models below 1,000W, although 2,500W and 4,000W are common larger sizes. Keep in mind that operating the house with loads in excess of the inverter rating will cause nuisance overload tripping and may shorten inverter life.

When trying to read the continuous rating of an inverter, you may be confronted with the term "VA" rather than watts. VA refers to the voltage multiplied by the current, which we have understood to be wattage. However, in an AC circuit the wattage is actually calculated by multiplying the VA times a function known as power factor. At the risk of trying to split hairs, when an AC motor load is operated, there is an effect known as *power factor* that has to be taken into consideration. In the average home, VA closely approximates watts and I generally advise that VA and watts can be used interchangeably. It is recommended that you reduce the rating of the inverter by as much as 20% if you are operating air conditioning units, pool and spa pumps, or other similar high-wattage *induction* motor loads that are subject to the effects of power factor. If you are operating a shop full of large woodworking tools with many of the motor loads running at the same time, it is best to allow for an inverter derating of approximately 25% or higher to allow for power factor. In practice there is little that you can do other than to be aware of the existence of power factor and reduce simultaneous usage of motor-driven appliances or purchase the next-larger inverter. An additional point is that universal motors that have brushes (sparks may be seen when the motor is running) can be run safely without any derating concerns as they are not affected by power factor issues.

Examples of universal motors include:

- regular and central vacuum cleaners
- food processors and mixers
- drills, routers, shop vacs, radial arm saws, and circular saws
- electric chain saws and hedge trimmers
- electric lawn mowers

If in doubt, contact an inverter dealer if you have unusual loads that must be used off-grid. Remember that if you operate a welding or woodworking shop and want to live off grid it may be better to separate home from shop and use a generator to power the big loads.

Inverter Surge Capacity

The surge capacity is an indication of how much short-term overload the inverter will be able to handle before it "trips." Surge capacity is necessary to allow some large loads to get started, particularly motorized loads requiring starting power two to three times their running power. Although this start period is very brief and lasts a fraction of a second, it should be considered. The main concern is whether or not you have any "unusual" electrical devices such as an arc welder in your home. In addition, it is wise to look at your electrical appliance usage list and see if there is a likelihood that several large motor loads may start at the same time. However, in the average home it is highly unlikely that surge capacity will become an issue.

Inverter Temperature Derating

The power protection circuitry for most inverters is temperature compensated, meaning that the maximum load that an inverter can run changes with the ambient temperature. As the temperature of the internal electronics of the power switching bridge increases, the allowable connected load current/power is reduced.

The graph in Figure 11-8 shows the effect of temperature on the capacity of a Xantrex model SW series 4,000W inverter to operate connected loads. Notice that the inverter reduces its capacity at temperatures above 77°F (25°C). The graph also assumes that the inverter is operating at sea level and without any restriction of the airflow around it.

Battery-Charging Features

Virtually all off-grid inverters come combined as inverter/charger units. Although you may save money purchasing an inverter without a battery charger, the charging unit must be purchased if a backup generator is installed. The sum of purchasing the two units separately will definitely be greater than the cost of purchasing a combined system.

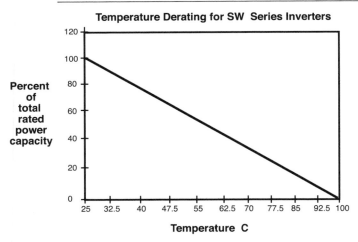

Figure 11-8. As ambient temperature increases, inverters must be derated as illustrated here or according to the manufacturer's data sheet. (Courtesy Xantrex Technology Inc.)

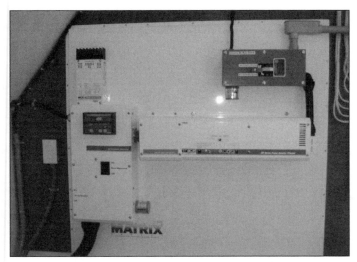

Figure 11-9. This modular, low-cost power panel features a Xantrex DR 2425, 2,500W modified square wave inverter, C-40 voltage regulator, battery, and PV panel DC disconnect switch as well as generator control, AC transfer switch, and metering, all mounted on a small 48-square-inch (1.2-square-meter) panel. This type of simple installation is perfect for an off-grid cottage or smaller home and will not break the budget.

Having said that, deals do come around (eBay/classified ads) and if you want to install a battery charger after the inverter is installed you can purchase a separate battery charger such as the TrueCharge™ series from Xantrex Technology Inc.

DC to DC Voltage Converters

In a number of cases, it may be necessary to operate small direct-current loads at a voltage rating different from that of the battery. For example, you may wish to operate a 12V car stereo or trouble light from the battery in the event of inverter failure. Another common use for 12V power is to operate a small micro-inverter which in turn powers small phantom loads such as cell phone and PDA battery chargers.

Most off-grid households will operate the battery voltage at 24V, with larger homes operating at 48V. There is a temptation to use the knowledge gained from series wiring to connect a "jumper" across the battery bank to capture 12V. **Never do this!** Such a connection will cause the cells providing the 12V tap circuit to be drained more quickly. This can lead to premature failure from inadvertent over-discharging of those cells.

A better approach is to purchase a DC to DC voltage converter that will efficiently drop the higher direct current voltage to the desired lower one. These units will ensure that electrical power drained from the batteries is applied evenly across the entire battery bank. Units are available from companies such as Solar Converters Inc. (www. solarconverters.com).

Summary

Inverters are so highly advanced and are available in so many sizes that every cabin, cottage, and off-grid home can afford one. This eliminates the need for following in the steps of renewable energy "old-timers" who had no choice but to use 12V recreational vehicle appliances.

Chapter 12
FOSSIL-FUEL BACKUP POWER SOURCES

If there is an Achilles' heel with off-grid energy systems, it might be related to the need for a backup power source. Not that there is any problem with the reliability of renewable energy equipment. In fact, a properly designed renewable energy system is far more reliable than the electrical grid. By way of example, over the last dozen or so years our neighbors have gone from being more than a little skeptical about our off-grid system to recognizing that while the grid has been down dozens of times as a result of ice storms, blackouts, and general problems, our home has kept on ticking no matter what.

No, the problem has more to do with the variability of the weather than with the equipment. During the dark months of November and December, we have what seems to be weeks without any sight of the sun or a puff of wind. Once your system meter tells you it's time to charge the batteries, you have to listen.

Folks who are using grid-tie systems do not fare much better. If your system design is grid-dependent, once the utility goes black so does your home. Grid-tie systems with battery backup are only slightly better off, as they generally have insufficient battery capacity to handle more than a few days of grid outage and dark skies.

Bring in the fossil-fueled backup generator, the antithesis of what renewable and clean energy technologies are all about. Life is always full of tradeoffs, and the backup generator is one that can't be ignored.

A well-designed and functional off-grid system will operate 90% or more of the time without the help of a backup generator. But when the weather decides to pull the clouds over the PV panel's eyes or when major system servicing is required you have no choice but to either shut down the entire

Figure 12-1. Fossil-fueled backup generators are the antithesis of what renewable energy stands for. Nevertheless, they are an integral part of any off-grid system and can also provide a bit of insurance for those people who are connected to the grid. (Courtesy Generac Power System Inc.)

system or start the generator. (Remember battery depth of discharge from Chapter 9?)

A backup generator is designed to perform one primary function: to charge the battery bank to a full state and then shut down as quickly as possible. Generator power is polluting, noisy, and expensive compared to grid or renewably produced electricity. The less generator running time that is required the better. Further, generators generally perform best when they are fully loaded; it is preferable to run them near full capacity. Consider for a moment that most generators consume similar amounts of fuel whether they are running a single light bulb or charging a battery bank.

To limit running time, the inverter's internal battery charger will "load" the generator to a maximum level during the bulk-charging step discussed in Chapter 10. If the battery is being charged "normally" the inverter will switch to absorption mode and then float mode to complete the cycle before turning itself off. If battery equalization is desired the charging unit (inverter) will complete the high-voltage equalization charge and return to float mode prior to shutting the generator down.

Inverters or charge controllers equipped with automatic controls are able to signal the generator when to start and stop, completing the charging cycle as efficiently and as quickly as possible. Depending on generator capacity as well as battery size and depth of discharge the entire process can take from five to ten hours to complete.

It is common for a renewable energy system to operate for nine or ten months of the year without the generator ever switching on. When the cold and dark months arrive, the unit has to start and be ready to go. There is no point in having a cheap unit falter at exactly the point when you must depend on it.

Generator Types

A backup generator with a reciprocating internal combustion engine is more correctly known as a genset (a generator and motor set). There are many shapes and sizes and a variety of fuel supply choices. Before we look at models suitable for emergency systems let's review what types not to buy.

Small generators such as the ultra-portable model shown in Figure 12-2 are not suited to full home-emergency or battery-charging applications. As a quick rule of thumb, if you can lift the generator it is probably too small for household backup use. The Honda model EM5000S shown in Figure 12-3 is about the smallest (and least expensive) generator recommended for this application. If you have an old 4,000W unit kicking around in the garage, by all means put it to use. Keep in mind that performance, fuel economy, and battery charging time will be compromised with undersized models.

Figure 12-2. This small 1,000W ultra-portable unit from Yamaha is fabulous for operating small power tools or emergency lights, but it is not suited for a house full of appliances or for battery charging.

Generator Rating

Generators, like inverters, are rated in watts (W) or more correctly volt-amps (VA). The reason for this is that when inductive loads such as electric motors or welding units are connected to an electrical circuit they behave differently from resistive loads such as lights and heating units. Although the generator may have sufficient nameplate rating capacity to operate the desired load, the effects of inductance (power factor) may necessitate that a larger-capacity unit be used.

Additionally, the inrush current required to operate some loads is so great that small generators will cough, sputter, and be brought to their knees as soon as a large inductive load is connected.

Inexpensive gensets will have weak engines, inexpensive generators, and support electronics. This is not a problem for many lightweight household applications, but as soon as big motor-driven devices are activated the trouble starts. Anyone familiar with gensets will know that air compressors, well pumps, and furnaces may have a hard time getting started even if the genset has a higher power rating than the connected load.

Figure 12-3. The Honda EM5000S or equivalent-sized models are the smallest units that should be considered for emergency backup power and small or seasonal off-grid systems.

Motor loads have high starting power requirements that can exceed their normal running requirements by three or four times.

Similarly, battery charging can be very hard on smaller, less expensive gensets because a battery charger consumes power only from the very peak of the alternating current waveform. Smaller units, typically less than 5,000W, have difficulty providing power in this mode, even though the charging power applied to the battery may be a fraction of the generator's rating.

A gas engine driving a generator is a reciprocating device. You will recall from earlier chapters that rotating generators produce an alternating current (AC) voltage which traces a sine wave pattern. An example of this wave form is shown in Figure 12-4, with voltage represented on the "y" axis and time on the horizontal "x" axis. Starting from the extreme left of the waveform, the voltage starts out with zero amplitude and slowly rises until it reaches a peak of 170, at which point the polarity starts to reverse and the voltage drops back to zero before starting on the negative half of the cycle.

You may be wondering why the genset output voltage is 170V as opposed to the 120V level we are accustomed to. Because any AC sine wave voltage changes with time, we are faced with the issue of determining where on the waveform the voltage measurement should take place. A mathematical formula called the "root mean square" can be applied to such a waveform to calculate the voltage present in the "area under the curve." Perhaps a simpler, less technical way of visualizing the voltage of a sine wave is to refer to it as having an "average" level of 120V.

The generator's peak output voltage must be high enough for battery charging to occur. The battery-charging unit inside an inverter contains a transformer that is capable of stepping the applied voltages up or down depending on which mode of operation is selected. For example, with a 12V battery connected the inverter can step this voltage up to 120VAC to operate connected loads. To charge a battery, the inverter can reverse this

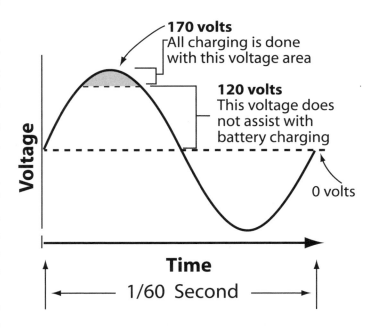

Figure 12-4. Battery-charging applications use only the top portion of the generator's sine wave voltage output. Small generators may not have sufficient capacity to prevent this area of the sine wave from collapsing and for this reason may not be suited for this application.

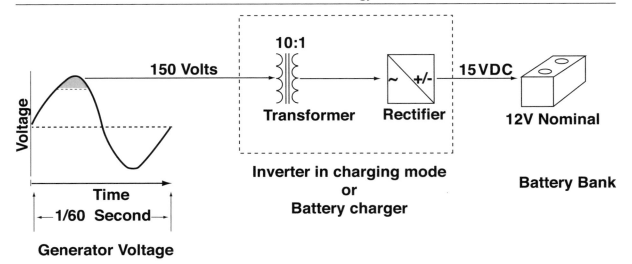

Figure 12-5. It takes a strong, high-quality generator to support battery charging and operate large motor loads such as furnaces and well pumps. Weak or inexpensive "utility" models are simply not suited to the job.

process by stepping the generator's 120VAC down to a DC voltage level sufficient to "push" current into the battery under charge. The waveform in Figure 12-4 shows that the generator's AC voltage starts at zero and climbs to a peak of 170. There will be a period when the generator's "stepped-down" voltage is less than that of the battery bank. No charging current will flow at this time.

Once the generator's instantaneous voltage exceeds the 120V peak on the sine waveform, the inverter's transformer will step it down by the appropriate ratio, convert the AC to DC, and feed the voltage into the battery. If the peak voltage is lower than 120V, no current will flow into the battery.

The simplified example shown in Figure 12-5 will help to clarify this point. The generator instantaneous voltage has risen to 150V. The applied voltage reaches the inverter's battery charging transformer where it is stepped down to the ratio determined by the nominal battery bank voltage, which is 10:1 in this example. The transformer output voltage is 15VAC. This voltage is applied to a rectifier that converts the alternating current to direct current. As the 15VDC output is greater than the battery's 12.19V (or 50% discharged) reading, current will flow into the

battery.

The "squashing" of the peak of the sine wave identifies gensets with weak or small generators that have insufficient power to support battery charging. Longer charging time and increased fuel consumption result in unnecessary genset wear and a higher cost for each watt of electricity stored in the battery.

By way of example, I recall one family that purchased an off-grid residence previously owned by an artist who used the location as a quiet summer residence. She occupied the place during the sunny spring and left before the snow flew in late fall. The new family purchased the spot after Labor Day and promptly moved in with their three children. One of the first things on their to-do list was to dispose of the old propane fridge and replace it with a "spare" electric model their sister-in-law just happened to have in her basement.

Needless to say, the little occasional-use off-grid system groaned, sputtered, and turned off the lights, all the while begging for mercy as the low-battery alarm beeped away.

I explained to these folks that the gift of a "free fridge" was no gift at all and suggested that they recycle it immediately. (No, don't donate it to the "poor"; they are already poor enough and can't afford your free fridge.) Next we looked at the off-grid system capacity and their personal load patterns and came to the conclusion that they were seriously short of renewable energy production capacity. It was pretty clear that a generator would be their only salvation until they could afford to increase their PV capacity to better match their energy consumption.

As it turned out, they had a fairly new 4,000W generator and after a bit of explanation on how to use a hydrometer I left them to see how they would cope. Within a few days they called back explaining that no matter how long they ran the generator they simply could not get the batteries to fully charge. I examined the system and found that as soon as the generator was "loaded up" to charge the battery bank, the charging current dropped to only 15A because of the "sine wave collapse" phenomenon discussed above. With a better quality generator, I would have expected a charging current of at least 120A, reducing run time significantly.

It was a very difficult discussion explaining to these folks that they had invested in a piece of junk for a generator and that they would have to commit to a better model for the job at hand. The happy ending is that they did end up purchasing an industrial unit and have reduced their charging time and fuel consumption several-fold since upgrading to a "real" 7,500W name-brand unit.

But I digress. It is highly recommended that you purchase a genset with a rating of 7,500W (120V/62A or 240V/31A) or higher. Check with the manufacturer to determine if the unit is equipped with an electronic voltage regulator module (sometimes referred to as an electric excitation module) within the generator. High-quality generators often have a peak voltage adjustment that ensures rapid battery charging and immediate starting of difficult motor loads. Lower-quality generators use passive excitation/voltage control, which is really not suited to battery-charging applications.

Discuss this issue with the generator sales staff. (I suggest a generator shop and not the local hardware store.) If they are not familiar with battery-charging applications for a particular model of generator, provide them with a copy of the above text and ask them to review the issue with the factory and provide a return guarantee if you are not satisfied. It is pretty tough to return a generator that is too small for your application.

As a quick rule of thumb, if you can lift the generator, it is probably too small for off-grid use.

Voltage Selection

The voltage selection of the genset will be determined by the connection method with the home and whether or not essential loads require 240V. It will be necessary to review voltage

Generator Type	Inverter Type	Typical Maximum Charging Current (Amps)	Approximate Charging Time in Hours (1000 Ah @50% Depth of Discharge)
Homelite 2500	DR 1512	11	52
Honda 3500	DR 1512	39	16
Westerbeke 12.5 kW	DR 1512	65	10

Table 12-1. Tests conducted by Xantrex Technology Inc. compare the charging current of several generator models when using their 12V inverter/charger. The Westerbeke 12.5kW model can charge a battery bank approximately six times faster than a Homelite 2500. The reduced generator running time saves unit wear, fuel consumption, noise, and money.

selection with your electrician during the planning phase.

See Chapter 13, "Putting It All Together Safely," for more information on generator voltage, selection, and connection.

Fuel Type and Economy

Gensets are available in several fuel choices including gasoline, natural gas, propane, and diesel. The less expensive units tend to be equipped with high-speed (3600 rpm) gasoline engines. Larger industrial-grade models are normally fueled using natural gas, propane, or diesel and have slower, more frugal 1800-rpm engines. The choice of model you select depends on several factors such as capital cost of the unit, proximity to a fuel source, desire for economy, and ease of use. From an environmental perspective, look for engines that have an EPA (Environmental Protection Agency) rating, are four-stroke, slow speed (typically 1800 rpm or slower), and, for urban applications, preferably burn natural gas. Fuel types that you can choose from are as follows:

- **Gasoline:** Everyone is familiar with the small gasoline engines that are ubiquitous throughout North America. Cheap and easily fueled, they have a very short life span when used in demanding applications. The majority of gasoline engines operate at 3,600 rpm, which results in rapid wear and high noise levels. For off-grid applications, expect a life span of five years or less before a major rebuild is required.
- **Natural Gas:** Natural gas engines are offered in two varieties: converted gasoline and full-size industrial. The converted engine is really no better than a gasoline engine, except that it offers the advantage of no fuel handling because it can be connected directly to the gas supply line. Industrial-sized natural gas engines are of a heavier design and operate more slowly, typically at 1800 rpm. This increases engine life and greatly reduces engine noise. However, natural gas is not available in

all areas because the fuel is transported via pipelines.

- **Propane:** Propane is similar to natural gas with the exception that this is the fuel of choice for off-grid and rural applications. Propane may already be the fuel source for other appliances in your home, making generator connection a breeze and eliminating an additional fuel source at the home.
- **Diesel:** The diesel engine has the best track record for longevity. Diesel units are heavy, long-lasting machines that generally operate at slow speeds. Fuel economy is highest with a diesel engine. Besides offering superior fuel economy, modern diesel engines have excellent cold-weather starting capabilities and are clean burning. The downside of diesel gensets is their higher capital cost and high operating noise level, necessitating a sound-damping enclosure.
- **Biodiesel:** As the name implies, biodiesel is a clean-burning, alternative fuel produced from domestic renewable resources. Biodiesel contains no petroleum but can be blended at any level with petroleum diesel to create a biodiesel blend (most often a blend called B20 with a ratio of 80% petroleum diesel to 20% biodiesel). It can be used in diesel engines with no major (or any) modifications. Biodiesel is simple to use, biodegradable, non-toxic, and essentially free of sulfur and aromatics.

The downside of biodiesel is availability, cost, and cold-weather operation, all of which must be considered before using this fuel source. (See further discussion of biodiesel fuel in Chapter 16).

A typical 8 kW propane genset consumes 1.93 gallons (7.3 liters) per hour when operated at 100% capacity. An equivalent genset in a natural gas-fueled model will consume 144 cubic feet per hour (4,077 lph). An equivalent diesel model from China Diesel Imports (www.chinadiesel.com) requires only 0.78 gallons per hour (3 lph). A diesel (or biodiesel) model thus requires 40%

less fuel than an equivalent propane model for the same amount of generator running time or battery charging. Over the considerable lifespan of a diesel model, this translates into a significant savings in operating cost. The fuel economy of a gasoline engine is comparatively poor.

Another argument against gasoline is the requirement to pay "road tax" when you fill up at the local gas pump. Less expensive colored or "off-road" gasoline can be purchased but may be difficult to locate. Even with the reduction in road taxes, a high-speed (3,600 rpm) gasoline model will not be as economical (or as quiet) as a diesel or low-speed natural gas or propane model.

Generator Noise and Heat

It's annoying enough to have to run a genset in the first place, but it's even more aggravating to have to listen to it running. The best way to eliminate this problem is not to operate one at all. The next best plan is to locate the unit a reasonable distance from the house and enclose it in a noise-reducing shed or chassis.

The Kohler natural gas genset shown in Figure 12-6 and the equivalent Generac model shown in Figure 12-1 are mounted in a noise-deadening, weatherproof chassis, which may be mounted on a cement pad in a similar manner to central air conditioning units.

You can either build a noise-reducing shed using common building materials or purchase a wood-framed tool shed building. The shed should be fabricated with a floating deck floor that does not contact the walls of the building. This construction prevents engine noise and vibration from radiating outside the building. The walls should be packed solidly with rock wool, fiberglass or, best of all, cellulose insulation. The insulation should then be covered with plywood or other finishing material, further deadening sound levels.

All internal combustion engines create an enormous amount of waste heat. A little dryer vent or hole in the wall won't cut it. The unit shown in Figure 12-7 is mounted so that the 18"

Figure 12-6. This Kohler natural gas genset is manufactured with a noise-deadening, waterproof chassis and provided with fully automatic controls. It will automatically start and provide power to the house as soon as the electrical grid fails. Once utility power returns, the unit will reconnect the house to the grid and shut itself down. Units can even be programmed to automatically start and stop periodically, ensuring that they will be ready at the next blackout or whenever the off-grid battery bank requires topping up.

x 18" (0.5 m x 0.5 m) radiator and fan assembly blows outside the building pointing away from the main house. This arrangement also requires an air intake, which is provided by air passing under the floating deck of the building.

The exhaust gas leaves the muffler vertically and passes overhead to a second automobile-style muffler before exiting the building, further reducing noise.

Generator Operation

A suitable power feed cable of sufficient capacity will have to be run either overhead or, more typically, in an underground trench, between the generator and the inverter or electrical transfer panel.

If the unit is equipped for manual starting, it will be necessary to go to the machine shed each time you wish to start and stop the unit. This is no problem on a nice summer day, but it can become a bit trying during winter storms when you need the darn thing the most. Automatic controls do not add an appreciable amount to the cost of the

Figure 12-7. Installing a genset in a well-insulated shed with a "floating" floor prevents motor noise and vibration from radiating outside the building. Include a large vent fan to remove waste heat, being sure to direct it away from the living area of the property.

genset and will greatly improve your relationship with the beast. Place a second direct-burial cable rated #14 AWG in the same trench as the power supply wiring. This second cable can be used to either manually turn the generator on and off or allow the inverter or battery voltage regulator to do the job automatically.

Locate the generator building so that prevailing winds blow exhaust gases away from the house. Also ensure that the unit's ventilation louvers or vents point away from the home to ensure a peaceful operating environment.

Other Considerations

A genset should have a long life, particularly if the unit is well maintained. Many underestimate how much television they watch, and I suspect that estimations of generator running time are a close second. (Incidentally, according to Nielsen Ratings, the average American watched 151 hours

of television per month in the last quarter of 2008!) But again I digress. Running time is not a problem for good-quality generators, but knowing when to perform periodic maintenance on the unit is. For example, the Lister-Petter diesel engine shown in Figure 12-7 requires various servicing functions at 125-, 250-, 500- and 1,000-hour intervals. Order the unit with a running time meter at a small extra cost.

Oil, air, and fuel filters must be changed periodically. Order a service manual and sufficient spare parts for your model and learn how to replace them yourself. The dealer will be able to recommend a suggested spare parts list.

Have engine oil on hand as well as the necessary tools to change filters and drain the engine crankcase. Inquire at your local garage about where used engine oil can be dropped off for recycling. Most garages will be happy to oblige, especially if you deal with them for automotive service. Never pour used motor oil into the ground or down a storm sewer. One quart (one liter) of oil can easily destroy 100,000 quarts of ground water, seriously damaging the environment.

To ease winter starting, use fully synthetic oil rated for operation with your generator model.

Further Notes on Biodiesel Fuel

In recent years, there have been remarkable strides in the "greening" of diesel engines. Anyone who thinks that diesel engines are slow, clunky, and smelly obviously hasn't been introduced to the new "common rail diesel engine" technologies of the past few years. Witness the Mercedes Benz E320 CDI series. This car is just as quiet as its gasoline counterpart and faster in both acceleration and top speed. Cold weather starting problems are also a thing of the past. Not only can you forget about block heaters, but the whole glow plug thing has gone the way of carburetors, muscle cars, and fuzzy dice. (OK, maybe not the fuzzy dice.) Look forward to a major resurgence of diesel engines in North America in the coming years.

As a further advantage for the diesel engine community, biodiesel fuel (manufactured using renewable soya, canola, and other grains as well as waste grease and animal fats) may be available in your area. Burning biodiesel is considered green since it has lower life-cycle carbon dioxide and smog-producing emissions than its fossil-fuel counterpart. There may be a slight cost penalty for its use, but considering the minimal running time of the genset in a well-balanced off-grid or emergency backup system and the symbiotic relationship this fuel offers with the renewable energy system, it is well worth the price. Biodiesel might even improve your relationship with your generator.

You will find more information on biodiesel in Chapter 16.

Renewable Energy Generators?

Although many folks think of renewable energy sources as wind, water, and sun, it is possible to fuel a generator with liquid renewable energy, a.k.a. biodiesel, and use the generator as the only source of electrical power generation.

Many systems have been developed using this perfectly acceptable method. It has the advantage of easing the cost of entry for people who might not be able to afford a roof full of PV panels. The concept is simple enough: if an off-grid system is going to have a backup generator in any event, why not simply start by having it charge the battery bank and forget about wind turbines and PV panels? Then, as finances allow, invest in PV, wind, or micro hydro, tying their energy into the system. Over time, the addition of these renewable sources will cause the generator's running time to drop, which will lower operating costs.

A small 12V off-grid system that consumes 2 kWh (2,000 Wh) per day would require a battery bank with a capacity of 1,000 amp-hours:

$$2,000 \ Wh/day \div 12V$$
$$= 167 \ Ah \ battery \ capacity \ per \ day$$

If you assume a minimum of three days between subsequent generator operations and a 50% battery depth of discharge, you arrive at 1,000 Ah capacity requirement:

$$167 \ Ah \ /day \ capacity \ required \ x \ 3 \ days)$$
$$\div 0.5 \ d.o.d.$$
$$= 1002 \ Ah$$

Referring to table 12-1, you will see that a Westerbeke 12.5kW generator will have to run approximately ten hours every third day to keep this size of battery bank fully charged. If the ongoing operational cost of running the generator is less than spending upfront capital on PV panels or wind turbines, then this is a perfectly reasonable way to get started.

Combined Heat and Power Generation

If you take a tour of the mechanical plant of any big hospital, the odds are quite good that they use a combined heat and power (CHP) generating system for electricity and steam production. On an industrial scale, most CHP systems use a gas turbine engine similar to those used in aircraft. These engines have been modified to burn natural gas to produce electricity and steam for process heat requirements within the establishment, including building heating, food preparation, and laundry.

Installing such a system makes economic sense because a large amount of energy in the primary fuel (natural gas, for example) can be captured and reused for heat after electricity has been produced. This process lowers operating costs for the facility by allowing one unit of fuel to perform two or more tasks. The ability of the CHP system to perform multiple processes at once increases the overall system efficiency to 80% or higher.

Although many folks think of renewable energy sources as wind, water, and sun, it is possible to fuel a generator with liquid renewable energy.

ECR International has performed a similar trick by reducing the size of CHP units to the home level with a system they call Freewatt. A small reciprocating engine equipped with an

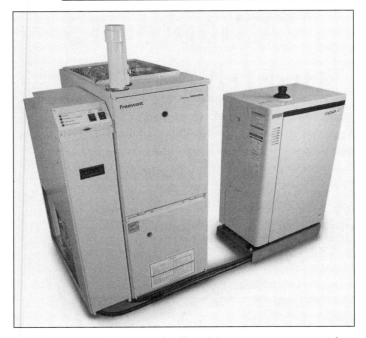

Figure 12-8. Both on- and off-grid homeowners may wish to consider the Freewatt® Micro Combined Heat and Power system from ECR International as an upgrade to a standard backup generator. The system runs quietly indoors (47 dBA) and generates up to 25 kWh of electricity per day (plenty for any off-grid household). At the same time, waste engine heat is captured and employed for space heating using either forced air or a hydronic system. (The Honda MCHP generator is to the right of the forced air furnace.) (Courtesy ECR International, Freewatt® is an innovation of Climate Energy LLC, a JV of ECR International and Yankee Scientific)

electrical generator and heat recovery system has been packaged with electrical controls in a chassis that is about the same size as a portable dishwasher. The unit is designed to produce domestic hot water and offset household electrical loads.

I suggest that people who wish to live off-grid consider such a unit for their home. The unit will sit quietly outdoors and generate up to 25 kWh of electricity per day (plenty for any off-grid household). At the same time, waste engine heat is captured and used for domestic hot water and/or space heating using either a forced-air or hydronic system. This unit can easily replace the dedicated backup generating unit that has become ubiquitous in most off-grid houses and should be able to provide a positive return on investment, something a stand-alone generator will never do.

For those looking for a unit to power their home during the next blackout, the Freewatt will not replace a backup generator straight out of the box. This is due to the limited power rating (1,000W) of the electrical generating unit, which is at a level far lower than even the most efficient home would require. However, over a twenty-four hour period, the Freewatt system can handle the backup energy load of almost any home. The trick is to store the energy produced over this period of time and to be able to dispense it as electrical demand dictates, possibly interconnecting the Freewatt unit as a battery charging system in place of the PV panel as shown in Figure 3-10.

Although the Freewatt system is not quite a renewable energy generator, with its very high efficiency rating it's the next best thing.

Chapter 13
PUTTING IT ALL TOGETHER SAFELY

We have finally made it to the point where we can stop talking about how all the bits and pieces that make up a renewable energy electrical system work and start putting it all together—*safely*. This chapter deals with interconnecting the various system components in a neat and effective arrangement.

If you are not familiar with electrical wiring, conduit, and general construction work, it's still well worth having a look at this section in order to understand what your electrician is talking about *before* you "throw the switch." This chapter and the relative appendices can also act as a reference should your electrician not be familiar with some of the details of working with direct current (DC), PV modules, and wind turbines.

WARNING!
Off-grid systems are a fairly rare phenomenon in North America and the electrical inspector in your jurisdiction may not be familiar with them. Be sure to check with your electrical inspection authority *before* you commit to such a system.

A Word or Two about Safety

Obviously, you want the installation work to be done correctly and safely. Owning and operating a renewable energy system is quite enjoyable, and you can almost forget that you have one at times. Although it is pretty cool stuff, it is not a toy and can cause electrocution or fire hazards if not respected. My first discussion with electrical contractors and inspection people left me bewildered. I clearly remember one person saying eleven years ago: "Why would you want one of those systems? You won't be able to run a toaster." Although I still chuckle at this while eating my morning toast, it goes to show that not everyone is up to speed with the technology.

Owning a renewable energy system is no different from owning and operating a standard electrical power station. Size doesn't matter. You can be killed or seriously injured with battery or inverter power, just as you can with energy from the grid.

Electrical Codes and Regulatory Issues

In North America, electrical installation work is authorized by local electrical safety inspection offices that issue work permits and review the work in accordance with national standards. In the United States, the National Electrical Code (NEC) has been developed over the last century to include almost all aspects of electrical wiring, PV, battery, and wind turbine installation. In Canada, the Canadian Electrical Code (CEC) performs the same function as the NEC.

The NEC and CEC comprise the Part 1 Installation Codes which regulate the interconnection and distribution of electricity to industrial, commercial, and residential buildings. These codes also deal directly with the internal wiring of your home.

Many people believe that because they have their own renewable energy system the code rules do not apply to them. This is wrong. With few exceptions, the installation of PV and wind systems must comply with the requirements of the code. In fact, way back in 1984 Article 690 was added to the NEC to deal specifically with the installation of PV systems.

In addition to the CEC/NEC rules, a Part 2 product standard is required to certify every electrical appliance that operates at 120/240 volts (VAC). Where safety concerns exist, this standard may be extended to lower voltage products such as battery-operated power tools. Many people are familiar with the Canadian Standards Association in Canada and Underwriters Laboratories in the United States. Working in conjunction with the CEC/NEC, these safety agencies are charged with the development of electrical and fire safety standards for household appliances. When a manufacturer develops a new product, the design must undergo extensive safety-related tests by these agencies. Products that meet the requirements are eligible to carry a "certification mark" which tells electrical inspectors that when properly installed they will be safe.

Legitimate manufacturers have their products undergo such testing and are eligible to use the UL, CSA, ETL or other authorized testing laboratory seal of approval. When comparing and purchasing products, look for this seal as a sign of a safe, quality design and be aware that electrical

devices without it will not be allowed to connect to the grid.

When you or your electrician is ready to begin wiring, it will be necessary to apply for an electrical permit. This permit will authorize you to:

- Perform all electrical wiring according to NEC/CEC codes and any local ordinances in effect at the time of installation.
- Install only electrical equipment that is properly certified. Each device must have a UL, CSA, or other approval agency certification "mark."
- Provide the inspector with copies of wiring plans, proof of certification, or other engineering or technical documentation to aid in his or her understanding of the renewable energy system.

Give the inspector written notification that the work is ready for inspection. You must not cover or hide any part of the wiring work, including backfilling of trenches, until the inspector has completed the inspection and provided an authorization certificate.

Your electrical inspector is not working against you. If he or she is asking a lot of questions it is to understand what you are doing. Renewable energy systems are not yet considered mainstream

Figure 13-1. Renewable energy electrical installation is not difficult, as this prewired, integrated panel from Outback Power System shows. Viewing the panel from left to right you see the AC circuit breakers, 2 x 120V inverters "stacked" for 240V output, DC circuit breakers for PV panel, and optional battery bank and MPPT series voltage regulator.

technology, and some inspectors may not be familiar with the specifics. On the other hand, your inspector will know an awful lot about wiring and installation details, and most professionals will be more than happy to assist with guidance and pointers.

Electrical inspectors will review the design and installation work and check for certification marks on the various appliances. Some system components on the market may not have test certification markings, but all electrical code rules require that products *must* have them. It is highly recommended that you check any products before you purchase them to ensure proper compliance. If you require a product and the manufacturer

has not had it tested, discuss this with your inspector before you buy. A certified product may be available, albeit at a higher cost. Alternatively, field inspection on site may be allowed in your jurisdiction for an additional fee.

> ## CAUTION!
>
> **As code rules are updated on a regular basis and may have subtle differences from one locale to another, use the information in this chapter as a guide but discuss the details with your electrician and inspector before proceeding with installation work.**

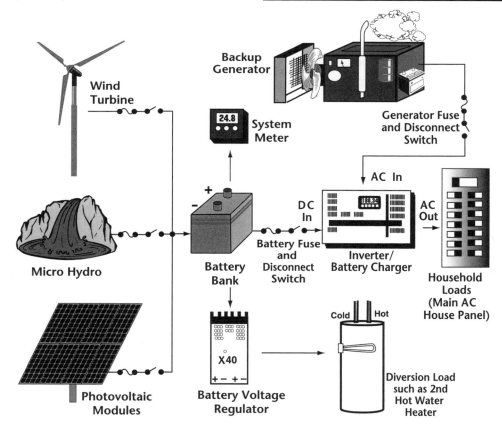

Figure 13-2. This off-grid system overview shows the placement and basic interconnection of each component. Although most systems have only one or two sources, in this case three renewable energy sources (wind, micro hydro, and solar electric PV) charge a battery bank. Power from the battery feeds an inverter which converts direct current to alternating current suitable for supplying household plug loads. A battery voltage regulator and diversion load ensure that the battery is not overcharged. System metering provides system status. A backup generator is used to recharge the battery bank should the renewable energy sources have insufficient output to maintain electrical loads.

What Goes Where?

In previous chapters we have dealt with each component as a separate piece of the pie. We are now ready to begin planning the wiring installation to interconnect all of the components.

The wiring overview in Figure 13-2 illustrates how interconnections are made between each component of the off-grid system. PV arrays always generate direct current (DC) and may be connected directly to the inverter for conversion to alternating current. Wind turbines may operate using direct or alternating current (AC), although in the case of the latter configuration a unit known as a *rectifier bridge* will convert the voltage to DC for supplying energy to the battery bank and inverter.

By contrast, a *grid-interactive* system which

sells excess electricity to the grid and provides emergency backup power during blackout periods is shown in Figure 13-3. The feed from the PV panel or wind turbine, battery bank, energy meter, charge control, diversion load, and inverter circuits will all be completed using direct current in the same manner as an off-grid system.

In either system the AC connections between the inverter and house supply panel follow standard household electrical wiring.

The distinction between DC and AC wiring is very profound. Wire, connectors, fuses, and switches are generally not interchangeable. Because the current on the DC side is very high (owing to the lower voltage), wire size tends to be much larger than on the AC side. For this reason we will review each wiring component separately.

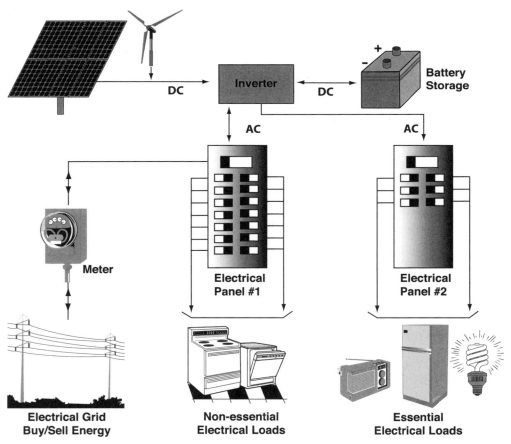

Figure 13-3. When the utility is operating, the grid-interactive configuration works in the same manner as the grid-dependent design. During a blackout, energy is supplied to essential house loads either from the renewable energy source or from a battery bank.

Direct Current (DC) Wiring Overview

Energy stored in a battery bank or supplied by a PV panel is typically supplied at a low DC voltage. When energy is required for our homes, the DC voltage is converted to AC by the inverter and stepped up to either 120V or 240V. If we assume that the inverter will supply a maximum house load of 1,500 watts (W) we know that 1,500W plus an allowance for inefficiencies has to flow out of the battery bank. The wattage, voltage, and current relationship for the household side of the inverter are:

1,500W load ÷ 120V house supply
= 12.5 amps (A) current flow

And on the low voltage input side of the inverter:

1,500W load ÷ 12V battery supply
= 125A current flow

A quick rule of thumb is to remember that the low-voltage side of the inverter current is 10 times greater than the 120VAC side when dealing with 12VDC configurations, 5 times greater for 24VDC, or 2.5 times greater for 48VDC.[1]

It stands to reason that, since the wattage is the product of the voltage and current, lowering the voltage will cause a corresponding rise in current and vice versa. It requires a wire the size of a Polish sausage to carry 125 amps of current in a 12V circuit yielding 1,500W. The same 1,500W can be supplied through a light-duty extension cord if the voltage is cranked up to 120V.

Large-gauge copper wire is expensive and difficult to work with, offering plenty of reason to stick with higher DC voltages where possible. This is also the reason that large AC energy consumers in the home such as electric stoves, furnaces, central air conditioning, and dryers are always rated 240V, allowing the use of smaller wire sizes.

Low-voltage electricity is also difficult to transmit any significant distance. As current flow increases to compensate for lower voltage, more of the power is lost due to wire resistance. The only way to address this problem is to raise the voltage, decrease the transmission distance, or increase the size of the wire. Often it is necessary to do all three. In short, keep DC wiring runs (between PV panels, inverter, and battery) as short as possible and increase the voltage where practical. Appendix 10 contains tables that relate the system voltage and load current to wire size and provide a maximum one-way cable length. This chart assumes a 1% electrical voltage loss, which in turn causes a corresponding loss in power. If you can tolerate a higher level of loss, then each of the applicable distances and losses may be doubled, tripled, etc.

For example, in the table for 24V systems in Appendix 10, assume that a PV array is delivering 40A of current and requires a cable run of 24 ft (7.3 m). The chart indicates that a #0 size of wire (some trades refer to this as "#1/0", while "#00" is referred to as "#2/0") will be required to maintain 1% electrical loss. Wire of this size is pretty big (about the diameter of a pencil), hard to work with, and fairly costly. Let's consider some alternatives:

1. Do nothing. Use the #0 wire and call it a day. There is nothing really *wrong* with using #0 wire; bigger wire is simply more

Figure 13-4. Direct-current wiring circuits such as the battery interconnection cables shown here are expensive and difficult to work with. Keep wiring runs as short as possible and purchase premanufactured cable such as this whenever possible.

expensive, harder to work with, and difficult to interconnect.

2. Increase the system voltage to 48V, thereby decreasing the cable size to #6 gauge. (Remember, doubling voltage halves the current to 20 amps.)

3. Allow an increase in voltage loss. Leave the system voltage at 24V, but allow the system voltage drop to increase to 4%. Use the 24V data table and reduce wire size as desired. For example, using a #6 wire will increase losses to 4%. (#6 wire will carry 40 amps with 1% loss over 6 feet. Increasing this length to 12 and 24 feet will increase losses to 2% and 4% respectively).

4. Check the table in Appendix 11 to be sure the selected wire size can carry the required current. Number 6 wire is rated for a maximum of 75 amps. Don't go overboard with losses to increase wire-run distances, as losses will reduce valuable power supplied to the inverter.

5. Run two sets of smaller wire. Suppose that we are connecting a PV array to a battery bank. It is possible to "split" the wiring of the array in two, reducing the current in each set to 20A. If each set of arrays is connected with #4 wire, we can still maintain our voltage drop at 1% and use smaller wire, which is easier to handle. The two arrays would be connected in parallel back to the inverter or battery bank, combining to produce our 40A supply.

The low-voltage side of an inverter requires an enormous amount of current, so the larger the cables connected to the inverter the better. Undersized cables result in additional stress on the inverter, lower efficiency, reduced surge power (required to start motor-operated loads) and lower peak output voltage. Don't use cables that are too small and degrade the efficiency that you have worked so hard to achieve.

In addition, keep the cable runs as short as possible. If necessary, rearrange your electrical panel to reduce the distance between the batteries and the inverter. The lower the DC system voltage, the shorter the cable run allowed. If long cables are required, either oversize them substantially or switch to a higher system voltage as discussed above.

Although large cable may seem expensive, spending an additional few dollars to ensure proper performance of your system will be well worth the investment. Xantrex Technology recommends that the positive and negative wires supplying an inverter be taped together to form a parallel set of leads. This reduces the inductance of the wires, resulting in better inverter performance.

AC/DC Disconnection and Over-Current Protection

For safety reasons, and to comply with local and national electrical codes, it is necessary to provide over-current protection and a disconnection means for all sources of voltage in the ungrounded conductor. This includes the connection between the PV array, wind turbine, and batteries as well as the connection between the batteries and inverter.

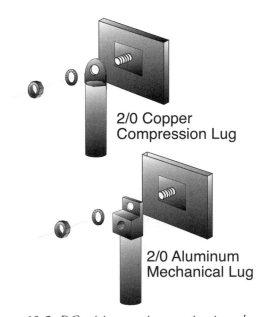

2/0 Copper
Compression Lug

2/0 Aluminum
Mechanical Lug

Figure 13-5. DC wiring requires terminations that are clean and electrically reliable. The copper compression lug on the left is a premanufactured cable for high-powered inverter and battery connections. The aluminum lug on the right can be assembled on-site without the need for special tools.

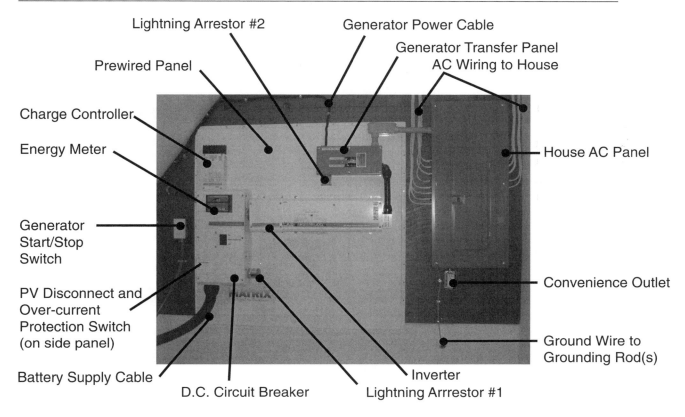

Lightning Arrestor #2
Generator Power Cable
Prewired Panel
Generator Transfer Panel
AC Wiring to House
Charge Controller
Energy Meter
House AC Panel
Generator Start/Stop Switch
PV Disconnect and Over-current Protection Switch (on side panel)
Convenience Outlet
Ground Wire to Grounding Rod(s)
Battery Supply Cable
D.C. Circuit Breaker
Inverter
Lightning Arrrestor #1

Figure 13-6. Renewable energy electrical installation is quite easy, as this pre-wired off-grid integrated panel from Xantrex Technology shows. The large gray panel to the right is where the standard 120V house wiring begins.

Most AC sources are already provided with a protection and disconnection device as an integral part of the house or appliance wiring. Generator and inverter units are generally provided with their own internal certified fuse or circuit breaker device.

Standard AC-rated circuit breakers and fuses will not work with DC circuits, and such an installation should never be attempted. Fuses and circuit breakers such as those shown in Figure 13-7 may be used, as they are rated for breaking a direct current electrical source.

Fuses and circuit breakers are similar to safety valves. When the flow of current through a conductor or appliance exceeds its specified rating in amps, a fuse which is wired in series will "blow," opening the electrical path. A circuit breaker works in the same manner, except that it

may be used as a temporary servicing switch and can also be reset after a trip condition.

Each circuit path will have a maximum current based on worst-case conditions. This current level must be calculated by reading the manufacturer's data sheets for the appliance, wire, inverter, PV panel, or other device. The current rating should be given a safety factor of 25%. Therefore, when sizing a cable for a run between a PV array and inverter, the cable-run distance, system voltage, and worst-case current and wire current-carrying capacity have to be considered. In addition, if the cable is contained in conduit that is exposed to the summer sun, the insulation temperature rating must also be considered. For example:

1. A PV array outputs a maximum of 30A under all conditions and the one-way wiring distance is 40 ft (12 m). The system voltage is 24VDC.

2. From the chart in Appendix 10, we see that a #00 (#2/0) wire is required to carry this current with a maximum loss of 1%.

3. From the chart in Appendix 11, we see that a #00 wire is capable of carrying a maximum current of 195A.

4. A safety factor of 25% is added to our maximum PV array current (30A x 1.25 = 37.5A).

5. The worst-case current calculated in #4 above is compared to the maximum current rating of the wire calculated in #3. If the worst-case current is less than the desired wire-size capacity, it is acceptable.

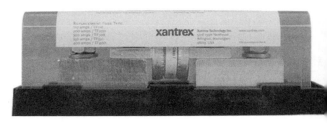

Figure 13-7a. "T" series DC fuse is shown above.

Cable Size Required	Rating in Conduit	Maximum Breaker Size	Wire Rating in Air	Maximum Fuse Size
#2 AWG	115 amps	125 amps	170 amps	175 amps
00 AWG	175 amps	175 amps	265 amps	300 amps
0000 AWG	250 amps	250 amps	360 amps	400 amps

Table 13-1. Battery and inverter cable sizing chart. Note that as a result of heat loss, wires run inside a conduit have a much lower rating than those exposed to air. Battery-to-inverter cable runs should not exceed ten feet in one direction. It is recommended that you use #0000 AWG-size wire for all inverter runs, regardless of length.

6. The circuit breaker or fuse size must be equal to or less than the maximum rating of the wire capacity as defined in the NEC/CEC. Note that electrical cables run in conduits or raceways must be derated because of heating effects.

7. The ambient temperature of the environment and other conditions affecting wire insulation temperature ratings are discussed in Chapter 2 of the NEC and Chapter 12 of the CEC.

These notes show you the complexity of wire selection, placement, and type. If you are not familiar with the above issues, purchase a copy of the latest NEC/CEC and review the appropriate standards. Alternatively, you can discuss these items with your electrical inspector at the planning stages.

Figure 13-7b. A direct current disconnection and over-current protection circuit breaker and mounting box.

To purchase the NEC contact:
National Fire Protection Association
1 Batterymarch Park,
Quincy, Massachusetts
USA 02169-7471
www.nfpa.org

To purchase the CEC contact:
Canadian Standards Association
5060 Spectrum Way
Mississauga, Ontario
L4W 5N6
ww.csa.ca

Battery Cables

According to the NEC, battery cables must not be made with arc welding wire or other non-approved wire types. Standard building-grade wire must be used. The CEC has no such restriction.

Figure 13-8. Battery cables for use with NEC-approved systems must be made with approved building wire. The CEC does not have this restriction.

Wiring Color Codes

Wiring color codes are an important part of keeping the interconnection circuits straight when installing, troubleshooting, or upgrading the system at a later date. The standard color schemes used are discussed below.

Bare copper, green, or green with a yellow stripe

Used to bond exposed bare metal in PV modules, frames, inverter chassis, control cabinets, and circuit breakers to a common ground connection (discussed later). The ground wire does not carry any electrical current except during times of electrical fault.

White, natural gray colored insulation

This cable wire may also be any color at all, other than green, provided the ends of the cable are wrapped with colored tape to clearly identify it as white. This wire carries current and is normally the negative conductor of the battery, PV, wind turbine, or inverter supply. The white wire is also bonded to the system ground connection, as detailed in Figure 13-8 and as discussed later in the text.

Red or other color

Convention requires that the red conductor of a two-wire system be the positive or ungrounded conductor of the electrical system. However, the ungrounded conductor may be any color except green or white.

The majority of DC systems are based on this standard color code. The AC side of the circuit uses a similar approach to color coding except that the ungrounded conductor(s) are generally black and red (for the second wire).

System Grounding

Grounding provides a method of safely dissipating electrical energy in a fault condition. Yes, that third pin you cut from your extension cord really does do something. It provides a path for electrical energy when the insulation system within an inverter or cable covering fails.

Imagine a teakettle for a moment. Two wires from the house supply enter the teakettle, plus a ground wire. During normal operation, the electricity flows from the house electrical panel via the ungrounded "hot" conductor to the kettle. Current flows through the heater element and back to the panel via the "neutral" conductor that is grounded. A separate ground wire connects the metal housing of the kettle (via the pesky third prong) to a large conductive stake driven into the earth just outside the house.

If the insulation or hot wire were to be damaged inside the kettle, it could touch the metal chassis. Because the chassis is bonded to ground (assuming you didn't cut the pin), electrical energy will travel from the chassis, through the ground wire to the conductive stake. This flow of current is unrestricted due to the bypassing of the element, causing overheating of the electrical wires. If it were not for the circuit breaker or fuse limiting this excessive current flow, a fire could start.

Electrical energy has an affinity for a grounded or "zero potential" object and will do whatever it takes to get to there. If there were no ground

Size of Largest Over-Current Device	Minimum Size of Ground Conductor
Up to 60A	#10 AWG
100A	#8 AWG
200A	#6 AWG
300A	#4 AWG
400A	#3 AWG

Table 13-2. This table shows the minimum size of the grounding conductor based on the largest over-current device supplying the DC side of the inverter.

connection on the defective kettle's chassis, electricity would simply stay there until an opportunity arose to jump to ground. If you were to touch the kettle and simultaneously touch the sink or be standing on a wet surface, the electricity would find its path through your body. This is not a good situation.

In a similar manner, the entire exposed metal and chassis of the system components are bonded to a *common* ground point as illustrated in Figure 13-9. The white or negative wire of the system is also bonded to this point, saving us from adding a second set of fuses and a disconnection device to satisfy NEC/CEC requirements.

The size of the ground wire is determined by the NEC/CEC and is based in part on the size of the main over-current protection device rating, as shown in Table 13-2.

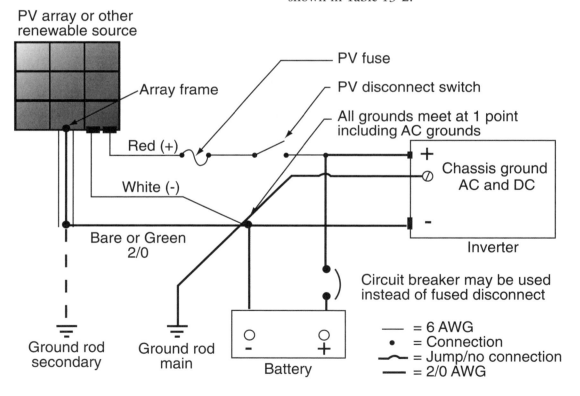

Figure 13-9. This view of a PV array, battery, and inverter DC interconnection shows the placement of over-current protection fuses and disconnect switch in the ungrounded (+) conductor. The negative (-) conductor (white) is bonded to ground and therefore requires no disconnection and over-current protection device. The ground wire (bare or green) connects at one point to a ground rod. A secondary ground rod may be used where the array or generator is a long distance from the main rod. Note the use of an interconnecting cable between both rods.

Lightning Protection

If you look carefully at Figure 13-6, you will notice two lightning arrestors attached to the main DC disconnection panel and the generator transfer panel. These devices contain an electronic gizmo known as a Metal Oxide Varistor (MOV), which connects between the DC +/- conductors and ground as well as the AC hot/neutral conductors and ground. A lightning arrestor is about the cheapest piece of insurance you can purchase to protect your power system. Connect one on every cable run that strings across your property. The system shown in Figure 13-6 has a roof-mounted PV panel and a generator located in a remote building. Cables running here and there can attract lightning on its way to ground potential. The arrestor "clamps" this voltage and passes it safely to the grounding conductor.

Interconnecting the Parts

A mechanical layout plan of your desired installation will help you determine the material required and assist you in visualizing the layout of the power station. It is also important to determine what functions you require from the overall design. Some configurations include:

- PV array only;
- PV and wind hybrid system;
- micro hydro only;
- any of the above with generator backup.

The number of configurations is extensive and it is not possible to cover every installation arrangement within these pages. Fortunately, the installation manuals for the inverter and equipment will assist with reconfiguring for custom-design requirements. By way of example, look at the schematic off-grid design using PV and wind power shown in Figure 13-10. For simplicity, no ground wiring is shown in this view. Refer to Figure 13-9 for grounding requirements.

PV Array

PV arrays may be mounted on a house roof, ground mount, or sun tracking unit. The interconnection of the modules is similar in each case, with one exception. The NEC requires that every roof-mounted PV array be equipped with a device known as a *Ground Fault Interrupter* (GFI), which is shown in Figure 13-11. It automatically disconnects the PV array in the event of an over-current, insulation, or water leakage fault that could cause overheating and a possible fire. A GFI is not required on ground, pole, or tracker mounts.

The first step in wiring your PV array is to determine the battery voltage you will be using. In Chapter 6, "Photovoltaic Electricity Production," we discussed the fact that standard PV modules are manufactured for 12 or 24V output. Figures 6-8 and 6-15 illustrate the steps required to increase PV array voltage by wiring modules in series as well as in parallel connection for increased current flow. Figure 13-12 shows a group of four 12V modules interconnected in series to form a 48V string. Each module is provided with a weatherproof junction box and knockout holes for liquid-tight strain-relief bushings. These bushings press into the hole in the junction box and are held in place by a retaining nut. The flexible cable is then passed between junction boxes and the series or parallel interconnection is completed. The schematic in Figure 13-10 shows two 24V PV panels wired in series to create 48VDC.

If a standard voltage regulator such as the Xantrex Model C60 is utilized, the PV array will be interconnected to supply an output voltage which matches the battery bank nominal voltage.

If a maximum power point tracking (MPPT) regulator such as the Outback MX60 is used, the array output voltage is wired for twice the nominal battery bank voltage.

Be careful to check for proper wire gauge and type to be sure it is suited for outdoor, wet installation. Review the wire choice with your electrical inspector or the NEC/CEC code rulebook.

The output from a grouping of several modules then meets at a combiner box mounted on the rear of a sun tracker unit, as shown in Figure 13-13. The array comprises a total of sixteen 12V modules which are connected in eight sets of two pairs. Each module is rated 75W. The two

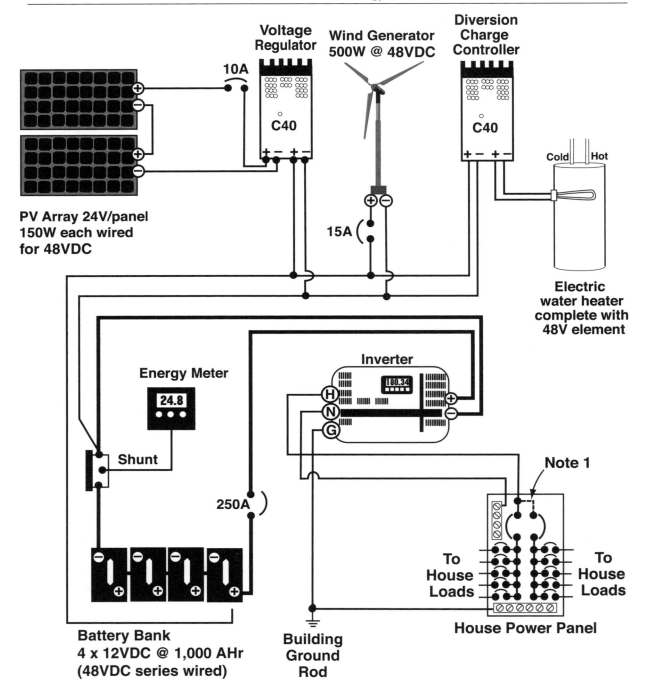

Figure 13-10a. This is one of the most complex versions of off-grid design using PV and wind energy sources, battery backup, series battery charge control, diversion load charge regulation (for the wind turbine), and energy metering.

*Note 1: A jumper wire may be fitted when the house panel is wired **ONLY** for 120VAC. Using this configuration, only one inverter is required. For 120/240 Volt operation, 2 inverters are required and this jumper wire is not installed.*

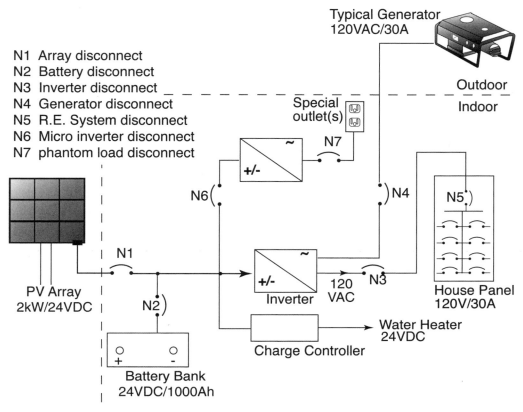

N1 Array disconnect
N2 Battery disconnect
N3 Inverter disconnect
N4 Generator disconnect
N5 R.E. System disconnect
N6 Micro inverter disconnect
N7 phantom load disconnect

Figure 13-10 b. A single line diagram is a simplified version of a full schematic. Your electrical inspector and/or electrician can help you configure the safety elements of the systems to meet your particular needs.

modules are wired in series forming a 24V set. The wires for the eight sets are then directed to the combiner box where all of the negative wires are connected together at one point. The positive leads are each directed to an individual fuse or circuit breaker (5A rating per set). The output of the circuit breaker is then fed to the common positive (+) terminal of the series-wired voltage controller, as shown in Figure 13-10. The parallel connection can be completed using terminal strips to connect all of the appropriately sized wires to the main DC supply cable.

Digressing for a moment, let's look at how this works when fully connected:

a) *16 modules x 75W per module = 1,200W peak*
for the array
b) *2 x 12V modules wired in series = 24V*
c) *Current from array = 1,200W peak ÷ 24V*
= 50A maximum

If we had simply left the array wired at 12V and paralleled the 16 modules we would have had to deal with twice the current (100A). Likewise, if the array were to be wired at 48V, the current would be 25A. Power in watts remains the same regardless of how we interconnect the modules. Higher voltage allows the use of a smaller conductor cable, reducing cost and energy losses.

Wind and Hydro Turbine Connection

Connection of a wind or hydro turbine is essentially the same as for the PV array. The installation manual for each model will provide a connection point or suggest a wiring interconnection method. For example, Bergey Windpower recommends strapping a weatherproof junction box near the top of the tower, drilled to allow the feed/connection wires from the turbine to enter using a liquid-tight fitting to provide strain relief for the

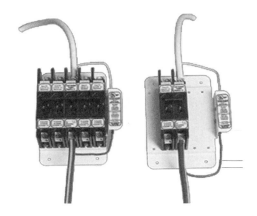

Figure 13-11. A ground fault interrupter provides protection against internal PV module faults that may cause overheating and fire. They are required when the PV array is roof mounted on a dwelling.

Figure 13-12. A PV array detail showing four 12V modules interconnected in series, forming a 48V string. The output of each of the four strings feeds into the combiner box shown at the right.

wires and keep the weather out.

A feed cable is then run down the tower to a second servicing junction box. The feed cable running the length of the tower (and finally to the house) should be installed in either a PVC plastic conduit or metal conduit secured to a tower leg with non-corrosive metal clamps. An equally acceptable alternative is a flexible cable/conduit combination known as teck cable. This material is more expensive than traditional conduit but installs in seconds and is just as strong. Use ultraviolet light-resistant cable ties to secure the teck lead wire to the tower leg.

Renewable Source to Battery Feed Cable

The electricity from the PV array (or wind turbine) must be routed to the battery room. Cables may be fed in conduit down the side of the house or wired overhead or underground if the generator is located some distance away. Underground cable connections tend to be the most common because of simplicity and lower installation cost. The NEC/CEC has provisions for both direct burial cable and cable protected by conduit. In either case, a trench is dug, typically 18" (0.5 m) deep, between the source and the inverter/battery room or house. Where the cable exits the ground to enter the house or travel up the array/turbine support leg, it is necessary to use a length of conduit to protect the wire from damage. The cable is placed in the trench and covered with 6" (15 cm) of soil. At this point, an "underground wire" marking tape is placed into the trench. The intent of the tape is to warn anyone digging in the area that a buried cable lies below. The trench is then filled in.

Over-Current Protection and Disconnect Devices

When the energy source is connected to the inverter (grid-dependent) or batteries (grid-interactive), the NEC/CEC requires that an over-current protection and disconnection device be installed. This may be done in one of two ways: you can purchase individual fused disconnect switches for each source, as shown in Figure 13-14 or you can have auxiliary circuit breakers added to the main battery/inverter disconnect box. The latter was chosen for the system outlined in Figure 13-1.

Every manufacturer of inverter-based systems provides some form of integrated panel as shown in Figure 13-1. Although initial cost will be higher for an integrated wiring panel, final cost will be the same or lower compared to purchasing the parts "a la carte" and wiring them together. In addition, any problems with incompatibility or "finger pointing" are eliminated when the complete system is provided by one manufacturer.

Battery Wiring

As you learned in Chapter 9, "Battery Selection and Design," wiring batteries in series increases the voltage (Figure 13-15A), and connecting them in parallel increases the capacity (Figure 13-15B). Battery manufacturers offer many voltages and capacity ratings to suit your specific requirements. The most important considerations for installation are voltage, capacity in amp-hours (Ah), and distance from the inverter. You will have to determine your essential load power requirements and the length of time you expect these items to operate during a blackout before selecting a battery configuration. Refer to Appendix 7, "Electrical Energy Consumption Worksheet," to calculate the backup energy requirement and resulting battery capacity.

As with integrated wiring panels, some manufacturers are offering inverter/battery combination packages to take the guesswork out of these calculations. Beacon Power offers its grid-interactive *Smart Power M5* and Outback Power Systems provides the *PS1-GVFX3648* system for outdoor installation.

As the size of the battery bank physically increases (providing power for long-duration blackouts) it becomes more difficult to use short wire lengths to feed the inverter. Plan the battery room layout using graph paper and cutouts of the selected battery model to determine the best physical layout, paying close attention to the

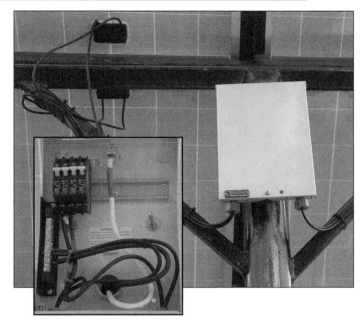

Figure 13-13. In a combiner and fuse box the feed wires from the array enter the box and are parallel connected, with each positive lead passing through a series-connected fuse or circuit breaker.

Figure 13-14. Individual fused disconnection boxes, such as the one shown here, may be used for each renewable source of energy or may be combined within the main disconnect chassis.

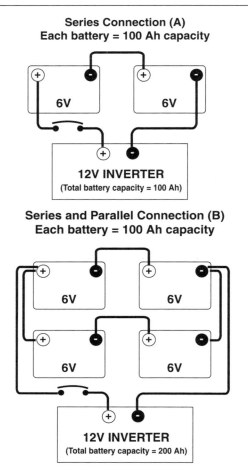

Figure 13-16a (above). This large battery bank comprises a series string of two shelves of eight 6V batteries generating 48V. The second identical string on top is wired in parallel with the first, increasing capacity.

Figure 13-15. Wiring batteries in series strings as in "A" increases the voltage. Placing two such groups in parallel as in "B" increases the capacity of the battery.

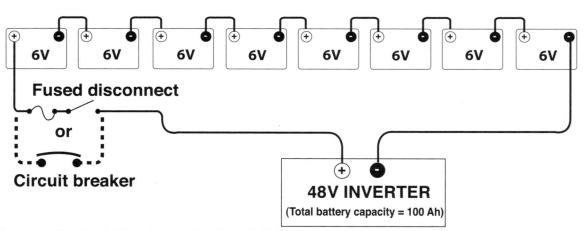

Figure 13-16b (above). This schematic details one shelf of batteries shown in Figure 13-16a. The second shelf of batteries would be wired in the same series string manner. The first and last batteries would then be parallel connected between shelves in the manner described.

cable connections and the length of the runs. Be sure to allow for battery disconnection and over-current protection within the wiring layout. Refer to Chapter 9 for further details regarding series/parallel battery theory.

Battery capacity and physical size can become so large that the length from one end of the battery bank to the inverter may exceed the manufacturer's recommended cable length for inverter operation. To circumvent this problem, interconnection wires should be bumped up to the largest size possible (#4/0) with the positive and negative leads taped together along their lengths to reduce cable inductance. Sometimes there is only so much you can do!

Battery Voltage Regulation

Battery voltage is regulated using a charge controller, which was discussed in Chapter 10, "DC Voltage Regulation." The charge controller may be connected as either a series regulator or a diversion regulator. The series regulator is the simplest arrangement, simply turning the PV input to the batteries on and off based on their state of charge. If your system includes a wind turbine, it's almost always necessary to maintain a load on the generator, requiring a shunt or diversion load arrangement.

Figure 13-10 illustrates a system designed with both a series controller equipped with a maximum power point tracking feature and a diversion charge controller with an electric water heater operating as the diversion load.

Series Regulator

Series regulation is by far the simplest and least expensive system to install. The downside with series regulation is that excess energy is wasted when the unit is regulating battery voltage. Morningstar Corporation indicates that series regulation adds less than 5% to the cost of a mid- to large-sized PV array and battery system. Maximum power point tracking options will increase this expense somewhat.

In Figure 13-10, a PV array has been configured by series wiring two 24V panels to provide a

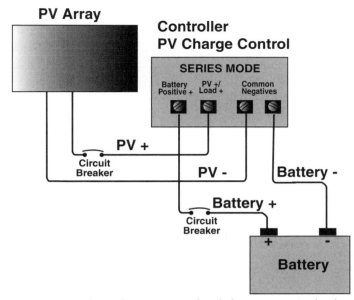

Figure 13-17 shows the connection details for a Xantrex Technology C-40 model charge controller which is not equipped with maximum power point tracking. The connection details can't get much simpler than this. Note that circuit breakers (or fused disconnect switches) must be used in each leg of an ungrounded wire (positive lead) fed by a source of voltage.

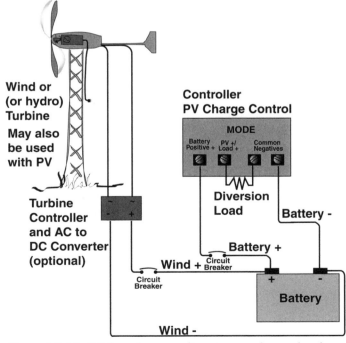

Figure 13-18. With a shunt or diversion regulator, the charge controller is connected to a diversion load such as a hot water heater or electric air-heating element.

nominal output of 48VDC. As you learned in Chapter 6, the open circuit voltage (PV panel in full sun without a connected load) will develop approximately 34V for a nominal 24V model. If this array were connected directly to a 48V battery bank, the output voltage of the PV array would match that of the battery. In this example, the battery bank is wired to provide 24VDC or twice the rated voltage of the PV array.

The PV array will generate maximum power at only one point on its current/voltage rating. It is certain that the battery voltage will not coincide with the maximum power point of the PV array. The series controller connected to the PV array contains a maximum power point tracking feature which allows the array voltage to track the maximum power point. The output of the controller "converts" this input power to a level that maximizes battery charging capability. In addition, the controller will automatically taper the charging current as illustrated in Figure 10 -7. Because of this MPPT function, the array voltage is wired higher than the battery bank nominal voltage.

Figure 13-17 shows the connection details for a Xantrex Technology C-40 model charge controller which is not equipped with maximum power point tracking. The connection details can't get much simpler than this. Note that circuit breakers (or fused disconnect switches) must be used in each leg of an ungrounded wire (positive lead) fed by a source of voltage.

Shunt or Diversion Regulator

Figure 13-17 shows a Xantrex model C40 charge controller configured for diversion load control. This configuration is required when grid failure occurs and the battery bank is fully charged. When these conditions are accompanied by strong wind, the wind generator will produce more energy than is required by the system. As discussed previously, wind and micro hydro turbines often require an electrical load to be connected at all times.

When the batteries are full their voltage will rise, telling the PV array series controller to stop

Figure 13-19. Absorbing large amounts of diverted electrical energy is easy if heat enters the equation. An air heating element such as the model shown here will absorb up to 1kW of excess energy and provide some home heating to boot.

charging (Figure 10-7). This releases the electrical load on the PV array. If the electrical load were released from the wind generator, the blades could possibly enter an over speed mode, destroying the unit. To prevent this from happening, the diversion charge controller "shunts" excess energy from the turbine to the diversion load (in this case a water heater), maintaining the required load.

Review Chapter 10 for details on plumbing and electric heating element swapping. The standard 120/240V elements of an electric water heater used as a diversion load **MUST** be replaced with new ones rated at the nominal wind turbine/battery bank voltage. Ensure that internal thermostats and over-temperature protection devices within the water heater are bypassed, either by removing them from the wiring circuit or installing a suitable heavy-gauge jumper wire. You do not want an internal thermostat opening, thereby disconnecting the load from the turbine.

Air-heating elements such as the model shown in Figure 13-19 can be used as a diversion load. However, air heating tends to be wasteful, as the majority of the heat may be generated outside the heating season. If an air-heating load is used and the byproduct heat is not required, consider mounting the heating element in a rainproof outdoor chassis.

The electrical rating of the diversion load should be 25% greater than the capacity of the devices feeding it. For example, the wind turbine shown in Figure 13-10 is rated 500W and the PV array does not contribute to the diversion energy because it is equipped with its own series charge controller. Therefore, the diversion load rating should be 500W plus a 25% safety factor, requiring an electric heating element of 625W capacity. The rating of the diversion controller must also be calculated to ensure that it is able to supply the maximum power to the diversion load. The charge controller rating is calculated by dividing the diversion heater rating (in watts) by the nominal system voltage:

625W diversion load ÷ 24V system voltage
= 26A diversion controller rating

This is well within the Xantrex C40 load rating capacity of 40 A.

Directing the excess heat to a hot tub or spa is another good place to "dump" excess energy. Use caution when connecting the diversion elements in this application. Both Underwriters Laboratories and Canadian Standards Association safety standards require the use of safety current collectors and GFI protection in spa systems[2].

The Inverter

DC Input Connection

The inverter is the heart and brains of the renewable energy system. Inverters are also pretty darned heavy, and if your system is designed to operate 240V house loads it may be necessary to have two inverters "stacked" together to generate this voltage. Figure 13-21 shows a prewired power panel from Xantrex Technology Inc. that contains:

- two 4,000W sine wave inverters
- two series-wired charge controllers

- dual 250A over-current disconnect units
- AC wiring chassis for generator and house panel connection

Purchasing your system prewired like the ones shown in Figures 13-1 and 13-21 is a wise decision. As the panels are assembled at the factory and approved to applicable safety standards, there is no problem satisfying your electrical inspector. In addition, costly errors in wiring are eliminated.

If you wish to complete your own wiring a la carte style, your electrical inspector will require a simplified wiring schematic such as the one shown in Figure 13-10. Note that this schematic drawing is in no way a full interpretation of the NEC/CEC code rules, but it does provide you with a basis for discussion with your inspector.

AC Output Connection

The AC side of the inverter will be well known to any electrician. Single-phase alternating current voltage at 120V potential is supplied by one black "hot" wire and a white "neutral" return wire. A safety ground is also required as discussed above. If the configuration requires a 240V supply, a second black "hot" wire is provided. The first and second hot wires are commonly referred to as line one and line two respectively. Systems that require a 240V supply will require two inverters connected in a "synchronized-stacked" arrangement providing additive (120 + 120V = 240V) supply.

Residential home wiring panels are designed exclusively for 240V supply. Where a home requires only 120V feeds, a jumper wire may be added to the panel, supplying a single 120V supply to both "legs" of the electrical panel. In this configuration, 240V loads may not be directly connected to the house supply panel. Note that it is possible to use a step-up transformer to convert 120V to 240V supply for the operation of specific appliances such as well pumps. It is generally less expensive to operate a single large 240V load from such a transformer than to purchase a second inverter and run it in "stacked" operation. However, if the load capacity of your home's electrical system is greater than a single large inverter can supply, the required

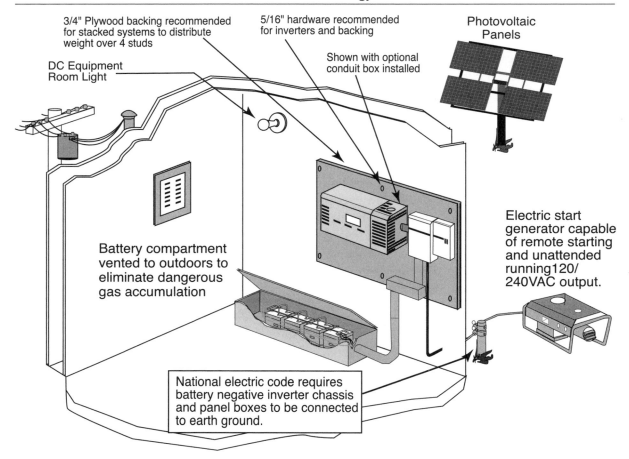

3/4" Plywood backing recommended for stacked systems to distribute weight over 4 studs

5/16" hardware recommended for inverters and backing

Photovoltaic Panels

Shown with optional conduit box installed

DC Equipment Room Light

Battery compartment vented to outdoors to eliminate dangerous gas accumulation

Electric start generator capable of remote starting and unattended running120/ 240VAC output.

National electric code requires battery negative inverter chassis and panel boxes to be connected to earth ground.

Figure 13-20. This drawing details many of the issues related to the DC connection of the batteries, inverter, charge controller, and disconnect and over-current protection system. Consult with the NEC/CEC and your electrical inspector to determine the details specific to your installation. (Drawing based on Xantrex Technology Inc. installation designs)

second transformer will automatically provide the required 240V output when connected with the first inverter.

The Generator

Generators supplied in North America are available with 120/240V split phase or 120V only output configurations. If your renewable energy system is wired to provide 120V only, having your supplier prewire the generator for 120V will make installation simpler. The same is true for 240V systems.

The requirement of having all ground and neutral points bonded at one location (see Figure 13-22) will necessitate the removal of the ground-

to-neutral connection inside the generator wiring chassis.

Energy Meters

Earlier we discussed how an electrical energy meter records the number of electrons flowing into or out of the battery (current exported or imported). A device known as a shunt (detailed in Figures 13-10 and 13-23) converts this current flow into a signal that can be measured by the meter, allowing the calculation of energy consumed and generated. The shunt is usually mounted inside the breaker chassis box, ready for wiring in series with the battery bank negative terminal.

If you are purchasing a prewired system it is

well worth a few extra dollars to have a shunt and energy meter prewired. An example of the energy meter is shown in Figure 9-17. These handy little devices record energy produced and consumed and, most importantly, battery state of charge, which will help prevent damaging over-discharge conditions.

If you are considering wiring your own system refer to Figure 13-23, which details the connection of the shunt. Refer to the specific energy-meter wiring-connection diagrams for the balance of the system connection. Note that where the small-gauge signal wire (typically #22 to #28 AWG) cable exits a hole in the breaker chassis, an anti-rubbing bushing must be added to prevent damage.

As with prewired panels it is wise to have the shunt installed by the manufacturer, as the large wire gauge and tight chassis conditions make installation after the fact almost impossible.

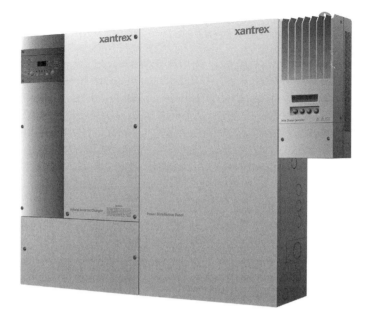

Figure 13-21. This prewired electrical system is compact, neat, properly installed and, most importantly, certified, which reduces problems with sometimes wary electrical inspectors. (Courtesy Xantrex Technology Inc.)

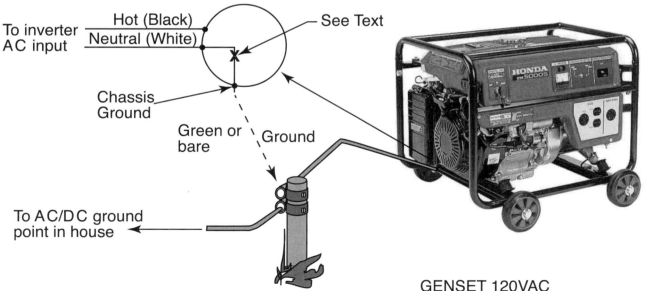

Figure 13-22. To ensure only one ground connection point in the system, the neutral-to-ground bond inside the generator will have to be removed as shown here.

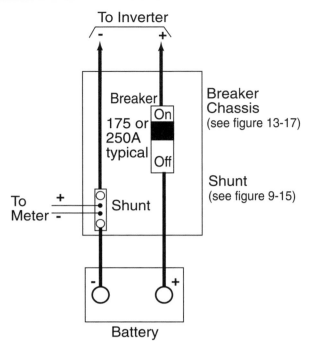

Figure 13-23. When ordering the main battery disconnect and circuit breaker, ensure that a 500A shunt is included. This will allow connection to an energy meter either now or in the future with a minimum of hassle.

Endnotes

1   Grid-dependent inverters such as the SMA Sunny Boy model accept high-voltage directly from PV panels. Wiring a sufficient number of PV panels in series raises the operating voltage on the DC side, eliminating the need for a large power transformer inside the inverter while reducing the wire size required for installation.

2   UL Standard UL 1795 and CSA standard C22.2 #218.1 require the use of current collectors to prevent shock hazard in the event of heater failure. 120V spa-heating elements may be purchased from your local pool and spa supply store. Consult with the applicable standards before connecting the diversion controller in this situation.

Chapter 14
POOLS, HOT TUBS AND SAUNAS POWERED BY SOL

Since the ancient Romans and Turks built their public bath houses, people have enjoyed the benefits and restorative powers of water and heat on the human body. Although times change, people are still flocking to pool, hot tub and sauna dealers in record numbers to purchase these wares. In many suburban homes, the property isn't considered complete unless one of these appliances graces the grounds.

Unfortunately, these much sought-after products are energy and water sink holes that are difficult to manage and typically aren't even considered for the off-grid home. As with all other construction techniques and building codes in North America, many of these products are not manufactured with energy and resource efficiency in mind. As a result, poorly insulated and designed hot tubs draw more energy than most off-grid homes will produce in a week.

If you are going to consider using these items, it will be necessary to understand how they consume energy and to develop ways to keep this consumption to a minimum.

Swimming Pools and Hot Tubs (Spas)

Swimming pools consume energy in a number of different ways. The primary consideration with any pool, whether above or below ground, indoors or out, is the requirement to keep the water clean and fresh looking. A pool that is cloudy or discolored will not only be a turnoff at your next summer block party; this condition is also an indication that things have gone very wrong with your water. Poor water quality can damage circulation equipment because of chemical imbalance, and it can also sting the eyes of swimmers or even cause unpleasantries such as the delicately named "beaver itch" and other equally distasteful or dangerous conditions.

Water purification in a pool or spa is achieved by the judicious balancing of water chemistry, which includes pH level, alkalinity, calcium hardness, and total dissolved solids. Purification involves the use of sanitizers such as chlorine, bromine, and ozone as well as the continuous, or nearly continuous, circulation of the pool or spa water.

Water is very heavy, and it requires a considerable amount of energy to move the large volumes of water from the pool through the plumbing and filter media and back to the pool. For homes connected to the utility grid, this is not a big deal; simply pay for more electricity each month. A typical pool will have a circulation pump of between ¾ and 1 horsepower (approximately 900 and 1,200W respectively), which you will be obligated to operate approximately 10 hours per day. On grid this will amount to approximately $1.00 per day of increased electrical cost to perform the required filtration.

Off grid, pool water circulation is more problematic as energy requirements can exceed sensible power generating system capabilities.

Pool circulation time may exceed this basic level as a result of excessive sunshine causing algae growth, large numbers of swimmers, or airborne debris, and dust blown into the pool. In addition, highly restrictive flow passages due to small diameter circulation pipes, sharp plumbing corners, excessive lengths of pipe runs between equipment and pool, as well as small surface area filters all work together to lower pumping efficiency and flow, requiring even more energy.

If you live in the northern United States or Canada, you might already be aware of the painfully short swimming pool season. The tomato-growing season appears to be a few weeks at best and the pool season isn't much different, unless you have a very high tolerance for cold water. As a result of the abysmal swimming season, many people resort to heating their pools with fossil fuels or electricity, pumping out on average 20 tons (18 tonnes) of greenhouse gases and other air pollutants per season. As discussed in Chapter 5.4, the Canadian Solar Industry Association states that "the average Canadian pool requires more energy and costs more money to heat than the average home." Given the level of energy required to heat a home, it's obvious that using *electricity* produced off-grid to heat a swimming pool is simply not going to work.

Off-grid, a pumping system that utilizes a 1,000W pump operating for 10 hours per day will require 10 kWh of energy just to circulate the pool water. This is more than twice the amount of energy we require to operate our entire home per day. The cost of the renewable-energy equipment necessary to produce this amount of power: at least 50 grand! Many off-gridders assume that they can simply circulate the pool water for an hour or two in an effort to save energy. This is simply not the case, as it results in a murky, unpleasant pond rather than the desired crystal-clear pool.

Ironically, a hot tub or spa requires a similar amount of energy even though the volume of water is greatly reduced. The reason: hot water. Although water circulation in a spa can be achieved with a smaller and more efficient circulation pump, enormous amounts of energy are required to heat and circulate the water in the spa vessel. As with a pool's circulation system, *electric* spa heating while operating an off-grid home or cottage is simply not practical.

Saunas (and Steam Rooms)

Although there is no denying the pleasure of a hot sauna or steam bath after a day's skiing or excessive Christmas shopping, the energy required to heat these units *electrically* off-grid is simply not feasible.

A common theme running through each of the above technologies is the requirement for heat. In each case, the assumption is that electricity produced using an off-grid system is too valuable to use for these luxuries, unless money is no object. Fortunately, there are ways to solve these problems, given a little bit of ingenuity.

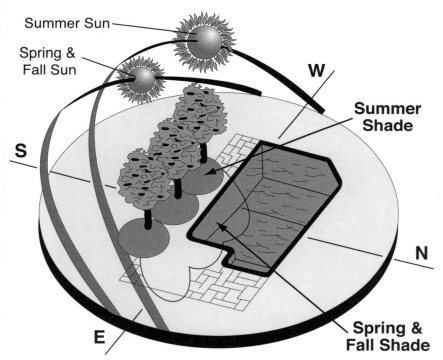

Figure 14.1-1. A home should be designed to incorporate passive solar technology, and so too should your swimming pool. Ensure the long axis of the pool is oriented east/west and that late spring, summer, and early fall sunlight can fall directly on the pool surface.

14.1 Conventional Swimming Pools Off-Grid

(including a discussion on spa circulation systems)

Regardless of whether you choose to install a simple above-ground pool or an elaborate in-ground model, site location is very important. A home should be designed to incorporate passive solar technology and so too should your swimming pool. Figure 14.1-1 shows the ideal layout for a pool, following the same rules used to develop a passive solar house, namely to ensure that the long axis of the pool is oriented east/west and that late spring, summer, and early fall sunlight can fall directly on the pool surface. Although the sun's ultraviolet light can contribute to the oxidation of pool sanitizers, resulting in increased chemical consumption, the added energy warming the pool will greatly offset this loss.

A swimming pool that absorbs the sun's energy during the day will quickly lose it at night unless there is a means of insulating the pool against this energy loss. A pool cover such as the model shown in Figure 14.1-2 will contribute greatly to retaining heat energy in the water, with many models providing security against children and animals falling in. Additionally, a cover that is used judiciously will help reduce water circulation time, as sanitizers will not oxidize as quickly and dust, leaves, and other debris will not enter the pool, further contaminating the water.

The cover, manufactured by Cover-Pools Inc. of Salt Lake City, Utah, can be automatically activated when the pool is not in use, possibly eliminating the requirement for a safety fence (depending on local codes) and helping to ensure that the cover is properly utilized.

Rick Clark, president of Cover-Pools, Inc., likes to refer consumers to the Web site for the U.S. Department of Energy (DOE), Office of Energy Efficiency and Renewable Energy (http://www.energysavers.gov/your_home/water_heating/index.cfm/mytopic=13140). According to the DOE, the most effective means of reducing

Figure 14.1-2. According to the Department Of Energy, the most effective means of reducing pool heating costs is to cover the pool when it is not in use, with potential savings of between 50 and 70%. (Courtesy Cover-Pools Inc. www.coverpools.com)

pool heating costs is to cover the pool when it is not in use, with potential savings of between 50 and 70%. "An automatic solid vinyl safety cover adds to the convenience of energy conservation. Since an automatic cover is easy to use, the pool owner is more likely to cover the pool every day during the swimming season," says Mr. Clark. Although the automatic pool cover offers excellent convenience, it remains a premium product. If you cannot afford such a model, consider purchasing a manually applied "solar blanket" available through your local swimming pool dealer. Just remember: regardless of which type or model of cover you purchase, if you don't install it, it won't work.

Figure 14.1-3 outlines a typical swimming pool (or spa) filtration system, shown with an optional solar thermal absorber attached. For readers interested in solar thermal pool heating, refer to Chapter 5.4 for further information.

All swimming pools and spas have some form of plumbing system which draws water from the surface of the pool (the skimmer intake) and secondary floor drain(s). Water is drawn into a pre-filter strainer basket, usually mounted on the intake of the pump, and then routed to a centrifugal pump where the water is pressurized. Water then flows into a cartridge or sand filter which removes fine particulate matter from the water before returning the water to the pool. Pumping time and energy are increased if the flow of water is impeded by long pipe runs, sharp angles and turns, or dirty strainer baskets and filters. Before installing your system, try to ensure that these conditions can be overcome through efficient site planning. Also consider installing the smallest pool or spa that will meet your needs rather than basing your decision on what your ego dictates. A small pool has a correspondingly small

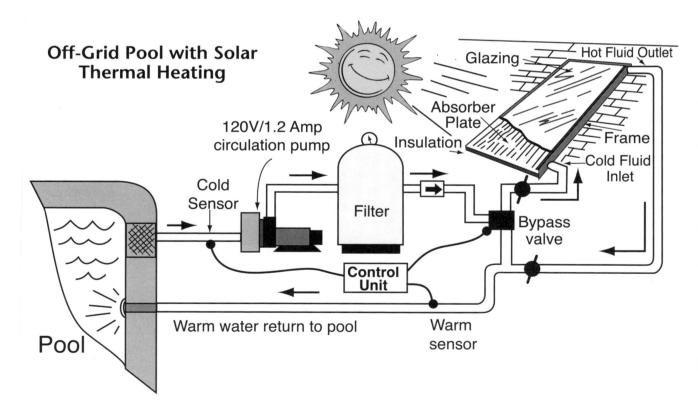

Figure 14.1-3. This figure outlines a typical swimming pool (or spa) filtration system, shown with an optional solar thermal absorber attached. For readers interested in solar thermal pool heating, refer to Chapter 5.4 for further information.

volume of water and will require less filtration time for a given amount of electrical pumping energy.

When developing a new pool, consider the following steps to ensure that plumbing efficiency has been increased to the highest level possible:

- Install the smallest pool that will suit your needs.
- Use oversized plumbing pipe and avoid sharp angles when changing pipe direction. Two 45° elbows spaced slightly apart are better than a single 90° elbow.
- Consider the use of oversized, flexible "spa hose" to further minimize plumbing friction.
- Ensure that the pump and plumbing lines are nominally horizontal (a slight downward angle to allow water to drain back to the pool is ideal). Having the pump and plumbing lines at different heights increases pumping "head pressure" and reduces water flow.
- Order a filter that has an internal surface area that is at least twice the size required for your pool size. A larger filter surface area reduces "backpressure" on the pump, increasing the flow through the filter and in turn reducing pumping time. You may have to explain to your skeptical pool dealer that you are off grid and that every drop of energy efficiency counts.
- Ensure that a solar blanket or other pool cover is installed whenever the pool is not in use.
- Maintain a strict pool cleaning and chemical maintenance schedule. Properly maintained water will require less filtration time.
- Encourage pool and spa users to have a shower *before* entering the pool. This will reduce the amount of deposits such as dead skin and deodorants entering the water and increasing filtration time.
- Install a battery-operated timer (one without a phantom load) to operate the filtration pump during the daylight hours in order to capture excess solar energy production. The timer may also be used to limit water circulation

Figure 14.1-4. Provided restrictions in the pool plumbing have been minimized, smaller, high-efficiency pumps can be used, resulting in energy savings of up to 10 times.

during the peak solar hours of the day and thus prevent pool overheating.

As noted above, standard swimming pools often use very large centrifugal pumps rated ¾ to 1 horsepower or larger. Provided the restrictions in the plumbing have been minimized, smaller, higher efficiency pumps can be employed. An example of a high-efficiency pump is shown in Figure 14.1-4. The "Circ-Master" is manufactured by Aqua-Flo Inc. of Chino, California and is designed to replace standard high-power water pumps where a smaller model may suffice. This model is rated 1/15th horsepower and draws 1.3A when connected to a standard 120V utility service connection or inverter. The total power required to operate this pump is 156W and, with carefully reduced filtration time based on the above rules, the pool may require as little as 900 watt-hours (0.9 kWh) of energy per day, with energy consumption reduced by 10 times compared to a standard pool.

Plumbing a pool for solar thermal heating is accomplished by adding a check valve and optional motorized bypass valve as shown in Figure 14.1-5. The schematic view indicates that pressurized water exiting the filter flows into a one-way check valve. (This prevents a thermo-siphon backflow condition that occurs at night when the circulation pump is turned off.) Pressurized water then flows to an optional thermostat-controlled, motorized bypass valve which directs water to return to the pool when it is sufficiently warm or through the solar collector panels when heat is required.

Pools installed with a small number of solar collectors may not require a motorized bypass valve, as these smaller installations are not likely to overheat the pool. It is also possible to control the pool's water temperature by limiting the pump operation through the peak daylight hours using a programmable timer.

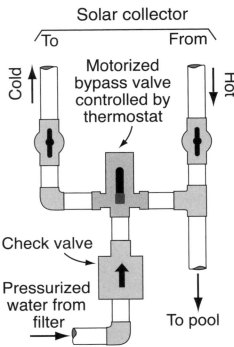

Figure 14.1-5. It is a very simple task to plumb a swimming pool to accept a solar thermal heating system. In this figure the plumbing photograph is illustrated in a schematic view shown on the right. Pressurized water from the filter enters a one-way check valve. An optional motorized bypass valve controlled by a thermostat directs water to the cold side of a solar thermal collector panel, with the return water directed back to the pool.

14.2

An Alternative Way Forward: The Eco-Pool Concept

Overview

The attraction of a swimming pool cannot be denied. For many children, visiting summer camp or going to a cottage and swimming away the hot summer days is an image that is burned into their collective memory and recalled in later years as those youthful summers are relived.

Figure 14.2-1. The Hubicki family eco-pool forms a lovely vista from the living room patio doors. This natural pool is built with recycled and local materials and is maintained by nature's rains and a pond ecosystem.

Figure 14.2-2. The eco-pool water is oxygenated as it spills over a waterfall and nourishes the water plants of the pond.

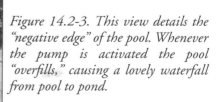

Figure 14.2-3. This view details the "negative edge" of the pool. Whenever the pump is activated the pool "overfills," causing a lovely waterfall from pool to pond.

Figure 14.2-4. The water plants making up the clarifying pond are clearly visible in this view.

Figure 14.2-5. Long after the swimming season has ended, the eco-pool and pond still provide a beautiful setting.

Figure 14.2-6. In this view, Mike is shown with the settling filter lid open. During operation, water is forced to swirl around inside the container, causing heavy sand or other particles from the pond to settle before the water is returned to the pool.

If it is your desire is to recreate the natural setting of past summers, then why start with a concrete and vinyl circular swimming pool when what you really want is a lake or "swimming hole"? In an effort to resolve the dichotomy of a modern pool and a natural setting, people working in the area of permaculture have developed the concept of the eco-pool.

Tina and Mike Hubicki are one couple who wished for the pleasures of a swimming pool but didn't want to stray from their approach of using natural materials and sustainable methods in their quest. "In 2004 we built our eco-pool based on work done in Germany and other locations around the world," explains Mike. "The pool is constructed in a 14 ft x 28 ft (4.3 m x 8.6 m) rectangle which feeds a 15 ft x 30 ft (4.6 m x 9.2 m) natural pond via a 12 in (0.3 m) high waterfall created by a "negative edge" between the pool and the pond."

"The basins for the pool and pond were constructed using recycled tires for earth retention and steps around the inside of the pool," Mike continues. "Old carpet was used as a base to protect the rubber pond liner. Water is circulated using an energy-efficient pump (discussed in Section 14.1 above) and water sterilization is completed without the need for chemicals by using a copper-silver ionizer unit. Total power requirements for the pool and pond system are approximately 1.5 kWh per day. Apart from the initial filling, the pool and pond surface form a catchment basin for rainwater, ensuring sufficient water level."

The eco-pool has minimal chemical requirements for water sterilization owing to the unique natural filtration system. I asked Mike to explain how the pool works as an integrated system. "When the pump is activated, water flows into the pool, filling it beyond its maximum level. This causes water to spill over the negative edge waterfall, aerating the water with oxygen. Oxygenated water spills into the stilling pond, where a natural selection of water plants and microorganisms purifies the water. Pond water is then directed to a settling filter which allows sand

and other particulate matter to separate. The water is then pressurized by the pump and returned to the pool. An electrically operated silver-copper water ionizer (available at most pool dealerships) provides final water purification to ensure that the water stays crystal clear."

Eco-Pool Construction

The best way to describe the construction of the eco-pool is to show you by walking step-by-step through the construction process.

Figure 14.2-7. No, the high hoe operator is not sleeping; he is studying the plans to ensure that the eco-pool elevation and site layout are within tolerance. Although it is possible to excavate by hand, a backhoe or "bobcat" unit will save a lot of wear on your back.

Figure 14.2-8. It is essential to have all the excavation level and set to the correct depth. As you will see in the following pictures, the entire pool structure sits on columns resting on grade. Careful attention to this detail will ensure that the finish decking materials line up with the entrance doorway and guarantees straight and true construction.

Figure 14.2-9. Once the excavation has been roughed in, the tires are placed around the perimeter of the pool and set back to form a series of steps, easing the entrance into and exit from the pool. Each tire is tamped with earth to make a solid, secure structure.

Figure 14.2-10. Hand-dug trenches hold the plumbing lines which run to and from the pump, pool, and sediment wells.

Figure 14.2-11. The pool floor drain is moved into position and a plastic bag is placed over the mouth of the inlet to prevent dirt from entering.

Figure 14.2-12. Gate valves are placed in the plumbing lines to facilitate servicing the pump and filter without having to drain the pool.

Figure 14.2-13. This view details the connections of the pool plumbing to the sediment wells.

Figure 14.2-14. A series of cement patio stones is placed over the tires that will form the negative edge waterfall seen at the top of the picture.

Figure 14.2-15. Once the tires and plumbing lines are roughed into position, a series of decking joists is temporarily placed in position to ensure correct alignment between the desired decking surface height and the pool. Note that the decking is a "floating" design that is not directly attached to the house structure.

Figure 14.2-16. Mike has used a piece of sewer pipe as the vessel for the hot tub planned for the future. The hot tub will share the pool's hydraulic system and decking.

Figure 14.2-17. This view shows the roughed-in clarifying pond, with one edge located directly under the negative edge waterfall of the pool. The pond has been deliberately designed to have a freeform, natural look.

Figure 14.2-18. Cement patio blocks form the foundation for the negative edge waterfall. It is imperative that this surface be level or the waterfall will not be uniform.

Figure 14.2-19. Deck support beams are made by laying pressure-treated 2" x 10" (51 mm x 255 mm) boards on their sides, providing a wide footing for the deck finish boards.

Figure 14.2-20. Standard concrete "cottage blocks" are installed along the finish grade, which in turn supports 2" x 10" pressure-treated joists. It is very important to ensure level orientation between the joists and tire support boards to ensure a quality deck finish.

Figure 14.2-21. The boys are smiling because they have completed the most difficult and heavy part of the construction process.

Figure 14.2-22. At this point, the entire construction enjoys a coffee break.

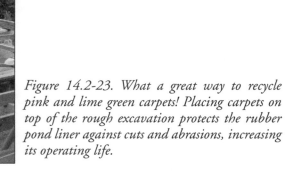

Figure 14.2-23. What a great way to recycle pink and lime green carpets! Placing carpets on top of the rough excavation protects the rubber pond liner against cuts and abrasions, increasing its operating life.

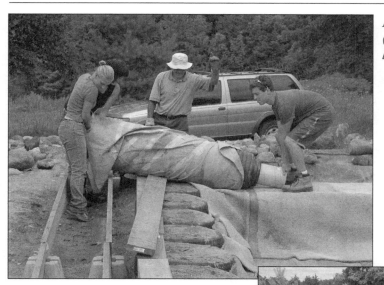

Figure 14.2-24. With a mighty heave, the crew (recruits of wary friends) carefully unrolls the pool liner into the eco-pool excavation.

Figure 14.2-25. The liner is pulled into position and aligned parallel to the pool sides.

Figure 14.2-26. The alignment continues, ensuring that folds and pressure points are eliminated.

Figure 14.2-27. Careful attention is paid to the corners as well as the "steps" formed by the tires. The pool liner must rest on a solid surface so as not to stretch when water is added.

Figure 14.2-28. Corner folding is perhaps the trickiest part of the installation.

Figure 14.2-29. Sufficient liner material must overlap the pool edges to be held in position by the deck support boards.

Figure 14.2-30. Once the liner is in position, the water delivery truck is called in.

Figure 14.2-31 Mike always did want to become a firefighter.

Figure 14.2-32. As water is added to the pool, the crew continues to check the liner and make final adjustments.

Figure 14.2-33. The liner will have a tendency to "pull away" from the walls of the pool and slide towards the bottom. Here the crew continues to press the liner into its proper position to ensure that it stays in place.

Figure 14.2-34. Once the pool is filled, the overlap of liner material is gently stretched into position and fixed under the deck support boards placed on the tires. The weight of the deck material will hold the liner in position.

Figure 14.2-35. The decking material is now nailed into position starting at the house end of the deck.

Figure 14.2-36. Joists in the vicinity of the patio doors run perpendicular to the main deck and necessitate that decking material be installed at right angles to the rest of the deck.

Figure 14.2-37. The pond liner is fitted into position. Less precision is required with this area as the "lines" of the pond should be more fluid. Note the use of stream stones to hold the liner in position and define the boundaries of the pond.

Figure 14.2-38. Stone dust walkways have been integrated into the perimeter of the pond and intersect with the main deck. This makes a perfect transition from the formal deck to the natural pond.

Figure 14.2-39. This view details the pool negative edge waterfall-to-pond transition area. The pool liner folds into the pond liner, ensuring that 100% of the circulation water stays within the confines of the hydraulic system.

Figure 14.2-40. Additional stream stones and perennial plants edge the completed deck and hide the support structure underneath.

Figure 14.2-41. The hot tub (concrete sewer pipe) is planked with vertical cedar boards held together with a cable band, giving the impression of a large coopered barrel.

Figure 14.2-42. The inside of the hot tub is lined with Styrofoam®-brand closed-cell insulation cut to size and held in position with compatible "insulation adhesive."

Figure 14.2-43. The insulation is clamped into position and allowed to set. Later the inside of the hot tub will be lined with waterproof and rot-resistant red cedar and plumbed into a new solar thermal heating system.

Figure 14.2-44. This is the boundary between the waterfall and clarifying pond.

Figure 14.2-45. Once the pool is filled, the black liner and pool water surface play tricks on the eye. The negative edge waterfall is a delight to see and hear.

Figure 14.2-46. Water plants coupled with the random placement of the rounded river rocks form a natural pond which starts the water purification process.

Figure 14.2-47. Viewed from the pond surface, the eco-pool almost disappears.

14.3

Hot Tubs and Solar Power: A Marriage Made in Heaven

We discussed in Section 14.1 how the circulation system of a typical spa operates. What we did not discuss is the enormous amount of energy that is required to heat a spa to its normal operating temperature of 104°F (40°C). You may remember from Chapter 1.3 that thermal energy is often measured in British Thermal Units (BTU), from which we can calculate the amount of electrical energy required to heat a hot tub or spa. (This calculation can also be completed using the metric system of joules or calories.)

A small three-person spa may have a water capacity of approximately 260 gallons (≈1,000 liters) which will have a mass of 2,200 pounds (1,000 kg). If we assume the initial temperature of the water used to fill the spa to be approximately 50°F (10°C), then the amount of heat energy required can be calculated by multiplying the mass of the water by the differential between its cold and desired water temperatures (an additional amount for efficiency factor and heating losses is also required, but this "efficiency factor" will depend on the boiler or heater system):

2,200 pounds of water x 54°F heat rise
= BTU of energy required
= 119,000 BTU

From Appendix 1, we can find the amount of energy contained in various energy sources. For example, each kWh of electricity has sufficient energy to provide 3,413 BTU of heat:

119,000 BTU to heat spa
÷ 3,413 BTU / kWh of electricity
= kWh of electricity required
= 34.8 kWh of electricity

This amount of energy is more than that required to run the typical off-grid household and approximately $30,000 worth of off-grid electrical generating equipment would be required. Furthermore, this calculation neither includes the energy required to circulate the spa water nor factors in the heating losses of the system. (The day-to-day average operating energy would be considerably less than the initial heat-up energy, but it would nonetheless be a considerable amount.) The point of this exercise is simply to say that heating a spa with off-grid-generated electricity is simply not in your best financial interests.

There are several alternatives to using electricity to heat a hot tub unit. The simplest and most common is to purchase a propane pool/spa heater (Figure 14-3.2). Based on the energy content of propane, approximately 1 quart (1 liter) will provide sufficient heat (23,000 BTU) for the spa discussed above, with less required to maintain the water temperature on a day-to-day basis. Of course propane is not a renewable fuel, so let's examine other options that use renewable energy heating methods.

Figure 14.3-1. If you are going to consider an electrically heated hot tub for off-grid use, consider Softub models such as the one pictured here. These units are manufactured with pool-grade vinyl and are heavily insulated with a closed-cell foam material. The hot tub captures the waste heat of the electric circulating pump and transfers this energy into the water. Although the Softub units require approximately the same amount of energy as other well-insulated units, they tend to be smaller (requiring less water and heating energy) as well as being more energy efficient. (Courtesy Softub Canada)

Figure 14.3-2. There are several alternatives to using electricity to heat a hot tub unit. The simplest and most common is to purchase a propane pool/spa heater.

Wood-Heated Hot Tubs

Wood-fired hot tubs are manufactured both by Snorkel and by Madawaska Millworks and are designed for use where electrical energy is either at a premium or nonexistent, making them perfect for the cottage or for occasional use. Many are designed for fill-use-drain operation where water is pumped from a lake, heated, and drained from the unit after a few days' use. Another big advantage of wood-firing is that the water heats very rapidly.

Wood is an environmentally sound energy source for hot tubs as the burner units consume enormous amounts of air, resulting in clean and low-particulate smoke output. One person who regularly uses one on winter trips to his cabin draws lake water from a hole drilled in the ice, gets the fire rolling, and hops in, all within just under two hours. When the fun is over, he pulls the plug and the water drains directly on the ground—a system that integrates very well with nature.

Chemical sterilizers can be used with these models, although the lack of a circulation pump and filtration system makes water purification very tricky. If you decide to add a filtration

Figure 14.3-3. The wood-heated hot tub may remind people of Gilligan's Island and the ever-present headhunters; however, these units are very enjoyable and extremely fast heating, which is a blessing for weekend or cottage use. (Courtesy Snorkel Stove Company)

Figure 14.3-4. Wood-heated hot tubs are perfect for the cottage or for occasional use. Many are designed for fill-use-drain operation where water is pumped from a lake, heated, and drained from the unit after a few days of use, possibly eliminating the need for chemical sanitizers. (Courtesy Madawaska Millworks)

system, ensure that the plumbing and electrical components are installed in accordance with electrical standards, that a ground fault circuit interrupter (GFCI) is connected to the electrical mains connection (yes, even off-grid homes require these safety devices), and that your electrical inspector approves the installation.

Solar-Heated Hot Tubs

Robert (Bob) Owens is a self-described solar-tinkerer who decided that if a solar thermal panel or two could provide enough hot water for the family home in Bradenton, Florida, he could probably heat his spa with the same sort of system.

Bob installed two flat plate solar thermal panels on the roof of his house and plumbed them into the standard spa circulation lines. A plywood box/seat was built to contain the temperature differential control and hot water circulation pump (Figure 14.3-7). Referring back to Chapter 5.4, you will recall that a typical solar thermal system operates on the basis of an electronic control module that measures the outlet or "hot side" of a solar thermal panel and compares this temperature to the coldest point in the circulation system, typically at the pump intake. When the temperature differential is greater than a preset number of degrees (typically 10°F or 6°C) *and* the spa temperature is lower than a preset upper limit (typically between 100°F and 104°F/38°C and 40°C for spas), the circulation pump activates, forcing spa water into the solar collector panel, extracting heat, and returning warm water to the spa.

Temperature measurement requires the use of devices known as "thermistors" which change their electrical resistance as a function of change in temperature. (Figure 14.3-8). Bob did not want to drill holes to mount the thermistors directly in the water path, so he opted to glue them to the side of the spa surface after carefully removing a patch of insulation (Figure 14.3-9). Once installed, the insulation was reapplied and the system was ready to run.

Figure 14.3-5. Robert (Bob) Owens of Florida decided that there was no need to pay to heat his spa when a little bit of ingenuity and solar thermal technology could do it for him. Here Bob toasts the successful marriage of sun and water. Note the water "spout" to Bob's right, which returns hot water to the spa from the solar thermal system. (Courtesy Robert Owens)

Figure 14.3-6. Using two flat plate solar collectors such as these, Bob is able to heat his large spa from a cold fill in less than three days. Requiring just a few running hours per week, the system maintains the desired water temperature of 100°F (38°C).

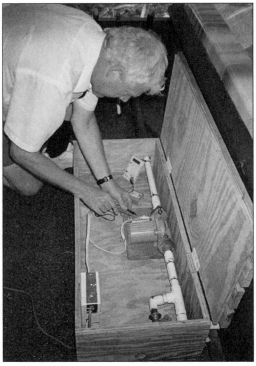

Figure 14.3-7. Unfortunately, commercial spa and hot tub manufacturers don't see the beauty or environmental advantage of running their units from solar energy. Bob Owens decided to take things into his own hands and make the necessary modifications himself. (Courtesy Robert Owens)

Figure 14.3-9. One thermistor is mounted on the "cold" point, which corresponds to the water temperature in the spa. The other sensor measures the temperature of the "hot side" or output of the solar thermal collector. (Courtesy Robert Owens)

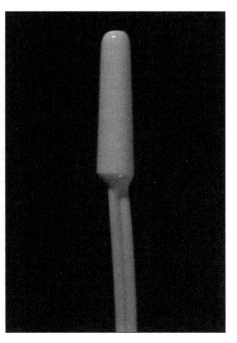

Figure 14.3-8. A thermistor device such as this model measures water temperature and sends an electric signal to a control module. When the spa temperature is below the desired setpoint and the solar thermal panels are a few degrees warmer than the spa water, the control module activates the pump. (Courtesy Robert Owens)

Regardless of what fuel you use to heat your hot tub, there is no point in wasting it. A solid thermal cover such as the model in Figure 14.3-10 will save at least 75% of your heating energy and considerably more in winter. Spa covers also provide a degree of safety, preventing small children from entering the spa while unattended. Many models come with locking hardware that makes unauthorized removal of the cover virtually impossible.

Figure 14.3-10. Regardless of what fuel you use to heat your hot tub, there is no point in wasting it. A solid thermal cover such as the model pictured will save at least 75% of your heating energy and considerably more in winter.

Figure 14.3-11. Unlike a hot tub, a sauna does not have any way of buffering or storing heat from one use to the next. (Courtesy Madawaska Millworks)

Figure 14.3-12. Since saunas are usually heated electrically, these models are not usually appropriate for off-grid living. This coopered barrel design can be supplied with an electric heater (which could be powered with a biodiesel or high-concentration ethanol gasoline generator to maximize environmental sustainability) or with a traditional wood stove heater operating from renewable energy sources. (Courtesy Madawaska Millworks)

Chapter 15
COMMUNICATIONS BEYOND THE "END OF THE LINE"

Living free of the electricity grid is extremely gratifying, but at the same time communicating with the outside world can be a challenge. This is because the utility poles that normally bring electricity into the home usually supply other services, including cable television and telephone. Fortunately there is no need to worry. Just because the telephone poles don't quite make it to your home you don't have to remain disconnected from the rest of society. Although your first reaction might be to call your local phone company to run the lines into your home, there are other options for you to consider.

If the phone lines don't reach your house and you want the convenience of telephone, Internet, or email service, the number of solutions on the market is growing by leaps and bounds. But do they really work as well as the slick ads and Web sites suggest? The technological forest of communications options requires a bit of thinning to find out what really works, what doesn't, and if these "solutions" are within the budget of real people.

The following discussion provides an overview of various methods of communicating over the last few miles (or more) of the rural "wire chasm" where the phone company may not extend service and also reviews the following alternate communication technologies:

- cellular phone service
- whole home (fixed) cell service
- point-to-point phone service extender
- fixed-point, broadband Code Division Multiple Access (CDMA) (digital) wireless service (provided by the phone company)
- radiotelephone service
- satellite phone service
- wide-area wireless high-speed Internet service
- satellite Internet service
- Voice over Internet Protocol (VoIP) service

Cellular Phone Service
With the spread of cellular service across North America, this should be the second option to consider after standard wireline telephone service. Most cellular providers try to "optimize" their investment in infrastructure, so they tend to concentrate service in high-density urban areas and along major highway corridors. Providing

Figure 15-1. With the spread of cellular phone service across North America, rural homeowners should consider this their second option after standard wireline telephone service.

cell service to a few dozen rural folks is clearly not a priority. However, cellular coverage is beginning to blanket the continent, and with the use of high-gain directional antennas service coverage may be extended well beyond what the carrier's "coverage maps" actually indicate.

When cellular (cell) service was first introduced, it was based on analog technology which required high-power and heavy transmitters to supply the desired 3W of radio signal. These early units were typically installed in automobiles and came equipped with a fixed antenna on the roof or window. Later models were introduced as a "bag phone" which was also equipped with a small antenna, battery, and charging device.

Figure 15-2. As cell technology improved and became ubiquitous, service providers began the transition to low-power digital handheld phones with limited transmitting range. High-frequency digital systems allow higher call density per cell transceiver tower as well as expanded features such as voicemail, email, and data transmission. Unfortunately, low-density rural would-be subscribers are often left out of the cell game.

As cell technology improved and became ubiquitous, service providers began the transition to low-power digital handheld phones with limited transmitting range. Since most users live in high-density urban areas and desire small, portable units with long battery life, limited radio signal strength is not a concern. High-frequency digital systems also allow higher call density per cell transceiver tower as

well as expanded features such as voicemail, email, and data transmission. Unfortunately, low-density rural would-be subscribers are often left out of the cell game.

The greater the power output of your phone, the more likely it is that you'll be able to link up with a cellular antenna tower in your area. Your first task will be to find which cellular provider offers the best service in your area and then try and track down the highest wattage phone, amplifier, and high-gain antenna you can. You may wish to inquire about older analog transceivers. Low-frequency high-power analog units have greater range than their higher-frequency, low-power digital counterparts. Keep in mind that the US Federal Communications Commission (FCC) is no longer requiring cell service providers to upgrade or even maintain analog service as the drive to a digital world continues.

Cellular service comprises a multiplicity of transceiver and tower units sprinkled across the countryside to provide service to a series of coverage areas or "cells." As you travel along a highway, you may notice your phone service becoming weaker and suddenly strong again. This is the process of one cell automatically "handing over" the call to the next cell.

In a rural location, you may find that some or all of your calls are long distance. This is because you may be quite a distance from a cell tower and the serving areas do not match those of the standard wireline service providers. This may also occur if your home is within range of more than one cell tower, in which case some calls are billed as local calls while others are charged a higher "roaming" rate.

This occurs as a result of the system scanning for the nearest cell tower that has both the capacity and signal strength to handle your call. Each time you call you may get bumped from one cell tower to another, as if the calls have originated from different locations.

Cell phones are normally equipped with omni-directional antennas that broadcast their signals equally and in all directions. A directional antenna, commonly called a "yagi" (Figure 15-5), can be mounted on any convenient elevated tower or on the side of a building. Once connected to the antenna jack of a cell phone unit, it will direct the transmission directly to the desired service provider cell tower, eliminating "tower bouncing." In addition to providing a directional signal, yagis also increase signal strength or "signal gain" by focusing the majority of the transmitted signal in one direction, thus providing improved clarity and transmission range. It is also possible to ask your service provider to select your local calling area to prevent unnecessary roaming charges.

Whole Home (Fixed) Cell Service

As an alternative to using portable cellular phone devices such as those shown in Figures 15-1 and 15-2, consider installing a whole home or fixed cell system such as the model shown in Figure 15-3. The Phonecell® SX5e can be thought of as a wireless version of a regular household phone system. The off-grid home is wired for telephone

Figure 15-3. As an alternative to using portable cellular phone devices, consider installing a whole home or fixed cell system such as the Phonecell® SX5e. It can be thought of as a wireless version of a regular household phone system. (Courtesy Telular Corporation)

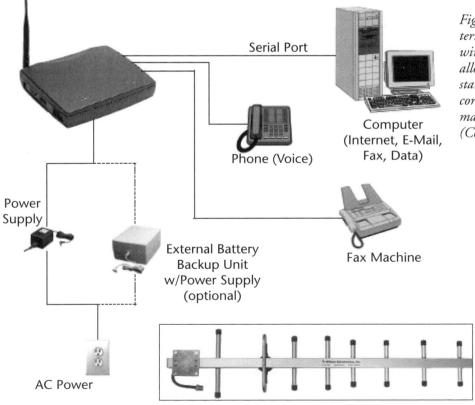

Figure 15-4. The Phonecell® terminal simulates the normal wireline phone service operation, allowing it to connect with standard house phones (including cordless models) as well as fax machines and the Internet. (Courtesy Telular Corporation)

Figure 15-5. A directional or "yagi" antenna will increase radio transmission range by directing the signal directly to the distant cell tower system. Directional antennas must be elevated and may be mounted on any convenient tower or on the side of a building structure. (Courtesy Wilson Electronics Inc.)

service in the normal manner, but rather than being connected to a landline system the wires are routed to the Phonecell® terminal.

The terminal simulates the normal wireline phone service operation, allowing it to connect with standard house phones (including cordless models) as well as fax machines. The device also provides Internet and email connectivity using the digital cellular phone system. The internal 2W radio transmitter and optional external directional antenna ensure maximum signal strength.

Costs for fixed cell terminals are in the range of $500 plus external antennas and installation (if required). Operating costs are the same as for regular cell phones, with many service providers offering "family plans" or bundles where a group of cell phones share a pool of minutes at a reduced rate

Point-to-Point Phone Service Extender

If you live in an area not covered by cellular service and you have access to a phone line within 20 miles (32 kilometers) of your home, you may want to consider a point-to-point phone service extender. This system will involve positioning a device at the end of the distant phone line that converts the telephone calls into a radio signal. A similar device located at your home will convert the radio signal back to a phone signal that is compatible with regular phone, fax, and computer modem connections. You will in fact be using two radio frequencies (duplex): one to send information (voice or data) and a second to receive it.

This type of system will require a broadcasting license since radio channels are considered public property. A yearly fee is collected by the Federal Communications Commission (FCC) in the United States and Industry Canada for the use of this radio "airspace."

Point-to-point service extenders have advantages and disadvantages over other communications technologies:

Advantages

- These systems mimic a real phone line, so you will be able to use your fax machine and

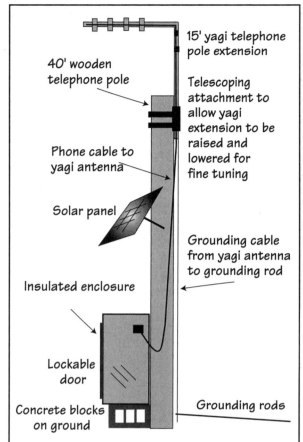

Figure 15-6. If you live in an area not covered by cellular service and have access to a phone line within 20 miles (32 kilometers) of your home, you may want to consider a point-to-point phone service extender. This system requires positioning a device at the end of a distant phone line that converts telephone calls into a radio signal. A similar device located at your home converts the radio signal back to a phone signal that is compatible with regular phone, fax, and computer modem connections.

computer modem to log onto the Internet and use your email.
- Local calls are not billed by time as they are with cell phones.
- You maintain the same access to long-distance service providers that you would normally have with a regular landline telephone.

Disadvantages

- The purchase price of these systems can be significant. Systems typically cost between $2,000 and $5,000+ to install. If you are running a business from your home and can amortize this over time, it may not be a concern, but it can hurt the wallet of mere mortals.

- The "bandwidth" on some (typically low-frequency) systems will be limited, which means that Internet service will be slower than with regular landline dialup service. If you want to download large files, these systems can be prohibitively slow.

- As you are using radio frequencies, connection and audio quality may be subject to the whims of the weather.

- You will need a location to install the transmission box and antenna at the distant end of the phone line. Your local phone company may allow you to use one of their poles, but it may not be able to assist in all instances. If the end-of-the-line telephone poles are on private property, you may be able to persuade the landowners to allow you to use their existing pole and grid electricity from an existing service to operate the radio transceiver unit. Alternatively, you may use a photovoltaic panel and battery bank to operate the unit.

- Depending on system frequency you may be subject to radio license fees.

Installation Issues

Point-to-point phone service extender systems will require an antenna similar to the model shown in Figure 15-5. All makes of antenna should clear the tree line and be able to "see" the distant antenna located at the wireline end of the system. The higher the frequency the more important it becomes to have line of sight between the two points, as signals become attenuated (diminished) by radio signal absorption from trees or other features of the landscape. Low-frequency units are affected less by local landscape issues and may be the only system that will work in a given area.

Figure 15-7. If the end-of-the-line telephone poles are on private property, you may be able persuade the landowners to allow you to use their existing utility pole and grid-supplied power to operate the radio transceiver unit. A bottle of wine at Christmas will more than cover the electricity charge. Alternatively, you can use a photovoltaic panel and battery bank as seen in this photograph.

Figure 15-8. Phone line extenders require a radio transceiver box, battery, charger, and antenna at the end-of-the-line phone service. Your local phone company may allow you to install the transceiver unit on one of its poles, with the electrical utility providing power to operate the unit.

Insulated enclosure for transmit unit and solar power/batteries for Optaphone

*Inside dimensions are 28" wide, 20" tall, and 9" deep.
Back panel should allow mounting of voltage regulator.
Unit has a locking handle with door to prevent vandalism.
Unit must be mouse/animal proof.*

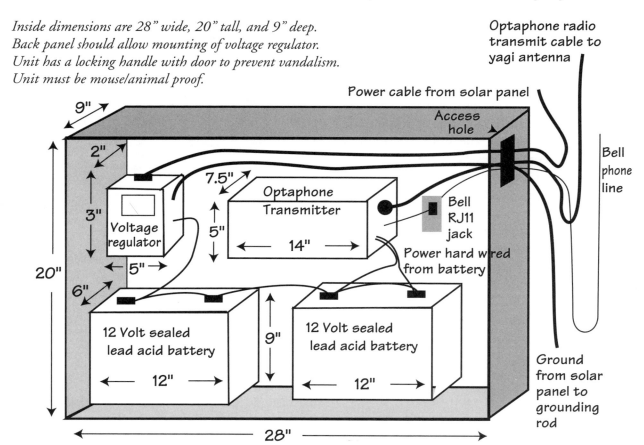

Figure 15-9. Mechanical view of the insulated enclosure for the radio transceiver unit and PV-charged battery pack.

At the home end, the transceiver base station is powered just like any AC or DC appliance. (This is an excellent example of a necessary phantom load.) As telephone service is normally desired on a 24/7 basis, it is wise to connect the base station directly to the battery bank through a dedicated DC circuit at the appropriate voltage. If the transceiver is not designed for direct connection to the house battery bank, a DC to DC voltage converter (see the final section of Chapter 11) may be installed to power the unit directly via its DC voltage input connection. Alternatively, a small phantom load inverter or even the regular house inverter can be used to power the unit.

At the remote location you will have to deal with how to provide power to the unit. If the unit is located at a neighbor's home, power should not be a problem provided a power line extension from their existing service is allowed.

If you decide to locate the unit in a remote utility box, chances are that you can purchase power from your local electrical utility, since where there are phone lines there *may* be electricity as well. You will have to contact the electric utility and arrange to pay the expense for this new service, requiring a meter and service outlet which could be similar to the cost of having electricity brought into a home. A drawback to this system is that phone service will also be affected by power outages and radio service interruptions.

An alternative to powering the remote phone transceiver with grid power is to make your own electricity, just as you are doing at your home. The communication company that sells you the phone system should be able to help you with this. In the installation shown in Figures 15-6 through 15-10, power is provided by one small photovoltaic panel sized to accommodate the system energy requirements. In this example, PV-supplied electricity charges two gel-cell sealed lead acid batteries. The batteries are surrounded with closed-cell insulation which allows them to remain as warm as possible.

This installation depends on the system being able to store enough DC power in the batteries

Figure 15-10. An alternative to powering the remote phone transceiver with grid power is to make your own electricity, just as you are doing at your home. A small photovoltaic panel and suitable battery bank will easily do the job. Your local radio system provider will be able to provide the necessary design (and possibly components) for this project.

to run the phone for four hours per day for two weeks with no sun, especially during the cold winter months (when the batteries will be cold and not perform as well). During the six years that this system has been in operation, sometimes well in excess of four hours per day, there has always been enough power.

The utility box also contains a series-type solar charger (Chapter 10) similar to those used in a home as well as the service box where the phone company supplies its connection. The directional yagi antenna pole is grounded to two 8-ft grounding rods, providing necessary lightning protection. The three lines that come into the box (the phone line, the power line from the solar panel, and the antenna feed) all form a "U" or drip line before they enter the box. This allows water running down the lines during a rainstorm to "drip" off rather than run into the electrical equipment. Keeping water out of this utility box is absolutely essential.

High-Frequency Phone Service Extender Systems

High-frequency (2.5 GHz or 2,500 MHz) phone service extender systems are similar to the low-frequency (approximately 400 MHz) models described above. Operating at a higher frequency provides greater bandwidth and permits higher-speed Internet and facsimile data connection.

In the 2.5 GHz range, radio signals are more easily absorbed than with low-frequency models. Obstacles as small as raindrops can interfere with the signal, so it is imperative to have good "line of sight" between the transmitting and receiving antennas. This means that if you're at the top of one of the antennas you should be able to see the top of the other antenna with nothing impeding the view.

Remember that the system will only be able to provide Internet speeds as fast as that provided by the phone company wireline at the receiving end of the system. Rural dialup service is often much slower than it is in an urban area. The further you are from a major telephone trunk line, the slower Internet access will be.

Fixed-Point Broadband CDMA (Digital) Wireless Service (provided by the phone company)

Your local phone company may actually offer a service similar to the phone extender technology described above but capable of servicing multiple customers. The advantage to the phone company is that it will save the cost of stringing copper phone cable along "the last mile" junction between high-speed phone and data trunk lines and the rural home. As the cost of CDMA wireless technology continues to drop and the cost of installing utility poles increases, you may find your phone company receptive to installing a system for you.

Radiotelephone Service

Yet another technology to consider is phone-company-supplied radiotelephone service, a glorified walkie-talkie system that allows connection between remote sites and a powerful radio tower operated by the phone company. These systems fall into two categories: operator-assisted or automatic.

Manual or operator-assisted systems require the remote location to press a "call" button, which alerts a mobile operator to acknowledge the request. As in the old days, the caller gives the desired number to the operator, who completes the connection. Automatic systems replace the operator with a touch-tone keypad, allowing direct access to the desired number.

In either system, the radio frequency is shared by a number of users in a county-wide party line system. Although radio reception quality ranges from "OK" to "Would you repeat that please?", for many people this may be the only economic means of communication.

Satellite Phone Service

If your dream home is in the wilds of Alaska, one further communication system to consider is the satellite phone. You may already be aware of these units installed in the seatbacks of planes or used by journalists covering news stories from remote reaches of the globe.

These remarkable phone units are capable of operating anywhere in the world and provide email and Internet connection as well. Another amazing feature of these systems is their ability to drain your bank account. One model was recently priced at US$700. The airtime fees range from $2 per minute for a 50-minute block to $6,000 per year for the 10-hour "economy" package.

I'm sure Bill Gates has a couple of these babies lying around the yacht!

Wide Area Wireless High-Speed Internet Service

With the demand for high-speed Internet access booming, many technology companies are scrambling to find ways to bridge the "last mile." This term was coined to reflect the difficulty of offering high-speed, or broadband, Internet data service using traditional telephone wires, many of which were designed over 100 years ago simply to carry the sound of a human voice.

Over the years, the major telephone companies have been upgrading their line circuits from copper wires to glass optical fibers and digital laser technology. Pumping light down an optical fiber is not only less expensive than using traditional copper, it is faster and less subject to signal degradation. Although the optical fiber network stretches across vast miles of land and sea, where it doesn't go is what counts to most of us: the last mile.

The last mile refers to the distance from your home to the "local central office" (often just a box along the highway), which contains the termination point between the high-speed optical fiber and the low-speed copper wire lines serving your house. The cost to upgrade the last mile to high-speed service is prohibitive.

Cable TV companies have already bridged the last mile, offering Internet service that dovetails nicely with their primary business of delivering television programming, which requires data bandwidth similar to that required for Internet data. Unfortunately for rural off-grid dwellers, they don't get cable service either.

Recognizing the last-mile dilemma, companies have developed wide-area wireless high-speed Internet that is designed to bridge the last mile—and then some. If you're lucky enough to be in the "and then some" category, you're in luck.

Wide-area wireless service requires the service provider to install a communication antenna and radio on a high point in the local area, usually a municipal water tower. Anyone within visual line of sight of the tower (and within a given signal propagation distance) can erect a small, low-cost

Figure 15-11. If your dream home is in the wilds of Alaska or beyond the range of cellular or extender technology, one further communication system to consider is the satellite phone. These remarkable phone units are capable of operating anywhere in the world and provide Internet connection as well. (Courtesy Globalstar USA, LLC)

Figure 15-12. With the demand for high-speed Internet access booming, many technology companies are scrambling to find ways to bridge the "last mile" between high-speed data trunks and your home. If you are lucky enough to be within the broadcast area of high-speed wireless service, you may have solved your phone and Internet problems in one shot. (Courtesy Storm Internet, www.storm.ca)

semi-parabolic dish such as the model shown in Figure 15-13.

The service provides exceptionally high-speed service (generally up to 3 megabits per second in both directions) for about the same cost as Internet over cable technology.

As the wave propagation is circular, signal strength may allow reliable connection within a 6-mile (10 km) radius or more from the transmitting tower.

Satellite Internet Service

Many rural homeowners receive outstanding television reception and channel selection through a satellite dish and receiver combination. Across North America, the cousin to satellite TV reception is a specialized high-speed satellite bi-directional Internet service provider. Most systems provide 0.5 megabits per second download speed with uploads operating at between 50 and 100 kilobits per second. This is about 15 times faster than dialup for the download direction and between 2 and 4 times faster during uploads.

Figure 15-13. Many rural homeowners receive outstanding television reception and channel selection through their satellite dish and receiver. Across North America, the cousin to satellite TV reception is a specialized high-speed satellite bi-directional service provider to keep you connected with the world.

The electrical requirements for the dish and support electronics are quite reasonable, typically in the order of 15W or about the same as a compact fluorescent lamp. This level of energy consumption fits well within the energy budget of most off-grid homes. As with any computer and monitor, ensure that the power supply feeding the dish is connected to a power bar, time-of-day timer, or outlet switch that will turn all the equipment off (no phantom loads) when your work is complete.

Voice over Internet Protocol (VoIP) Service

The concept of Voice over Internet Protocol (VoIP) is easy enough to understand. It's unraveling the hype from the facts that takes a bit of work. A microphone digitizes a person's voice, converting it into a series of digital data in much the same manner as compact discs or MP3 players convert music into digital data. Using a reasonably fast Internet connection, the voice data is transmitted to the receiving PC and converted back into sound so that it can be played over a sound card, dedicated Internet phone, or headset.

A system that relies on communication between two computers as in the above example is known as *peer-to-peer* communication. Currently, the hands-down winner in the race to replace the telephone with VoIP technology belongs to Skype (www.skype.com). This European-based company offers a free software package by the same name which allows free PC-to-PC calls to anywhere in the world. Skype is able to provide this software without charge because there is no required infrastructure other than the existing Internet connection, a PC, and either a headset and microphone or other speaker/microphone package.

But it doesn't stop there. Peer-to-peer calling is fine if both users are online and running the appropriate software. But what if you are in Vancouver and you want to call Mom who is holidaying in Florida? Skype is currently running another program known as SkypeOut that allows

Figure 15-14. Deseray O'Rielly makes Voice over Internet Protocol (VoIP) look like child's play. A system that relies on communication between two computers is known as peer-to-peer communication, with Skype (www.skype.com) the hands-down winner in the race to replace the telephone.

the calling PC to route the call through the traditional wireline phone system. Because Skype is "renting" the phone system to complete the call, users are charged a flat rate of approximately two cents per minute for calls to North America and Western Europe. Other areas are being added to the system, albeit at differing and higher rates.

Skype has also developed a product known as SkypeIn, which provides a phone number that people can call from any standard telephone. As with regular Skype, the long distance carrier is the Internet, allowing the company to offer this subscription service for a very reasonable fee of $5.00 per month for unlimited inbound calls in North America. (As of August 2009, SkypeIn is not available in Canada.)

Skype stresses that its service is not designed to replace a regular landline service, as it is unable to provide 911 emergency calling features.

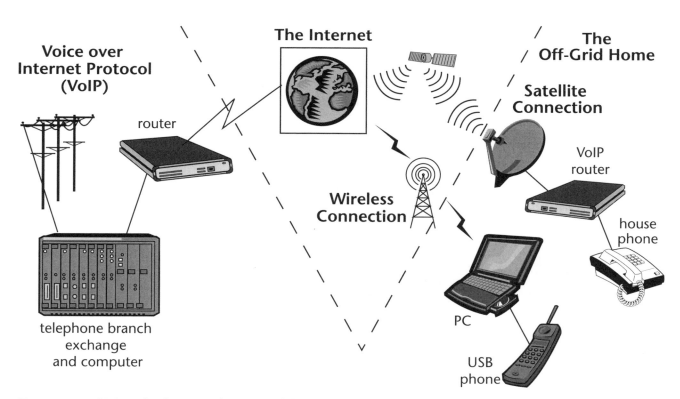

Figure 15-15. VoIP technology uses the power of the Internet to digitize the human voice in the same manner as MP3 music is recorded. The voice data is then routed in real time to another computer user (peer-to-peer) or through a service provider to any telephone on the planet.

Figure 15-16. Vonage offers a VoIP box which connects directly to the Internet via a cable/digital subscriber line (DSL) modem and router. This tiny unit then connects to the house phone line and provides an invisible connection to the outside world.

Other firms such as Vonage and Net2Phone provide VoIP services using a free interface box (Figure 15-16) which connects your house telephone directly to the Internet via a cable/DSL modem and router. Unlike Skype, the Vonage and Net2Phone units allow calls to be made with any touch-tone phone to any phone on Earth. The party you are calling doesn't need to use the VoIP service and does not require an Internet connection. If both parties use the VoIP service, there is no charge to either party.

Both companies have virtually duplicated the features of regular landline and cellular calling by providing a host of features such as "911" emergency service, call waiting, call forwarding, etc. And with low minute rates for long distance (non-Vonage/Net2Phone system calls), the off-grid homeowner will never need to be attached to society through a physical umbilical cord again.

Does VoIP Technology Work?

Because people have always assumed telephone service to be an essential part of life, it may seem strange to have to think about alternate ways of getting a dial tone into the off-grid home. But technology has a way of shaking things up, and VoIP technology is no different.

To test the suitability of VoIP technology in an off-grid home, I decided to recruit some serious technical help from a computer guru friend, Ken O'Rielly. Ken is a software/hardware developer at a technology company where he is affectionately known as "The IT Ninja." "It is pretty clear that testing VoIP technology for the off-grid home is going to be slightly different than for a typical urban home," remarked Ken, as he started thinking about the project. "Firstly, an off-grid home is not likely to have access to high-performance DSL or cable service. Secondly, if an off-grid home has a radio-based phone line extender or cell service, there is not much point in using VoIP technology since an extender would have insufficient speed and cell service would be more expensive than the VoIP technology it would interact with."

Ken decided that since some off-grid residents might have access to high-speed wireless service it would make sense to test this technology to provide a benchmark for sound quality and ease of operation.

"When using a peer-to-peer connection, regardless of which broadband technology was used the end result was that Skype and other VoIP technologies were no different than speaking with someone on a regular telephone," Ken explained. "I tried the technology with people all over the United States and Canada and found no flaws or problems with the technology. The USB phone that Bill supplied for testing was a nice touch. There were some compatibility issues with the computer motherboards I tested (or more correctly, with the USB chipsets used to communicate with the telephone). Windows was unable to recognize the device and considerable tinkering was required to get it going." Ken also noted that Skype works on both PCs and Macs.

When Ken was able to get the USB phone working, it did provide a sense of familiarity for non-technical people. It also provided a very clear voice for both users.

The story changed considerably when we switched the Internet link so that one user was on cable (urban) and the second user was on satellite high-speed (off grid). "This is a more realistic connection for people living off grid," Ken said. "The reason people are off grid in the first place is because of the distance from the power lines, and cable suppliers aren't likely to extend their lines for a few bucks per month."

Ken connected to Cam Mather, who lives off grid about 15 miles (24 km) from the nearest power line. Cam uses a satellite service from Xplornet (www.xplornet.com) which provides him with 500 kbps download and 100 kbps upload speed, using the large-format 39" (98 cm) dish on the Ku-band.

"While speaking with Cam, roughly 75% of the conversation I heard was perfect. The 25% that was unacceptable was either garbled or choppy and I had to constantly ask Cam to repeat himself," said Ken. "At Cam's end, I was coming in perfectly clearly and he had no problems at all. This makes sense because of the download speed of the satellite. One problem that did occur with the satellite system was an approximately three-second delay after one of us spoke before the other would hear it. We opted to play "trucker" and use the old CB radio slang, saying 'over' each time one of us finished talking."

Cam was unable to get SkypeOut operating at all, so we opted to test the Vonage technology and phone adaptor shown in Figure 15-16. The purpose of this test was to see if this North American technology would operate from Cam's satellite Internet system when he dialed out to a regular landline (see Figure 15-15 for details of Vonage connection to landline service). As Ken had experienced with the peer-to-peer technology, Cam sounded like R2-D2 about 25% of the time. On the other hand, Cam could hear me as plain as day.

Cam discussed this matter with Xplornet customer service, and after he had explained that he had the "upgraded" dish and enhanced data package the response finally was: "Unfortunately we don't support the Vonage VoIP because of the latency and upload speeds."

I repeated this same test with another off-gridder by the name of Terry Tufts. Terry and his wife Catherine live several miles from the nearest utility poles and are out of range of high-speed wireless Internet service. When faced with the dilemma of how to get phone and Internet service, they settled on Xplornet's Ka-band satellite service by Telesat. The Telesat Ka-band uses 15 regional "spot beams" to provide national broadband coverage. The effect of using spot beams is a more concentrated signal, which results in smaller dish sizes and affordable service package pricing for the end user.

Figure 15-17. VoIP technology has vastly improved over the past few years and it appears now that anyone living off-grid can still enjoy the benefits of telephone and Internet service from a variety of providers. Lorraine Kemp is seen in this photograph making a phone call using the Skype VoIP service. A USB phone is connected to an HP Netbook computer, which is in turn connected to the household wireless network. Internet service is provided by a 3-Mbps wide-area wireless broadband system, transmitting approximately 5 miles (8 km) to a repeater tower.

This time, the results were much more promising. Although the conversation was interrupted by latency (delay) issues, Terry and I were able to chat extensively over a period of days. Terry reports that the system continues to work well months after its initial installation. Although the service is not perfect, Terry explains that "it sure beats the hell out of being disconnected from the world."

Conclusion

If you are looking to move away from utility lines, make sure you have a phone strategy and ensure that it will work before you take the plunge. More and more people work out of their homes or are dependent on phone and Internet service for many of their daily activities such as banking and doing research. While the dream to "drop out" to Walden Pond remains an admirable one, most of us still require some interaction with the industrial economy, and the phone and Internet system is an integral part of that link.

Chapter 16
BIOFUELS

Zero-Carbon Liquid Fuels

While the world waits breathlessly for hydrogen-powered fuel cells, developments in liquid biofuels are continuing at a surprising, perhaps reckless rate aided by pork-barrel politics and a general lack of common sense.

Almost everyone has heard of ethanol and some enlightened souls may be familiar with biodiesel. While these technologies are not nearly as sexy as hydrogen and fuel cells, they are available now and can offer numerous advantages over oil while also making it unnecessary to replace trillions of dollars' worth of existing fuel infrastructure systems.

At the same time, if biofuel supplies and regulations are not properly developed, which appears to be the direction society is currently taking, they can also provide an equally bewildering series of problems ranging from increased carbon emissions to creating massive damage to the world's rain forests, not to mention wreaking havoc with food supplies and prices.

This rather extensive chapter will attempt to provide an overview of the issues, pro and con, related to biofuels. It will also show what place there is for a sustainable biofuel industry when it is used in conjunction with home heating, transportation and even off-grid power generation.

Finally, I will provide an overview of the small-scale production of zero-carbon biodiesel that can be used as a liquid fuel source in any of these applications.

An Introduction to Biofuels

Modern internal combustion engines including those used in transportation vehicles do not have to run on gasoline. In fact, early automotive pioneers did not have access to refined gasoline and consequently used peanut oil and alcohol for fuel. However, the discovery of large volumes of easy-to-access and cheap fossil fuels spelled the end of these early alternatives. Likewise, heating systems including boilers, hot water and forced air furnaces are equally indiscriminate when it comes to their choice of fuel sources.

Ethanol from Food – A Non-Starter

The consumption of fossilized fuels is not the only means of extracting energy from plant life. Fermentation of grapes and apples has been fuelling binges for as long as man can remember. The naturally occurring sugars in the fruit produce wine and cider with a maximum alcohol content of approximately 14%. Applying heat to these beverages, in a process known as distillation, allows extraction of the ethanol alcohol at up to 100% concentration.

Figure 16-1. Plant grains and fiber can be converted to sugar, fermented into ethyl alcohol (ethanol), and used as a blending ingredient with gasoline or as the main fuel. High concentrations of ethanol can reduce greenhouse gas emissions by up to 80% relative to gasoline. (Courtesy: Iogen Corporation)

The primary source of plant sugars can vary, as automotive-grade ethanol can be produced from food grains such as corn, wheat, and barley. Recent advances in enzymatic processes even allow the conversion of plant waste in the form of straw and agricultural residue into sugars which can in turn be fermented into ethanol.

Conventional grain-derived and cellulose-based ethanols (those made from farming and timber byproducts) are the same and can be easily integrated into the existing gasoline supply chain. Ethanol may be used as a blending agent or as the main fuel source. Gasoline blends with up to 10% ethanol can be used in any vehicle manufactured after 1970.[1] High-level ethanol concentrations of between 60% and 85% can be used in special "flex-fuel vehicles."[2] At the time of writing Ford and General Motors warranties allowed up to 10% ethanol blends in their North American vehicles and have begun to introduce models that operate on high-ethanol blends as well.[3]

Adding ethanol to gasoline increases octane, reducing engine knock and providing cleaner and more complete combustion, which is good for the environment. Ethanol produces lower greenhouse gas emissions than gasoline: a 10% ethanol blend with gasoline (known as E10) may reduce GHG emissions by 4% for grain-produced ethanol and 8% for cellulose-based feedstocks. At concentrations of E85, GHG emissions are reduced by up to 80% when using cellulosic ethanol.[4]

Figure 16-2. The majority of ethanol is derived from corn. The production of ethanol fuels strengthens agricultural regions and creates new and more stable markets for the corn-cropper provided that government subsidies remain in place. At the same time, market pressure for more ethanol plants has caused large swings in the price of corn and other grains, causing hardship in poor regions of Mexico as well as domestically. In addition, the environmental benefits of growing crops to produce road fuels are negligible at best. The long-term prospects for food-crop-based ethanol are very poor and will immediately fade as soon as carbon taxation starts or government subsidies end.

Proponents argue that ethanol is produced from domestic renewable agricultural resources, thereby reducing our dependence on imported oil and providing a major source of economic diversity for rural farming economies.

In the United States, 77 ethanol plants produce over 3.3 billion gallons (12.5 billion liters) of ethanol per year. Canadian production currently stands at 62.6 million US gallons (237 million liters) per year. Of the 77 plants in the United States, 62 use corn as the feedstock. The

Figure 16-3. North American ethanol production currently stands at 3.36 billion gallons (12.72 billion liters) and is expected to increase by 28% before the end of the decade. This is just a drop in the bucket compared with fossil-fuel gasoline usage, but with good demand management (through CAFE and properly managed carbon taxation) it could go a long way toward reducing greenhouse gas emissions. (Courtesy: Iogen Corporation)

remainder use a variety of seed corn, corn and barley, corn and beverage waste, cheese whey, brewery waste, corn and wheat starch, sugars, corn and milo, and potato waste. In the United States there are 55 proposed new plants and 11 currently under construction. Canada currently has 6 plants with 1 under construction and 8 new proposals on the drawing board.[5]

There is no question that using properly managed, domestically grown

At current rates of gasoline consumption it will be impossible to use ethanol as a replacement.

feedstocks from renewable fuel crops such as switchgrass or from waste biomass sources is better than importing fossil fuel from the Middle East. However, there is a great deal of controversy about diverting food grain stocks into fuel feedstocks.

At current rates of gasoline consumption it will be impossible to use ethanol as a full replacement for gasoline used in internal combustion engines; there is simply not enough farmland, irrigation water, or petrochemical feedstocks for fertilizer and pesticides available. Even if 100% of the US corn supply were distilled into ethanol, it would supply only a small fraction of total domestic gasoline demand. In his 2007 State of the Union address, President George Bush called for 20% biofuel content in gasoline and diesel fuel. This will require an eightfold increase in US biorefinery capacity and will supply approximately 15% of US motor fuel requirements. In addition, a massive increase in natural gas imports into the United States will be required to distil the corn into alcohol. North American natural gas supplies are in a free fall, so this option will be costly if not impossible.

According to Alberta economics professor Kurt Klein, "this level of biofuel production would be devastating for the livestock industry and the food supply. Costs of all forms of livestock feed would increase in price with a horrendous dislocation of agricultural production. There would be no more exports of corn and wheat from the U.S., and remember that Canada actually imports about 20 percent of its corn requirements from the U.S. now."[6]

I also worry about the long-term effects on the farming sector directly. During this massive boom in ethanol production, corncroppers are ramping up production as never before. This effort requires increased land use for monocropping corn, which comes at the expense of all other crops. Farmers must invest more aggressively in land leases and equipment purchases to ensure that they get their share of the rewards, all the while levering assets to make this happen.

During the boom cycle, ethanol and hence corn prices will rise, affecting everyone who buys corn and corn products. This is, indeed, everyone. The corn cropper is hurting his dairy neighbor down the road because protein meal prices are rising as a result of the increased pressure placed on all animal feeds. And everyone has no doubt heard of the "tortilla riots" in Mexico when tens of thousands of Mexicans swarmed the streets of Mexico City in a protest against the rising price of tortillas. These flat corn breads are the main source of calories for many of Mexico's poor, who saw prices rise by over 400% in early 2007 partially as a result of US-based ethanol production.

Figure 16-4. Iogen Corporation founder Brian Foody Sr. stands in a field of straw grass, which may become the twenty-first-century equivalent of a Saudi Arabian oil field. (Courtesy: Iogen Corporation)

Is it acceptable to cause financial hardship or starve people in distant lands in order to allow the middle and upper classes to drive around in their SUVs with a supposedly clear conscience?

And what happens when the subsidies stop flowing and people come to their senses? Corn ethanol will cease to exist as a transportation fuel and farmers and related suppliers will go bankrupt by the thousands.

Add to this the dubious greenhouse gas emission reductions and negligible forward energy-balance issues and food-grade ethanol will not be a viable long-term solution to our energy woes.

Cellulosic Ethanol – The Better Choice

Cellulose-based ethanol eliminates the diversion of food crop to fuel conversion and offers an advanced new transportation fuel feedstock which has some advantages over grain-based ethanol:

- Unlike grain-based ethanol production, the manufacturing process does not consume fossil fuels for distillation and thus reduces greenhouse gas emissions.
- Cellulose ethanol is derived from nonfood renewable sources such as straw and corn stover. Fast-growing perennial plants such as switchgrass can be grown with little assistance from petroleum-based pesticides and fertilizers and can be cultivated on marginal lands.
- There is a potential for large-scale production since it is made from agricultural residues which are produced in large quantities and would otherwise be destroyed by burning.

Cellulose-based ethanol has, in the past, been very expensive as a result of the inefficient processes required to produce it. The Iogen Corporation, working in conjunction with the Government of Canada and Shell Corporation, has recently launched its most cost-effective method for producing what the company refers to as EcoEthanol™. Once the first phase of production, using straw and corn stover, is under way it won't be long before other feedstocks can be used, including hay, fast-growing switchgrass, and wood-processing byproducts.

The Downside of Ethanol

There are very few downsides to ethanol as a fuel. Because ethanol contains oxygen it permits cleaner and more complete combustion. Ethanol contains approximately 47% less energy than gasoline on a volume basis. At high-concentration levels, it takes more ethanol than gasoline to drive a given distance; however, this will be much less of a problem when society begins driving highly efficient plug-in electric hybrid vehicles that use very little liquid fuel in normal driving. In any event, the current higher price per unit volume of ethanol compared to gasoline will eventually be offset by carbon valuation policies that will favor sustainably produced renewable fuels.

Biodiesel as a Source of Green Fuel

Internal combustion engines ignite fuel using one of two methods. A spark ignition engine produces power through the combustion of the gasoline and air mixture contained within the cylinders. An electric spark which jumps across the gap of an electrode ignites this volatile mixture.

The compression ignition or diesel engine, as it is more commonly known, uses heat developed during the compression cycle. (Have you ever noticed how hot a bicycle air pump becomes after a few strokes?) With a compression ratio

Figure 16-5. Loading straw into an ethanol plant is a lot cleaner and much more sustainable than extracting oil from the North Sea. Waste grasses, straw, and wood products are "carbon-neutral" fuels that are available domestically and in endless supply. Think of these sources as "stored sunshine." (Courtesy: Iogen Corporation)

Fuel Source	Volumetric Energy Density
Ethanol	89 MJ/Gallon 23.4 MJ/Liter
Gasoline	131 MJ/Gallon 34.6 MJ/Liter

Table 16-1. This table illustrates the volumetric energy density of ethanol compared to gasoline. A typical gasoline-powered car will travel approximately one-third further than a vehicle using the same volume of ethanol. The current higher price per unit volume of ethanol compared to gasoline will eventually be offset by carbon valuation policies that will favor sustainably produced renewable fuels.

of 18:1 or higher, sufficient heat is developed to cause diesel fuel sprayed into the cylinder to self-ignite. The higher energy content of diesel fuel and the oxygen-rich combustion process of the compression ignition engine contribute to improved fuel economy, power, and reduced CO_2 emissions, all desirable features for fleet and other high-mileage vehicles.

Biodiesel, like its cousin ethanol, is a domestic, relatively clean-burning renewable fuel source for diesel engines and oil-based heating appliances. It is derived from virgin or recycled vegetable oils and animal-fat residues from rendering and fish-processing facilities. Biodiesel is produced by chemically reacting these vegetable oils or animal fats with alcohol and a catalyst to produce compounds known as fatty acid methyl esters (FAME or biodiesel) and the coproduct of the chemical production process, glycerine.

> *Biodiesel yields 3.2 units of energy output for every unit of energy input.*

Biodiesel that is destined for use as transportation fuel must meet the requirements of the American Society of Testing and Materials (ASTM) Standard D6751 for pure or "neat" fuel graded B100.[7] Fossil-fuel diesel (or petrodiesel) must meet its own similar requirements within the ASTM standard. Biodiesel and petrodiesel can be blended at any desired rate, with a blend of 5% biodiesel and 95% petrodiesel denoted as B5, for example. The testing and certification of any transportation fuel is a requirement of automotive and engine manufacturers and is implemented to minimize the risk of damage and related warranty costs.

From an agricultural viewpoint, biodiesel offers many of the same advantages and disadvantages to the farming community as ethanol feedstocks. In addition, the process of making biodiesel is *relatively* simple and low cost, which may lead to cooperative and rural ownership of processing and production facilities, further increasing farming income and risk diversity. Even the coproduct of the biodiesel manufacturing process, glycerine, has a market in food, cosmetic, and other industries.

A primary advantage of the biodiesel production process is that two streams of product are generated from the oilseed input. After crushing, the protein meal is sold for livestock feed as well as all manner of products from tofu to baking additives. The oil stream can likewise be used in the food industry or provide the input feedstock for biodiesel.

From an energy and therefore carbon balance standpoint, biodiesel produced from soybeans is superior to corn-based ethanol, yielding 3.2 units of energy output for every unit of energy used in the growing and production cycle.[8]

In Canada, it is estimated that some 0.5 million tonnes of canola seed become distressed as a result of weather or storage problems. Although the oil that is extracted from these seeds is not suitable for a food product, there is no reason why it could not be used as a fuel input.

Likewise, the restaurant industry disposes of or recycles millions of tonnes of used cooking oils known as yellow grease, which can also be diverted into the biodiesel production process as is beginning to happen in North America.

At the back end of the restaurant industry are millions of grease traps that capture cooking oils and food products contained in the wash water. Rather than allowing them to travel into municipal sewage facilities, these grease traps are evacuated by

Figure 16-6. Biodiesel fuel, a renewable, clean-burning fuel source for diesel engines and oil-based heating appliances, is produced from virgin vegetable oils as well as waste vegetable or animal fats. Dr. Martin Reaney, Chair of Lipid Utilization and Quality at the University of Saskatchewan, studies the canola oilseed plant used extensively in the production of biodiesel. He is also researching ways of using so-called "industrial mustards" that may produce high oil yields and grow on marginal soils.

recovery companies and the grease is disposed of. The extracted effluent is known as brown grease and has a very high energy content. Because of nonexistent carbon regulations, this material has traditionally been landfilled, allowing the material to decay and release immense quantities of carbon dioxide and methane into the atmosphere. Current environmental regulations are tightening up and are beginning to prevent the material from being landfilled. Studies have shown that this material can be converted into biodiesel, although it would not be competitively priced with gasoline. Allow carbon taxation to step in and the inequity in pricing would be solved in an instant.

Figure 16-7. Every year approximately 5% or more of the total Canadian canola crop is declared "off-specification" and in effect wasted. Generating a special commodity price for "fuel-grade" canola (or any oilseed) would put more money into farmers' pockets and help alleviate the losses incurred in a horticultural roll of the dice with nature.

Figure 16-8. As the price of petrodiesel continues to rise and biodiesel production costs fall, rural communities will produce their own democratic energy to fuel their part of the economy. (Courtesy: Lyle Estill/Piedmont Biofuels)

Figure 16-9. The simplicity of a biodiesel production facility is shown in this aerial view of a continuous-production plant. Soybeans are delivered to the processing plant for conversion to food-grade oil. Alcohol and a catalyst are added to the oil to produce biodiesel and glycerine. The biodiesel is stored in the tank farm. The glycerine byproduct is sold and the process water and alcohol are recycled, creating an environmentally friendly processing cycle, with the biodiesel having a much higher energy balance than corn-based ethanol. (Courtesy: West Central)

Biodiesel in the Transportation Sector

The modern diesel engine is a far cry from the smoky, anaemic model of the 1970s. Fast forward to today. Gone are the smelly, smoky, lumbering diesels of old. Witness the new Mercedes E320 family of "common rail, turbo-diesel" engines that offer no "dieseling" noise, smoke, or vibration, achieve superb mileage, and have better acceleration than the same model car equipped with a gasoline engine.[9]

Biodiesel offers some distinct advantages as an automotive fuel:[10]

- It can be substituted (according to vehicle manufacturer blending limits) for diesel fuel in all modern automobiles. Although B100 may cause failure of fuel system components such as hoses, o-rings, and gaskets that are made with natural rubber, most manufacturers stopped using natural rubber in favour of synthetic materials in the early 1990s. According to the US Department of Energy, B20 blends minimize all of these problems.
- Performance is not compromised using biodiesel. According to a 3 ½-year test conducted by the US Department of Energy in 1998, using low blends of canola-based biodiesel provides a small increase in fuel economy. Numerous lab and field trials have shown that biodiesel offers the same horsepower, torque, and haulage rates as petrodiesel.
- Lubricity (the capacity to reduce engine wear from friction) is considerably higher with biodiesel. Even at very low concentration levels, lubricity is markedly improved. Reductions in the sulfur levels of petrodiesel to meet new, stringent emissions regulations have, at the same time, reduced lubricity levels in petrodiesel dramatically. Biodiesel blending is currently being considered by the petrodiesel industry as a means of circumventing this problem.
- Because of biodiesel's higher cetane (ignition) rating, engine noise and ignition knocking (the broken motor sound when older diesels are idling) are reduced.
- Carbon lifecycle emissions can be dramatically reduced or even eliminated based on careful selection of oil feedstocks and process chemicals.

Biodiesel Composition

The concept of using plant matter to operate internal combustion engines is older than the gasoline and diesel fuels that are so ubiquitous in our lives today. Rudolf Diesel developed the *compression ignition* engine and demonstrated it at the Paris World Exhibition in 1900. His fuel of choice for powering the new engine: peanut oil.

Although the concept of using plant matter to operate internal combustion engines has been revisited numerous times since Diesel's early experiments, the discovery of cheap fossil oils delayed any significant development of biofuels.

The development of the diesel engine and fuel system progressed very quickly after its first demonstrations to the public owing to its increased efficiency compared with that of the steam engine, its relative portability (paving the way for personal

Figure 16-10. Over 200 public and private fleets in the United States and Canada currently use biodiesel, and the number is increasing rapidly. Environmental stewardship regarding climate change, urban smog, and air quality is creating mass-market acceptance of biodiesel fuel. (Courtesy: Saskatchewan Canola Development Commission)

transportation, farming, and industrial uses), and access to cheap and convenient diesel fuel oil. Engine development continued for the next 80 years using low viscosity petrodiesel fuel, while the much higher viscosity plant oils were left behind on the grocers' shelves for baking, salad dressing, and French fries.

With the first worldwide oil "shortages" in the 1970s, researchers began working in earnest in an attempt to develop the biofuel market. The many shortcomings related to the direct use of plant oils and their **total** incompatibility with petrodiesel fuel and existing distribution infrastructure[11] pushed researchers in the direction of chemically modified forms of plant oils and animal fats known as biodiesel.

Biodiesel is a renewable, relatively clean-burning, carbon-neutral fuel that can be obtained from a variety of oilseed plants, waste oils, and rendered animal fats. These unprocessed materials (collectively referred to as feedstock "oils") can be converted into a petrodiesel-compatible fuel using a process known as chemical transesterification.

The properties of rendered animal fats and plant oil vary widely from those of petroleum diesel fuel, primarily in the areas of *viscosity*, *atomization*, and the *coking* of engine components. All plant and animal oils have essentially the same chemical structure, consisting of *triglycerides*, which are chemical compounds formed from one molecule of glycerol and three fatty acids.

Glycerol (common name glycerin) is an alcohol that can combine with up to three fatty acids to form mono-, di-, and triglycerides.

Fatty acids are chains of *hydrocarbons* that vary in carbon length depending on the oil feedstock. If each carbon atom has two associated hydrogen atoms, the fatty acid is known to be *saturated*. If two carbon atoms are double bonded, having less hydrogen, the fatty acid is *unsaturated*. Likewise if more than two carbon atoms are unsaturated, the fatty acid is said to be *polyunsaturated*.

Triglycerides are the main compounds or components of animal fat and vegetable oils. They have a lower density than water and will therefore

Figure 16-11. The discovery of cheap fossil oils delayed any significant development of biofuels. Given the massive financial subsidies and military support lavished on the fossil fuel industry as well as the environmental damage it has caused, perhaps biofuels should have been with us from the beginning.

Figure 16-12. Biodiesel is a renewable, relatively clean-burning, carbon-neutral fuel that can be obtained from a variety of oilseed plants, waste oils, and rendered animal fats.

float on it. If the oil is solid at room temperature the triglycerides are known as "fats"; if they are liquid they are called "oils." As a general rule, triglycerides that are liquid at room temperature are unsaturated, which is a desirable property for both human consumption and engine fuels.

Plant oils have viscosities that can be as much as 20 times higher than that of fossil diesel fuel, while chicken fat, yellow grease, lard, and tallow remain stubbornly solid and unusable in their

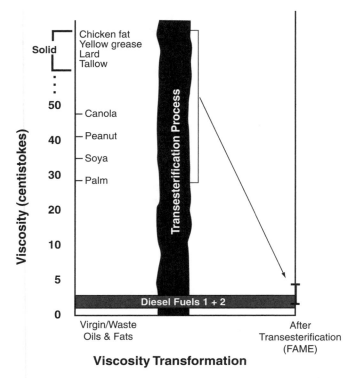

Viscosity Transformation

Figure 16-13. Plant oils have viscosities that can be as much as 20 times higher than that of fossil diesel fuel, while chicken fat, yellow grease, lard, and tallow remain stubbornly solid and unusable in their unaltered state. The problem of the high and variable viscosity of feedstock oils can be corrected by adapting the engine to the fuel or vice versa.

unaltered state. The problem of the high and variable viscosity of feedstock oils can be corrected by adapting the engine to the fuel or vice versa. This chapter focuses on the latter concept: adapting the fuel to the millions of engines, furnaces and diesel generators that are now operating or will be produced for many years to come.

Chemical transesterification of feedstock oils is a well-known process which solves the problem of feedstock viscosity as demonstrated in Figure 16-13. The process was first described in 1852[12] when it was originally used as a means of producing high-quality soaps, and with a bit of retooling it was found to work wonders in the production of biodiesel. Simply stated, biodiesel is produced by the reaction of feedstock oils with an alcohol in the presence of a catalyst to produce

fatty acid methyl esters (FAME) or biodiesel. The typical process is:

100 kg feedstock oil + 10 kg methanol
= 100 kg FAME + 10 kg glycerol

The resulting FAME is known to be chemically contaminated with numerous compounds resulting from the esterification process, requiring further downstream processing to ensure a fuel quality compatible with ASTM Standard D6751 for biodiesel, a fuel comprised of the mono-alkyl esters of fatty acids derived from vegetable oils or animal fats. Hereinafter, FAME that is directly taken from the reaction process will be referred to as "*raw FAME*" and when it meets the fuel quality standard it will be referred to as *FAME*.

The first reference to FAME production was in 1937,[13] and within the next year a bus fuelled with palm-oil-based biodiesel ran between Brussels and Louvain.[14] However, at that point further scientific research and production ground to a halt.

Figure 16-14. Professor Martin Mittelbach, Ph.D., University of Graz, Austria, and his team of researchers have long been acknowledged as the founders of the modern biodiesel industry. Dr. Mittelbach is the author of numerous technical papers on the subject and the author of Biodiesel: The Comprehensive Handbook.

Some forty years later, Professor Martin Mittelbach, Ph.D., University of Graz, Austria, and his team of researchers were producing rapeseed-oil-based biodiesel and testing its feasibility as a diesel fuel substitute.

When prodded to discuss the origins of the modern biodiesel industry, Dr. Mittelbach modestly admits that he has been "involved since the beginning. We (University of Graz) were the first to produce biodiesel in Europe, more than 20 years ago, although I am not exactly sure what moved me into this program!...Because of my research on carbon-based compounds, a discussion ensued with an agricultural group in Austria that had some experience using straight vegetable oils mixed with 50% fossil fuels in tractors. The farmers found that after a period of time running on this mixture, total engine breakdowns would occur and they had to stop this practice. We looked into the problems, examined the prior research, and I guess you could say the rest is history."[15]

Within the next decade hundreds of research programs sprang up around the world as interest in clean and renewable fuels started to take hold.

In the United States, the demand for soybean meal (the residual husk of the bean after crushing) was greater than the demand for oil, causing an imbalance in supply and depressed soy oil prices and galvanizing the United Soybean Board into action. It voted to promote biodiesel production using soy oil, which ultimately led to the current National Biodiesel Board (NBB).

The Canadian Renewable Fuels Association (CRFA), which merged with the Biodiesel Association of Canada, develops market strategy and educational data for both the ethanol and the biodiesel industries.

As a result of lobbying efforts and continued research, biodiesel has received the support of numerous federal, state, provincial, and local governments which see it as a means of reducing greenhouse gas and smog-forming emissions, supporting local agribusiness, and helping to reduce North American dependency on foreign oil.

The Pros of Biodiesel

Blending

One primary advantage of biodiesel is its ability to fit almost seamlessly into the existing fuel distribution and retail sales system while other alternative "fuels" such as hydrogen require the complete rebuilding of distribution technology at a cost of trillions of dollars.

Biodiesel can be used in all modern diesel engines and oil-fired heating systems with minor (if any) modifications, with the following note of caution: FAME *may* cause long-term degradation of natural rubber hoses and gaskets and some paints, and replacing natural rubber hoses, "O" rings, and gaskets with polymeric (synthetic) versions such as Viton® may be necessary. However, experience has shown that blends of 20% biodiesel or less seldom cause any problems at all, and in any event late-model vehicles seldom use natural rubber components.

Biodiesel fuels are in commercial use in many European countries including Austria, the Czech Republic, Germany, France, Italy, and Spain. Biodiesel can be used in either its pure or neat form or as blends mixed with fossil diesel fuels.

Figure 16-15. A primary advantage of biodiesel is its ability to fit almost seamlessly into the existing fuel distribution and retail sales system, while other alternative "fuels" such as hydrogen require the complete rebuilding of distribution technology at a cost of trillions of dollars.

Neat biodiesel is designated B100, while blends are marked "BXX" where "XX" represents the percentage of biodiesel in the fuel mixture. Further, because biodiesel is stable at any concentration, users are free to choose the blending level they prefer based on availability, desired operating temperature, or price.[16]

Biodiesel Concentration

Germany, Austria, and Sweden market neat biodiesel, although blends of 5%-20% are the preferred concentration. The "European Directive for the Promotion of the Use of Biofuels," published in 2003, mandates that all member states ensure minimum market shares of biofuels (ethanol, ethanol derivatives, and biodiesel). Market share of biofuels is to reach 5.75% by 2010.[17]

For several reasons, such as the maximum available production of biodiesel, fuel quality and stability, and the political realities of displacing fossil diesel from the market, North American industry proponents consider a blend level of B20 to be the upper limit concentration. There is also a popular myth that 100% concentrations of biodiesel will damage engines, leading to expensive repairs. This is simply nonsense and is nothing more than posturing by engine and automobile manufacturers for two very obvious reasons:

Biodiesel is readily biodegradable and nontoxic

1. Approving a new fuel type will not increase vehicle sales but will add complexity to servicing and warranty issues. From the auto industry's point of view, why take on additional risk if there is no downstream financial benefit?
2. Approving the use of 100% biodiesel will give legitimacy to the multitude of biodiesel home-brewers, who might use substandard fuel in the vehicle.

Figure 16-16. Biodiesel is readily biodegradable and nontoxic, making it the ideal fuel choice when used in environmentally sensitive areas such as parklands, marine habitats, or ski resorts. It is known to be less toxic than table salt and is as biodegradable as sugar.

Biodegradability and Nontoxicity

Biodiesel is readily biodegradable and nontoxic, making it the ideal fuel choice when used in environmentally sensitive areas such as parklands or marine habitats. It is known to be less toxic than table salt and is as biodegradable as sugar.[18]

High Cetane Value

The cetane value is a rating of the relative ignition quality of diesel and biodiesel fuels, with higher ratings offering improved ignition performance. As the cetane value increases, fuel ignition will be smoother and more complete, improving combustion and reducing emissions from unburned fuel. Virtually all biodiesel fuels have cetane values several percentage points higher than that of petroleum diesel fuel.

High Lubricity

Biodiesel, which contains no sulfur, has excellent lubricating properties, far in excess of those of petrodiesel, which help to reduce fuel system and engine wear. As petroleum diesel fuel sulfur levels continue to be legislated downwards, its lubricity will decline to the point where additives will be necessary. The addition of 1% biodiesel to low-sulfur petrodiesel will improve the fuel blend lubricity to within specification.

Low Emissions

As a renewable fuel source, biodiesel operates on a closed-carbon cycle, which reduces CO_2 production by 2.2 kg for every liter of fossil fuel displaced.[19] This is because of the regenerative (biological) nature of all energy sources that absorb CO_2 from the atmosphere during their growing phase, only to release the same compound during fuel combustion. Additionally, the FAME molecule contains 11% oxygen, which leads to improved combustion and significant reductions in particulate matter (PM) or soot.

As part of the Clean Air Act Amendments enacted by the US Congress, the Environmental Protection Agency (EPA) was directed to ensure that any new commercially available motor vehicle fuel or fuel additive would not present an increased health risk to the public. Under this directive, the EPA established a registration program and testing protocols which are outlined in CFR Title 40 Part 79 as part of Tier I and Tier II emissions testing.

The EPA completed a major study of the impact of various concentrations of soybean-based biodiesel in the operation of heavy-duty highway-based vehicles. The results of the study are shown in Figure 16-17 and clearly demonstrate the superior emissions reductions of FAME fuels.

Nitrogen oxides (NOx) do increase as a result of high engine combustion temperatures and a variety of new technologies have been developed to reduce this air toxin.

Renewability

The US Department of Energy (DOE) and the US Department of Agriculture (USDA) in 1998 completed a thorough study of the energy balance of biodiesel and found that for every unit of fossil energy used in the entire biodiesel production cycle, 3.2 units of energy were delivered when the fuel was consumed.[20]

Readers who are interested in learning more about the life cycle energy requirements for the production of soybean-based biodiesel are encouraged to read the entire 286 pages of analysis completed by the US Departments of Energy and Agriculture.[21]

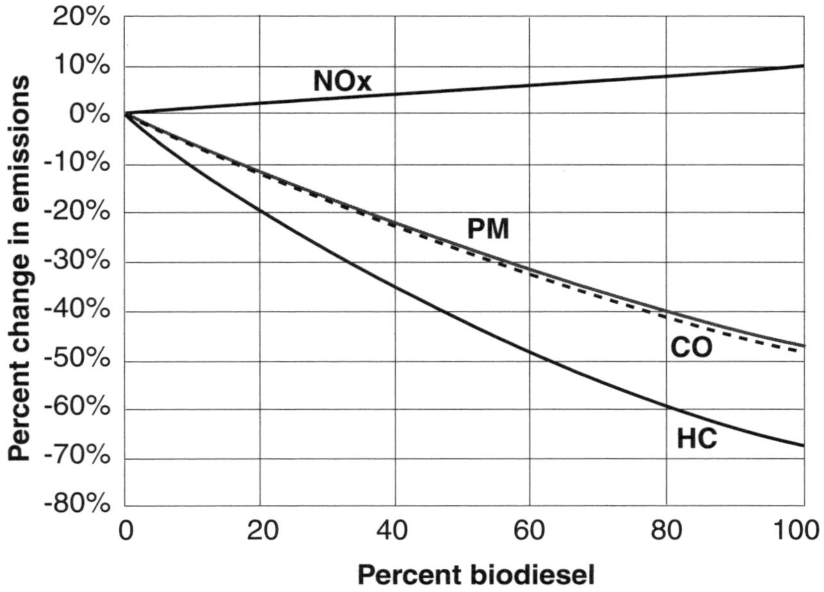

Figure 16-17 Average emission impacts of soybean-based biodiesel for heavy-duty highway engines. NOx = Nitrogen Oxides; PM = Particulate Matter; CO = Carbon Monoxide; HC = Hydrocarbons. (Source: U.S. EPA Report EPA420-P-02-001, October, 2002)

Low Sulfur

In order for petroleum diesel fuel to be given a rating of "low" or "ultra-

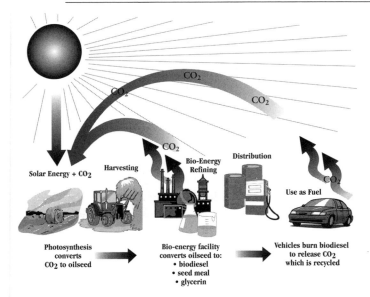

Solar Energy + CO₂ Harvesting   Bio-Energy Refining   Distribution   Use as Fuel

Photosynthesis
converts
CO₂ to oilseed

Bio-energy facility
converts oilseed to:
• biodiesel
• seed meal
• glycerin

Vehicles burn biodiesel
to release CO₂
which is recycled

Figure 16-18. Biodiesel is able to provide a positive net energy balance because growing plants absorb massive amounts of energy from the sun. Driving this energy absorption is the process of photosynthesis, which recycles atmospheric carbon dioxide, making biodiesel a carbon-neutral fuel source.

low" sulfur content, it is necessary to subject the fuel to an energy-intensive refining process that generates additional carbon dioxide emissions. The resulting fuel will have reduced lubricity levels that must be supplemented with lubricating additives.

Biodiesel by contrast retains its excellent lubricity while being intrinsically free of sulfur. Having a virtually zero-sulfur level allows the optimum use of oxidation catalytic converters in the exhaust system.

The Cons of Biodiesel

So far, biodiesel sounds like the perfect fuel. Unfortunately, there are a number of issues which must be considered on the negative side of the balance sheet.

Oxidation and Bacteriological Stability

Biodiesel is biodegradable, which is an excellent environmental benefit, but it creates long-term storage and fuel stability issues. When it is exposed to high temperatures, oxygen, or sunlight, or

placed in contact with nonferrous metals, FAME will deteriorate, resulting in polymerisation (fuel thickening) which leads to plugged filters and "glazing" within the fuel injection system. To prevent storage degradation, antioxidant additives may be added to extend storage life. Of course, the simplest corrective action is to simply limit the storage time of FAME fuels through rapid consumption and rotation.

Although one study says there is no problem with bacteria,[22] nevertheless there is evidence that there is a problem in that water content can destabilize FAME as well as create an active growing medium for microorganisms. Biodiesel manufacturers produce fuels that have very low water content, but biodiesel is hygroscopic and actually attracts water much more readily than petrodiesel does.

Biodiesel will become saturated at water levels above approximately 1,000 ppm, and if water ingress continues unchecked, water no longer remains bonded and collects at the bottom of the storage tank, leading to the very condition that promotes microbial growth.

Edward English II, Vice President and Technical Director of Fuel Quality Services, comments that "Biodiesel is a very different product from fossil diesel fuel and there needs to be better education surrounding handling and storage of these fuels. B100 may well leave the factory meeting ASTM standards, but each time it is pumped, transferred, stored, blended, and dispensed, it will pick up ever-increasing amounts of water. This stuff is like a sponge, and the only way neat or blended biodiesel will remain stable and microbial free is through proper handling, monitoring, and remedial procedures."[23]

Nitrogen Oxide Emissions

Numerous studies have confirmed that overall emissions from the combustion of biodiesel are low but show slightly elevated levels of nitrogen oxides (NOx). This increase is regarded as a problem related to higher combustion cylinder temperatures and is not inherently a fuel-related issue. Manufacturers believe that improvements in engine sensor and management technology as well as NOx catalytic reduction are just around the corner.

Unfortunately, this issue of NOx emissions is placing a damper on the entire diesel engine industry. Diesel proponents believe that gasoline-based exhaust emission strategies are strangling the potential of the diesel market and that each engine technology should receive its own emission profile, as is done in Europe, in light of the considerable fuel and greenhouse gas emissions savings of diesel engines.

Not every automotive manufacturer is worried. Mercedes-Benz is well known for its innovative, quality automobiles featuring state-of-the-art engineering. The company marked the epitome of its technological prowess by showcasing its leading edge BLUETEC technology, launching the diesel power train of the future. "Vision has therefore become reality as the extremely economical Mercedes-Benz CDI models are the cleanest diesel in the world in every category and consume between 20 and 40 percent less fuel than the gasoline counterparts," states a January 2006 corporate press release. Mercedes-Benz certainly recognizes the importance of emissions reduction technology, as over 50% of their total production volume is now captured by diesel engines.

Cold Flow Issues

No. 2 diesel fuel suffers from a thickening condition known as "waxing" or "gelling" when temperatures drop below the cloud point of the fuel. Should this occur within the fuel system of a vehicle, expensive cleaning becomes necessary.

By contrast, biodiesel suffers from a similar but reversible problem. Should low-temperature fuel gelling occur, causing loss of engine power or complete fuel starvation, the problem can be remedied by simply moving the vehicle to a warm location such as a parking garage until fuel temperatures moderate.

Both No. 2 diesel fuel and biodiesel can be "winterized" by the addition of so called "pour

Figure 16-19. The release of Mercedes-Benz BLUETEC technology coincided with the US release of low-sulfur diesel fuel in the autumn of 2006. Restricting sulfur content to a maximum of 15 ppm permits the use of particulate filters and efficient nitrogen oxide exhaust treatment. But the question remains: will the vehicles using BLUETEC technology be certified for use with biodiesel fuels? (Courtesy Mercedes-Benz Canada)

point enhancers," which may be as simple as blending No. 1-D into the fuel blend.

No. 2 diesel fuel is treated by fuel refineries to meet the expected minimum temperatures within a given geographical location. To a retail consumer,

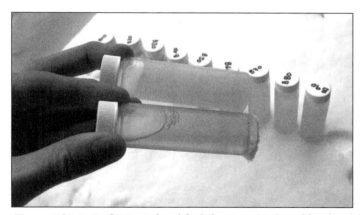

Figure 16-20. Both No. 2 diesel fuel (bottom view) and biodiesel can be "winterized" through the addition of so-called "pour point enhancers," which may be as simple as blending No. 1-D into the fuel blend. In this image, one vial of No. 2-D as well as vials of biodiesel in concentrations ranging from B10 through B100 are allowed to sit outdoors. B100 (top vial) is butter-solid at 14°F (-10°C), while No. 2-D as well as B10, 20, 30, and 40 all remain functional, above their respective cloud point ratings. Using this simple arrangement of test vials will provide the biodiesel user with a method of determining maximum winter blending limits.

the transition from "summer" to "winter" diesel fuel is completely transparent.

To counter the problem of cold flow, fuel-additive companies have developed a range of products to improve the cold weather performance of petrodiesel and biodiesel fuels. Primrose Oil Company Inc. offers the Flow-Master® winter diesel fuel treatment product, claiming it is more cost effective than using No. 1-D blended fuels.

The reason fuel gels at cold temperatures is that waxes inherent in the fuel begin to form microscopic crystals. If untreated, these crystals will immediately agglomerate (combine) with one another to form a gel and eventually solidify, blocking fuel lines and filters. Pour point enhancers limit the ability of wax crystals to grow large enough to agglomerate. Primrose indicates that its product will improve the cold flow rating of any untreated fuel by a minimum of 20°F to 30°F (11°C to 17°C).

It is virtually impossible to determine the exact concentration of biodiesel that can be used at a given geographical location without knowledge of the fuel's cloud point temperature. There is considerable variability in biodiesel cloud point temperature resulting from the inconsistency of the feedstock oil saturation level, with long-chain compounds displaying poor cold weather

Figure 16-21. Volkswagen of America has determined that diesel fuel containing up to 5% biodiesel certified to ASTM Standard D6751 meets the technical specifications for Volkswagen vehicles equipped with TDI engines imported into the United States.

properties. Tallow, lard, palm oil, and yellow greases used to make biodiesel may cause the fuel to remain solid or semisolid just below room temperature, requiring great care in storage and use.

Some researchers have taken a different approach to the problem by attempting to modify the FAME chemical structure through the use of alternate alcohols which have shown improved cold weather performance. Unfortunately, these alcohols have a higher cost and this process is of limited value in the current marketplace.

Repeated cooling and filtration of crystal growth within the FAME has also been attempted with varying rates of effectiveness. However, this process requires considerable amounts of energy and removes valuable esters that are lost during the filtration process, lowering overall biodiesel yields.

As carbon emissions become valued, there will be more impetus for researchers to solve this issue, using the most effective technological means at their disposal.

OEM Warranty Issues

One of the popular misconceptions about biodiesel is that it will not affect engine and fuel system warranties provided the fuel meets applicable specifications.[24] Statements such as these are not only misleading; they are simply wrong.

All major engine, vehicle, and fuel injection equipment manufacturers have clearly stated guidelines regarding the use of biodiesel fuels. Without hesitation, all manufacturers state that biodiesel that is used within the blend limits of their warranty statements must meet the appropriate national and/or international fuel standards.

The Volkswagen policy statement stipulates that "vehicle damage that results from misfueling or from the usage of substandard or unapproved fuels cannot be covered under our vehicle warranties," clearly placing the burden of proof regarding **any** fuel-related damage, fossil- or FAME-based, on the customer.

The Diesel Engine

There are two types of engine predominantly used to power road-based vehicles. The spark ignition engine operates on gasoline or less frequently on liquid petroleum gas (LPG). The compression ignition engine uses diesel fuel and is named after Rudolf Diesel, who patented his heavy oil engine in 1892.

Diesel engines are very popular in Europe, with demand expected to rise from a current level of approximately 30% of total car sales to an anticipated 40% of the market. Surprisingly, the greatest demand for diesel technology comes not from the penny-pinching small-car market but from the luxury car buyers, with 44% of all luxury cars sold in Europe powered by diesel engines. Luxury diesel sales represent a very large percentage of specific markets: Belgium, 87%; France, 82%; Austria, 77%; and Italy, 70%.[25]

Most North Americans have a complete disdain for diesel engines, thinking of them as slow, noisy, polluting, and generally uninspired. While this may have been the case with grandpa's old smoker, advances in technology have placed the diesel *ahead* of the gasoline-powered engine in several key areas:[26]

Fuel Economy: Because of their high volumetric efficiency, diesel engines use 30% to 60% less fuel than gasoline engines of similar power.

Power: Diesels produce more torque and power at lower engine speed than gasoline engines of similar displacement.

Durability: Diesel engines are designed to last well in excess of 300,000 miles (500,000 km) and require less maintenance than gasoline engines.

Greenhouse Gas Emissions: Diesel engines have higher thermal efficiency cycles than gasoline engines and diesel fuel contains more energy per gallon than gasoline, allowing a diesel engine to burn less fuel for a given power output and produce significantly lower CO_2 emissions.

Noise and Smoke: Using the latest "Common Rail Direct Injection" (CDI) and lean-burn technology as well as particulate traps and catalytic

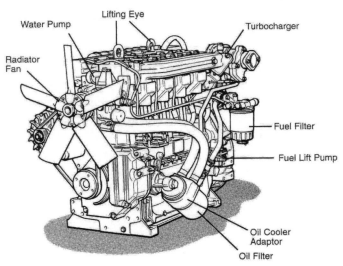

Figure 16-22. A typical light-duty diesel engine has between 4 and 8 cylinders, arranged with a series of intake and exhaust valves which control the air admission, compression, ignition, power, and exhaust phases. Because of their high volumetric efficiency, diesel engines use 30% to 60% less fuel than gasoline engines of similar power. (Courtesy: Lister Petter)

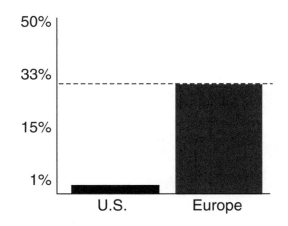

Figure 16-23. Current sales of diesel-powered vehicles have soared in the European Union owing to the numerous advantages diesels have over gasoline engines. In the luxury car market, France has seen 82% of market share go to diesel engines.

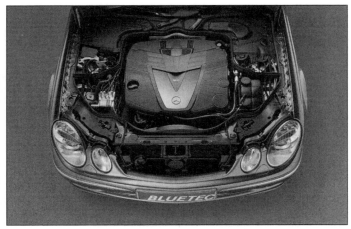

Figure 16-24. The prestigious Mercedes-Benz E320 CDI offers a 25% improvement in fuel efficiency compared to its gasoline-powered counterpart, while outperforming it by a full second in 0 to 60 mph acceleration, according to company data. Motor Trend's Frank Markus is even more direct: "Your eyes, ears, and nose will have trouble detecting that this is a diesel, while your backside will detect some serious pressure just at idle." In 2006 the company introduced its BLUETEC powertrain system, which meets all North American emission standards. (Courtesy: Mercedes-Benz Canada)

Figure 16-25. With "lean-burn" technology as well as particulate traps and catalytic converters, there is no need for modern diesel-powered vehicles to have any of the smoke or noise common in older designs. The City of Ottawa, Canada "Ecobus" features these technological advances.

converters, today's diesel vehicles have none of the smoke or noise common with older designs.

As a result, diesel-powered vehicles offer a cleaner, quieter, and more powerful alternative to identical automobiles and trucks equipped with less efficient gasoline-powered engines. Taking this one step further, diesel technology is an ideal candidate for advanced PHEV vehicles with their long life, fuel efficiency, and ability to operate on zero-carbon biodiesel.

The light-duty diesel market in North America is practically nonexistent, while the Europeans have created huge fuel and GHG savings by developing advanced engine technologies and emissions standards. Concurrently, the United States is fostering the Tier 2 emissions regulations that are a detriment to the development of diesel engine technology here, denying consumers access to the obvious benefits and national fuel consumption reductions that diesel technology would bring.

In order to get the benefits of diesel technology in North America, regulators may have to develop a compromise and recognize that different engine technologies have different emission profiles. US emissions standards accommodate gasoline engines which emit relatively little NOx but much more CO, HC, and CO_2 than their diesel counterparts. By placing the emphasis on smog-forming NOx and particulate matter (of greater concern in diesels) instead of CO_2, the US standards discourage the development and marketing of diesel engines.

A more balanced approach to matching technical developments with emissions standards would encourage the development and marketing of diesels in North America.[27]

Engine Technology Overview

All internal combustion engines include a cylinder block which houses the major components of the running gear. An eccentric crankshaft runs along the length of the engine, providing support and a means of transferring the linear motion of the piston to the rotary motion required to turn the

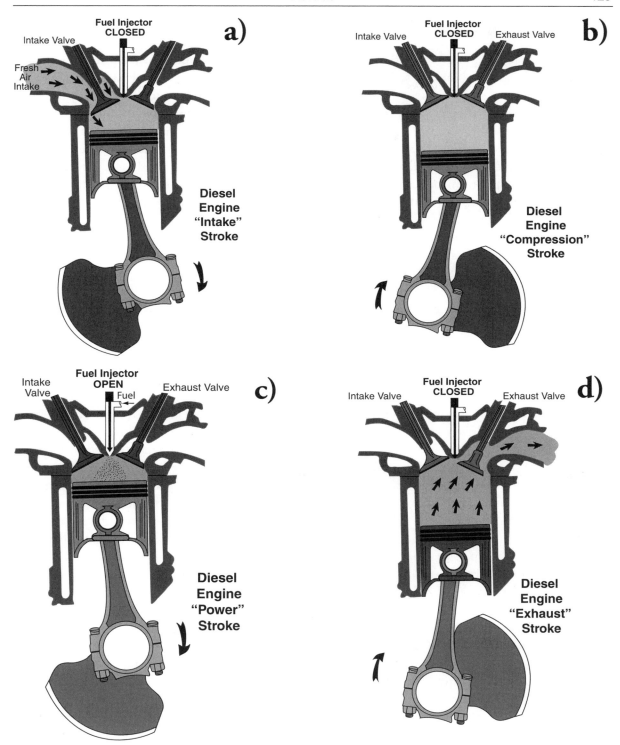

Figures 16-26a/b/c/d. An intake and exhaust valve (or, more commonly, valves) are operated from the camshaft, which forms part of the engine running gear and is synchronized with the rotation of the crankshaft. As the crankshaft rotates, the camshaft sequentially opens and closes the intake and exhaust valves in accordance with the four-stroke operating principle.

vehicle's drive wheels, or electrical generating unit.

Each piston is fitted into the cylinder machined in the engine block. A series of piston rings installed on the outer diameter of the piston and placed in contact with the cylinder wall ensures a gas-tight seal.

An intake and exhaust valve (or more commonly multiple intake and exhaust valves per cylinder) are operated from the camshaft, which forms part of the running gear of the engine and is synchronized with the rotation of the crankshaft. As the crankshaft rotates, the camshaft sequentially opens and closes the intake and exhaust valves in accordance with the four-stroke operating theory described below.

In the spark ignition engine, a properly balanced mixture of fuel and air is admitted into the cylinder and compressed, creating a volatile, explosive mixture. An electric spark is activated at the correct timing sequence, igniting the mixture and providing a downward power force on the piston which is then transferred to the crankshaft and ultimately to the drive wheels.

In contrast, the diesel engine eliminates the spark plug and related components and instead uses the heat of compression to perform the ignition function. The four-stroke timing sequence diagrams shown below illustrate this process.

In the diesel engine, air alone is admitted into the cylinder during the first 180° rotation of the crankshaft, creating the intake cycle shown in Figure 16-26a. The intake valve (left side) is opened by the camshaft (not shown) as the crankshaft rotates in a clockwise direction, forcing the piston down. The downward motion creates a vacuum which draws air into the cylinder. At the lowest point of piston travel in the cylinder, the intake valve closes, creating an airtight chamber.

The next 180° rotation cycle is known as the compression stroke, detailed in Figure 16-26b. The piston is now travelling upwards, compressing the air inside the cylinder. In diesel engines, the compression ratio (ratio of cylinder volume at bottom compared to volume at top) may range as high as 21:1, causing pressures in excess of 500 pounds per square inch (3.4 MPa) and temperatures in excess of 1,000° F (538° C).

When the piston reaches the top of the cylinder at the conclusion of the intake cycle, the power stroke begins, as shown in Figure 16-26c. Diesel fuel is supplied to the fuel injector under very high pressure. Using mechanical or, more commonly, computer/electronic control, the injector is opened and fuel is sprayed in a fine mist (*atomized*) into the cylinder. A fraction of a second later, the fuel will ignite, causing massive expansion of the burning gases and forcing the piston downward, applying power to the piston rod, crankshaft, and vehicle drive train.

The time delay between the opening of the fuel injector (signifying the initial fuel spray into the cylinder) and ignition of the fuel is known as *ignition delay* and is determined by the *cetane* rating of the diesel fuel. The ability of a fuel to combust in the presence of heat is known as the auto-ignition ability of fuel and is a key property of diesel fuel.

The final 180° cycle is the exhaust stroke, when the camshaft opens the exhaust valve as seen in Figure 16-26d. As the piston sweeps upward, combusted fuel (exhaust) components are forced out of the cylinder and sent on their way to the exhaust system, noise reduction muffler, and tailpipe. Advanced engines may also contain particulate matter traps to capture the dusty particles caused by the incomplete combustion of diesel fuel, NOx adsorbers to capture and treat nitrogen oxides in the exhaust stream, and catalytic converters which utilize the high temperatures to catalyze or neutralize exhaust gas components.

It requires two full rotations of the crankshaft to obtain one half-cycle (180°) power stroke from the engine. In order to improve the smoothness and power of the engine, multiple pistons are fitted into the cylinder block and connected to the crankshaft. Each piston is "timed" to produce its power stroke at slightly different times, allowing almost continuous overlap in the power-generation cycle of the engine, in turn reducing mechanical vibration and noise.

Automobiles such as the Mercedes-Benz Smart™ car have only three cylinders and a total engine displacement (net cylinder volume x number of cylinders) of only 52 cubic inches (850 cc). By contrast, a large diesel-powered pickup truck may have an engine with eight cylinders and a displacement of 458 cubic inches (7,500 cc). Of course industrial engines can have much larger displacements than this.

Fuel Injection Systems

At the risk of oversimplifying diesel engine fuel injection technology, I am going to make the generalization that there are two important classes of diesel engine that are relevant to our discussion: engines using "basic" fuel injection and those using "common rail" fuel injection.

Basic Fuel Injection

Figure 16-27 shows an overview of a typical fuel injection system used in many current and all older diesel vehicles. Fuel is drawn from the vehicle's storage tank and filtered to remove any debris and water that may be present in the tank. The filter is manufactured from tightly woven cellulose material which is able to stop particles larger than 20 microns (0.0008 inches) in size, preventing abrasive material from damaging the fuel system components.

Filtered fuel enters a high-pressure injection pump which pressurizes the fuel, feeding it into the fuel injector located in the top end of the cylinder. The injection pump is provided with multiple fuel supply lines, one for each cylinder and fuel injector. As diesel fuel is incompressible,

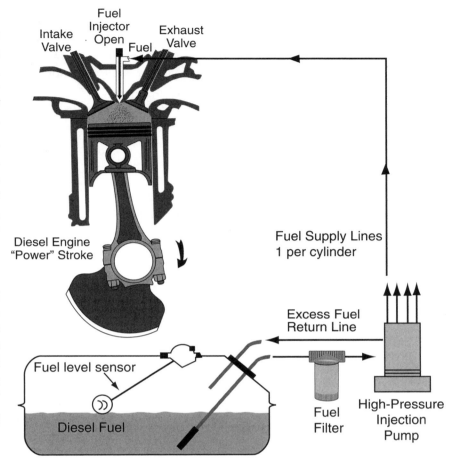

Basic Fuel Injection System

Figure 16-27. This drawing details a typical fuel injection system used in many current and all older diesel engines.

an excess fuel return line constantly recirculates fuel when an internal pressure relief valve reaches fuel operating conditions. Pressurizing any liquid will cause its temperature to rise, thereby increasing the temperature of fuel in the tank and distribution system.

When the piston reaches the top of the cylinder during the compression stroke, air in the cylinder will be heated above the ignition temperature of the fuel. Depending on the model and age of the engine, fuel will be atomized by either mechanical or electrical control of the injector nozzle. The "fineness" of the resulting fuel spray will determine

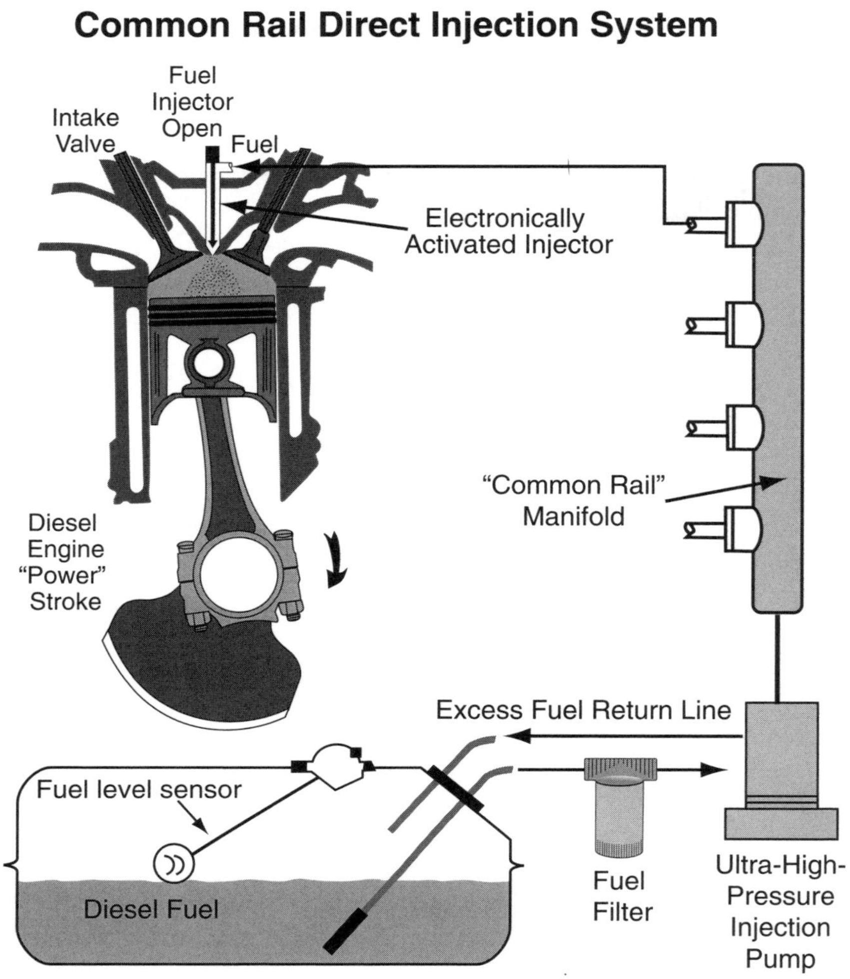

Common Rail Direct Injection System

Fuel
Intake   Injector
Valve   Open   Fuel

Electronically
Activated Injector

"Common Rail"
Manifold

Diesel
Engine
"Power"
Stroke

Excess Fuel Return Line

Fuel level sensor

Diesel Fuel

Fuel
Filter

Ultra-High-
Pressure
Injection
Pump

Figure 16-28. Common Rail Direct Injection (CDI) systems use an ultrahigh pressure injection pump and a "common rail" or pressure manifold to ensure high, even fuel pressures, ensuring precise control of the combustion process.

Common Rail Direct Injection

Common Rail Direct Injection (CDI) systems (Figure 16-28) use an ultra-high pressure injection pump and a "common rail" or pressure manifold to ensure high, even fuel pressure. CDI systems pressurize the diesel fuel to enormous levels, often in excess of 22,000 pounds per square inch (psi) ($\approx$150 Mpa). The common rail manifold is able to act as a pressure storage reservoir and ensure that fuel is instantly available when the fuel injectors require it. Note that higher fuel pressures generate higher excess fuel temperatures as compared with basic fuel injection systems.

Special piezoelectric fuel injectors can open and close thousands of times per second, allowing the fuel control computer to provide multiple "bursts" of fuel, which offers precise control of the combustion process. Pilot injection of minute amounts of diesel fuel prior to the main combustion injection initiation virtually eliminates diesel engine "clacking" noises.

A 4-valve-per-cylinder, 2-intake and 2-exhaust arrangement increases power, fuel efficiency, and responsiveness by allowing the fuel injector to be placed in a central location, creating a symmetrical fuel spray pattern and best fuel/air mixing.

combustion completeness, engine noise, and efficiency. A disadvantage of this system is that fuel pump pressure varies according to engine speed, resulting in varying fuel spray patterns.

Researchers have found that this problem can be minimized with the use of increased and constant fuel pressure and by "timing" the spray pattern by modulating the fuel injector to create "microbursts" of fuel in the cylinder as the piston is sweeping through the end of the compression stroke and the beginning of the power stroke. In order to accomplish this, much higher pressures and faster fuel injectors are required.

The Biodiesel Production Process

The simplest method used to produce FAME is based on the single reaction tank batch process. An updated version using two reaction tanks is used in many smaller commercial facilities (< 4 million liters per year / ≈ 1 million gallons per year). The advantage of the two-tank system that it ensures completeness of the reaction process and guarantees that *total glycerin* levels are within the ASTM specification of less than 0.24%. An example of such a system is shown in Figure 16-29.

- Vegetable oil or animal fat is loaded into a storage tank and heated.
- An alcohol and catalyst (methoxide) reactant is loaded into a second tank.
- The feedstock oil and reactants are pumped into Batch Tank #1, where they are mixed and heated until the conclusion of the reaction process.
- The materials in Batch Tank #1 separate into two layers or phases: raw FAME and glycerin.
- Glycerin is pumped to a processing section which neutralizes the catalyst, recovers and recycles the methanol, and provides unrefined glycerin (glycerol) for sale.

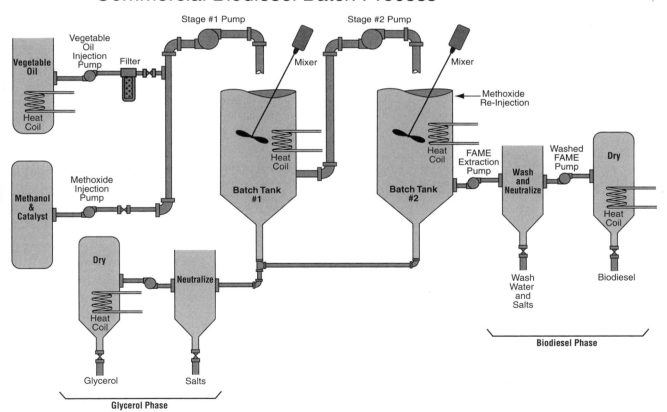

Figure 16-29. The simplest method used to produce FAME is based on the single reaction tank batch process. An updated version using two reaction tanks is used to ensure completeness of the reaction process and to guarantee that total glycerin levels are within the ASTM specification of less than 0.24%.

- The raw FAME is pumped to Batch Tank #2 where it is reprocessed with methoxide reactant to remove trace amounts of bonded glycerin.
- Excess glycerin is pumped off and recovered as in the first reaction step.
- Raw FAME is sent to a washing and catalyst neutralizing tank. Wash waters are processed for disposal.
- The high-quality FAME is then dried and readied for testing and sale.

Vegetable oil or animal fats having a free fatty acid content of less than 2% and preferably less than 1% are loaded into the feedstock tank shown at upper left. A steam-heated exchanger coil heats the oil feedstock to approximately 60°C. It does not make any difference which parent feedstock oil or fat is used to produce the FAME provided that it meets all of the requirements of ASTM D6751 at the completion of the production process. The reader is again reminded that fats and oils that are fully or partially hydrogenated as well as highly saturated will be affected by cold flow issues.

A second storage tank is filled with a commercial premixed (methoxide) reactant solution comprising methanol (alcohol) and a catalyst of either sodium hydroxide (NaOH) or potassium hydroxide (KOH) which, when blended, is known as sodium or potassium methoxide depending on the chosen catalyst. The sodium or potassium methoxide can also be produced at the factory.

Methanol is the most commonly used alcohol, although other alcohols may be interchanged. Alcohol selection is determined by several factors including cost, toxicity, ease of recycling, and quantities required to complete the reaction process. Although methanol is principally derived from natural gas and is highly toxic in the environment, which lowers the overall "green value" of biodiesel, it meets the litmus test of low cost, ease of use, and ability to be recycled. Cellulosic ethanol is an obvious zero-carbon contender but is currently not used because of the lack of any carbon valuation policy.

Because oil and alcohol do not have an affinity for each other, a catalyst is used to initiate the transesterification reaction. A catalyst may be either alkaline (base or basic) or acidic in nature, with experience showing that acid-catalyzed reactions operate too slowly for cost-effective biodiesel production purposes.

The selection of base catalyst is also a matter of choice, although sodium hydroxide is the most commonly selected compound. The importance of the feedstock selection and free fatty acid level must be considered along with catalyst selection. The combination of base catalysts and FFA levels exceeding approximately 2% will cause the formation of soaps that can hinder the reaction process and create a contaminant in the glycerin phase.

Once the feedstock oil has reached operating temperature, it is pumped and filtered into the batch reactor tank, where it is continuously stirred and heated. The methoxide reactant is then pumped into the reactor tank and vigorous mixing is continued for a period ranging from 30 minutes to one hour, with the close contact of the feedstock oil, alcohol, and catalyst ensuring the completeness of the reaction.

At the end of the mixing period, the mixer and heating coils are deactivated and the solution is allowed to settle, causing it to separate into two phases with the raw glycerin (glycerol) sinking to the bottom of the tank and the raw FAME floating on top as the second phase. Reaction completeness is reportedly in the range of 85% to 94%, which is below the requirements necessary to ensure that bonded glycerin meets the ASTM standard.

The glycerol phase is pumped to a processing unit where methanol is captured and recycled for further use. The glycerol is also washed with acidic water to neutralize the basic catalyst. Wash water is processed and either discarded after meeting required environmental standards or recycled for further use in the plant. The glycerol is then dried and sold to a refining company.

The raw FAME that was produced in the first reaction is sent to a second reaction tank where it is reprocessed with fresh methoxide reactant in a process similar to that of the first reaction. Introducing a second reaction drives the transesterification to 95%+ completeness, ensuring that bonded glycerin levels will now fall within the required standards.

Glycerol is again pumped off and processed as in the first reaction stage.

The raw FAME is subjected to methanol capture before being washed in the same manner as in the glycerol phase. A drying step ensures that the high-quality, washed FAME contains less than the prescribed amount of suspended water before it is sent to storage.

Small-Scale Biodiesel Production System

When Lorraine and I designed our off-grid home we decided to add a small workshop area in the detached garage to house the biodiesel production facility that would be used to fuel our backup generator and car. This extra space is located behind the door shown in Figure 16-32 and consists of an area of approximately 6 x 18 feet (1.8 x 5.5 meters). The room is well insulated and tightly sealed with standard construction-grade vapor barrier and acoustical gap sealant and is fitted with a ventilation fan that is capable of drawing approximately 500 cubic feet (14 cubic meters) of air from one end of the room to the other per minute of operation. The high-volume air turnover rate is required to keep methanol vapors to an absolute minimum, certainly below the *lower explosive limit* (*LEL*), as well as to ensure that air quality remains high and nonpoisonous.

The production lab is shown in an overview image in Figure 16-33. All of the production equipment is mounted along the left wall, while the "wet laboratory," safety equipment, and storage area are located along the right wall. This image shows, starting at the bottom left corner, the first tank which receives the WVO and is used to filter, dry, and deacidify the oil to desired standards. The second tank is the main reaction tank which performs the chemical conversion (transesterification) of WVO into biodiesel. It is also used to separate the raw biodiesel from glycerol and to recycle the excess methanol prior to washing the biodiesel.

The next small white conical tank is the chemical mixer which combines the methanol stored in the adjacent tank with a powdered catalyst to form sodium (or potassium) methoxide, which is pumped into the main reaction tank to start the conversion of WVO into biodiesel.

The small white tank beside the methanol is the biodiesel wash water storage tank. It contains a submersible heater which warms the wash water to 60°C before it is sprayed into the raw biodiesel which is stored in the large white conical wash tank. The wash tank is fitted with both water spray nozzles and an air bubbling system which are used to remove contaminants from the raw biodiesel.

Figure 16-32. When Lorraine and I designed our off-grid home we decided to add a small workshop area in the detached garage to house the biodiesel production facility that would be used to provide fuel for our backup generator and car.

Figure 16-33. The production lab is shown in an overview. All of the production equipment is mounted along the left wall, while the "wet laboratory," safety equipment, and storage area are located along the right wall.

Production Lab Overview - *Waste Oil Drying and Reaction System*

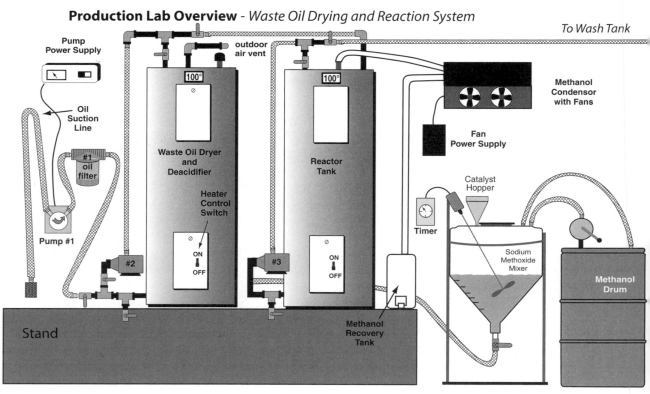

Figure 16-30. Biodiesel Production Lab Overview – Waste Oil Dryer and Reaction System

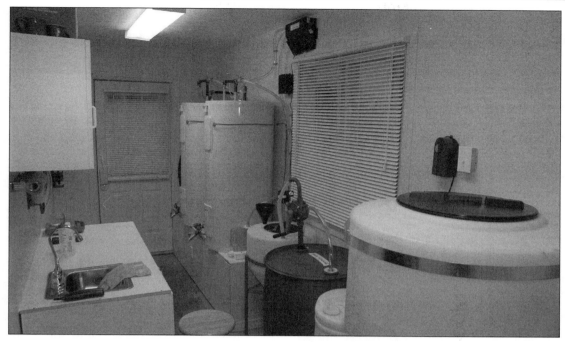

Figure 16-34. This view of the production lab was taken from the rear wall facing the front door. Note that the WVO and main reaction tanks are mounted on a cabinet which raises them off the floor. This arrangement provides additional storage space and makes accessing the valves and controls of the reaction tanks much easier on one's back.

Production Lab Overview - *FAME Wash and Drying System*

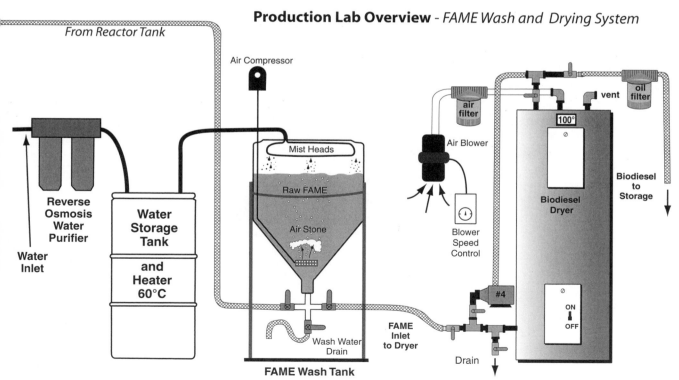

Figure 16-31. Biodiesel Production Lab Overview – FAME Wash and Drying System

Figure 16-35. This view of the wet lab shows the ample storage space for process chemicals, gloves, safety equipment, and other materials required in the biodiesel production process.

Figure 16-36. Safety is of primary importance when working with any chemicals. The wet lab area has storage for work smocks, gloves, and face shields as well as a telephone, eyewash station, and Class A, B, and C and foam fire extinguishers. Antistatic, ground-connected wrist straps are also provided to prevent the accidental ignition of methanol. The safe production of biodiesel does not happen by chance!

The last tank visible at the far rear of the picture (located behind the large conical wash tank) is the biodiesel dryer. This unit removes any remaining water from the biodiesel before final filtration and storage.

Also visible in this image are the ventilation fan located along the rear wall and the electrical subpanel which provides electrical circuit control and protection for each of the process tanks, pumps, and heaters used in the system.

Running along the right sidewall is the wet lab area which is used to analyze both the WVO and the biodiesel produced by the system. The storage cabinets include all of the necessary process chemicals as well as test equipment, scales, and measuring beakers.

Safety is of primary importance when handling any chemicals, and a variety of smocks, eye protection, rubber gloves, and spill cleanup materials as well as an eyewash station and multiple fire extinguishers are provided.

The photograph in Figure 16-34 was taken from the rear of the production lab and shows how all of the equipment and lab facilities fit neatly into this compact area. Note that the WVO receiver tank and the main reaction tank are lifted off the floor on a horizontal storage cabinet. In addition to providing additional storage space, the cabinet also houses a small reverse osmosis water filtration unit which feeds mineral-free water to the biodiesel wash tank.

The small black box mounted above the main reaction tank is the air-to-liquid heat exchanger which is used to condense methanol vapors driven off from the biodiesel reaction process. The unit operates as a fan-driven cooling unit, causing methanol vapor to condense and drip into the storage tank located next to the reaction tank. Condensed methanol is returned to the methanol storage tank for future use.

Because of the inline design of the production system, it is possible to simultaneously process approximately four 40-gallon (≈ 150 liter) batches of biodiesel at one time, with one batch in each stage:

- drying and deacidification of WVO
- transesterification of WVO into biodiesel along with methanol recovery
- washing of raw biodiesel
- drying of biodiesel prior to testing and storage

Processing of waste stream glycerol and wash water is handled "offline."

Figure 16-37. WVO is delivered to the facility and immediately transferred to the receiver/dryer tank.

The WVO Receiver /Dryer

WVO is delivered to the facility and immediately transferred to the receiver/dryer tank shown in Figure 16-37. This tank is made from a 60-gallon (227-liter) electric water heater although larger, commercially available process tanks can be used if the system is to be scaled up for cooperative or small-scale commercial use.

Fellow biodiesel enthusiast Geoff Shewfelt is shown in Figures 16-38 and 16-39 with a load of WVO recently received from a local restaurant and transferred immediately to the receiver/dryer tank, where a fuel transfer pump and suction line draw the WVO from the storage pails (Figure 16-40). The WVO is analyzed to determine its free fatty acid composition and may be subjected to deacidification and heating to remove excess water. WVO absorbs water from the foods that are fried in it, and if sufficient water remains in high-FFA oil the transesterification process may fail, producing a jelly-like gravy rather than biodiesel. This is an important step that most biodiesel processors tend to skip.

Upon completion of drying and/or deacidification, the WVO is transferred to the main biodiesel reaction tank using the circulation pump fitted to the receiver/dryer unit.

Figure 16-38. A load of WVO recently delivered from a restaurant will be transferred immediately to the receiver/dryer tank.

Figure 16-39. Geoff Shewfelt is shown inserting the suction hose of the receiver/dryer tank into a pail of WVO. The suction hose is equipped with a particle strainer at one end and a 20-micron filter at the outlet of the transfer pump, ensuring that food particles are filtered out and only WVO is sent to the receiver/dryer tank.

Figure 16-40. This detail view shows the suction hose and particle strainer inserted into a pail of cold-pressed canola oil that was recovered from a batch of off-specification oilseeds.

The Biodiesel Reaction Tank

The biodiesel reaction tank is configured in a similar manner to the receiver/dryer tank as shown in Figure 16-41. The reaction tank is fitted with a circulation pump and a "sight glass" (Figure 16-42) created from reinforced, braided plastic tubing. This sight glass permits the filling of the reaction tank with an exact amount of WVO and reaction chemicals and also provides a way of monitoring the completeness of the reaction process.

A vapor recovery unit is also installed above the tank to capture the excess methanol driven

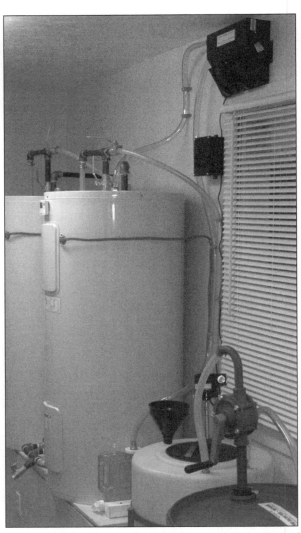

Figure 16-41. The biodiesel reaction tank is configured in a similar manner to the receiver/dryer tank. Both are visible in this view.

off during the transesterification process. Excess methanol is used to ensure that the conversion of waste oil to biodiesel is driven to completion, although a large volume of methanol is not required for the transesterification process, thus allowing recycling of the excess. It is much simpler and safer and requires less energy to capture the methanol at the reaction stage than to try and recover it from the wash water.

Leaving methanol in the biodiesel is simply not an option.

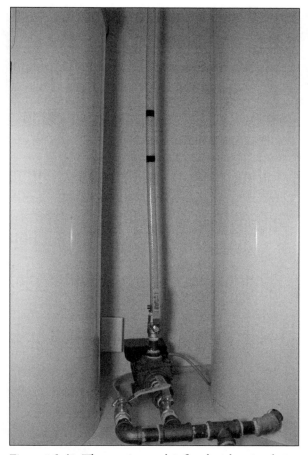

Figure 16-42. The reaction tank is fitted with a circulation pump and a "sight glass" created from reinforced, braided plastic tubing. This sight glass permits the filling of the reaction tank with an exact amount of WVO and reaction chemicals and also provides a way of monitoring the completeness of the reaction process.

Sodium Methoxide System

In order to "crack" WVO into biodiesel and its coproduct glycerol, it is necessary to use an alcohol and catalyst solution such as sodium (or potassium) methoxide. This solution is created by the careful measurement and mixing of methyl alcohol (methanol) and sodium (or potassium) hydroxide. Figure 16-43 shows a 55-gallon (208-liter) drum of methanol that has been delivered by a local fuel supply company. The drum is fitted with a hand-operated chemical pump suitable for methyl alcohol and a vapor recovery tube, both of which are fitted into the bung connections of the drum.

The outlet of the hand pump is connected to the white conical bottom tank (left), and the vapor recovery line is fitted to ensure that methanol vapors are returned to the storage tank. The conical tank is fitted with a screw-top sealing lid and a small funnel and stopper. The funnel is used as a hopper, which allows the addition of sodium hydroxide catalyst to the previously added methanol.

A small, spark-proof mixer is fitted to the tank as shown in Figures 16-43 and 16-44. The mixer is also connected to an electrically operated mechanical timer that is used to control the mix timing of the solution, ensuring that the sodium hydroxide completely dissolves in the methanol.

The outlet at the bottom end of the conical tank is connected to the suction side of the reaction tank through a shutoff valve. Opening this valve causes the sodium methoxide solution to be drawn into the reaction tank containing the WVO, starting the transesterification process.

Upon completion of the transesterification process, two liquid components or *phases* are created. A glycerol phase sinks to the bottom of the reaction tank, while the lower-density raw biodiesel phase floats on top. The glycerol is removed by draining it from the tank and transferring it to a separate refining station.

Upon removal of the glycerol phase, the raw biodiesel is reheated to cause excess methanol to

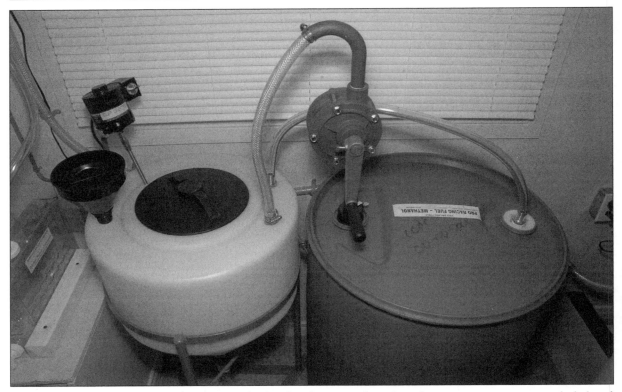

Figure 16-43. In order to "crack" WVO into biodiesel and its coproduct glycerol, it is necessary to use an alcohol and catalyst solution such as sodium methoxide. This solution is created by the careful measurement and mixing of methyl alcohol (methanol) and sodium hydroxide in the mixing tank shown above. The barrel to the right is filled with methyl alcohol.

Figure 16-44. A small, spark-proof mixer is fitted to the tank and is also connected to an electrically operated mechanical timer that is used to control the mix timing of the solution, ensuring that the sodium hydroxide completely dissolves in the methanol.

boil off. The methanol may be vented into the atmosphere or preferably directed to a reflux condenser that converts the vapors to liquid. The captured methanol may then be reused, lowering production costs.

After glycerol and methanol have been recovered from the reaction tank, the raw biodiesel is transferred to the wash tank for final processing.

Biodiesel Washing System

Raw biodiesel is transferred from the main reaction tank to the large white conical-bottom washing tank shown in Figure 16-45. This tank is fitted with two washing systems known as mist and bubble wash technologies. Regardless of which washing procedure is used, a small water storage tank located to the immediate left of the washing tank is required. This 10.5-gallon (40-liter) tank receives potable water from a reverse osmosis filtration unit located in the storage cabinet under the reaction tank, which is in turn fed by the household potable water supply. The purpose of the reverse osmosis system is to remove dissolved minerals such as calcium and iron that are contained in the well water in our geographical location. The water storage tank is fitted with a submersible water heating element that heats the wash water to between 120°F and 140°F (50°C and 60°C), greatly improving the wash speed and quality. Prior to starting the wash cycle, a small amount of acetic acid is added to the wash water and a submersible pump is activated.

If the mist washing process is used, the wash water is pumped to a series of mist heads mounted around the perimeter of the wash tank lid, causing a gentle shower of slightly acidic (softened) water

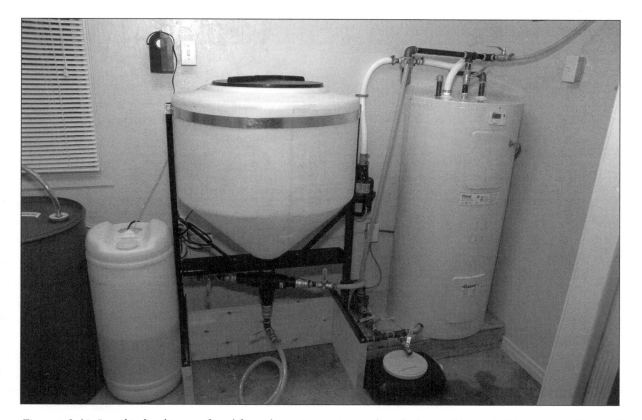

Figure 16-45. Raw biodiesel is transferred from the main reaction tank to the large white conical-bottom washing tank shown above. This tank is fitted with two washing systems known as mist and bubble wash technologies. The small white tank to the left of the wash tank is the wash water storage tank.

to spray over the biodiesel surface (see Figure 16-46). Water has a higher density than oil or biodiesel and therefore falls to the bottom of the tank, absorbing any free glycerin and excess catalyst (sodium hydroxide) along the way.

Washing may be completed using the spray mist method described above or using a "bubble washing" technique in which water is added to the biodiesel directly from the wash tank.

It is necessary to wash biodiesel a number of times in order to ensure compliance with fuel quality standards. At each successive washing stage the wash water contains fewer contaminants, allowing it to be reused for earlier, more heavily contaminated wash stages. This process is known as counter-current washing and greatly reduces the amount of water used in the production process. When the wash water is saturated with contaminants, it is drained off and temporarily stored before final treatment and drainage into the environment. This contaminated wash water is known to be toxic and must be treated prior to release.

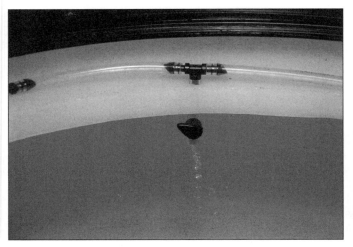

Figure 16-46. Prior to starting the wash cycle, a small amount of acetic acid is added to the biodiesel wash water and a submersible pump is activated, sending water to a series of mist heads mounted around the perimeter of the wash tank lid and causing a mist of slightly acidic (softened) water to spray over the biodiesel surface.

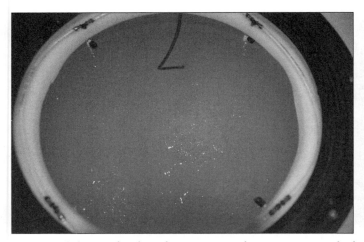

Figure 16-47. Biodiesel washing may use the spray mist method or a "bubble washing" technique (shown here) in which water is added to the biodiesel and tiny air bubbles are blown through the water/biodiesel mixture.

Figure 16-48. After the final wash water has been removed from the washing tank, the biodiesel is pumped into the dryer tank shown above.

Biodiesel Drying and Final Filtration

After the final wash water has been removed from the washing tank, the biodiesel is pumped into the dryer tank shown in Figure 16-48. The dryer uses an electric water heater arrangement and is equipped with a circulating pump. It is also fitted with a heated air blower and filter arrangement which blows hot air through the biodiesel (Figure 16-49) before venting outside to the atmosphere. This arrangement removes excess water from the fuel, aids in ensuring compliance with the ASTM limits for water and sediment, improves the biodiesel oxidation stability, and reduces the chances of microbial growth during storage.

Once the biodiesel has been heat-treated, it is pumped through a 20-micron fuel filter and is ready for quality testing and storage (Figure 16-50).

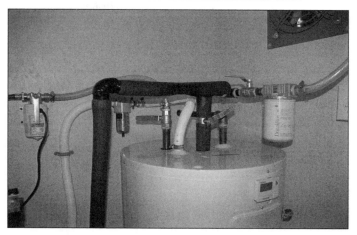

Figure 16-49. The fuel dryer is fitted with a heated air blower and filter arrangement which blows hot air through the biodiesel before venting outside to the atmosphere. This arrangement removes excess water from the fuel, ensures compliance with the ASTM limits for water and sediment, improves the biodiesel oxidation stability, and reduces the chances of microbial growth during storage.

Figure 16-50. Once the biodiesel has been heat-treated, it is pumped through a 20-micron fuel filter and is ready for quality testing and storage.

Figure 16-51. Simple five-gallon fuel totes are inexpensive and work well but increase the amount of handling labor as well as the frequency of spills. An electric fuel dispensing system such as this model keeps the dispensing method in familiar territory.

Figure 16-52. With the biodiesel processor operating at full capacity, the system can process approximately 45 to 50 gallons (≈ 170 to 190 liters) of biodiesel per day, which is far in excess of our total fuel requirements. However, processing, storing and handling that much fuel does take a fair bit of labor, space, and care.

Fuel Dispensing and Storage

With the biodiesel processor operating at full capacity, the system can process approximately 45 to 50 gallons (≈170 to 190 liters) of biodiesel per day, which is far in excess of our total requirements. However, processing, storing, and handling that much fuel does take a fair amount of space, labor, and care.

The Fuel Dispensing Unit

A fuel dispensing system can be as simple or as complex as your needs dictate. Simple five-gallon (20-liter) fuel totes are inexpensive and work well but increase the amount of handling labor as well as the frequency of spills. An electric fuel dispensing system such as the model shown in Figures 16-51 and 16-52 keeps the dispensing method in familiar territory. This dispensing unit is available at most farm and auto supply stores and comprises an explosion-proof fuel pump that is fitted to a standard 55-gallon fuel drum. The pump may be driven by either 120V household power or a 12V supply connection for in-vehicle use. A special water-absorbing filter known as an *agglomerator* is connected to the pump discharge prior to feeding the fuel nozzle.

Figure 16-53 shows a detail view of a typical fuel-dispensing unit. A 55-gallon fuel drum must be purchased or "rented" by paying a drum deposit charge. It is strongly recommended that only a new or recently used diesel fuel drum be adopted for your fuel dispensing system. The drum can be mounted on a drum dolly or a drum cart. A fully loaded fuel drum is very heavy and is difficult to move using standard drum carts; ideally the drum can be placed in a convenient central location and not moved.

A commercial fuel dispensing pump is mounted to the drum using the threaded bung fitting. A suction pipe is lowered into the tank to draw fuel into the pump intake as shown.

The agglomerator filter is fitted on the

discharge side of the pump as shown. Standard black pipefittings of the same type used in the biodiesel facility are required. The agglomerator filter is a disposable filter that is designed to agglomerate (group together or coalesce) water droplets suspended in the fuel and cause them to fall to the bottom of the sight glass bowl. A small drain valve is located at the bottom of the glass bowl which allows for drainage of any accumulated water. (Do not believe for one second that having an agglomerator will lessen the need for diligent fuel handling and storage. These filters will only remove relatively large water droplets and are therefore intended as precautionary devices only.)

Cold Weather Issues

Biodiesel is subject to a gelling or freezing condition at its cold-temperature limits as discussed earlier. If you live in an area where cold weather is the norm, it will be necessary to adjust the concentration of biodiesel stored in the fuel dispensing system as a function of temperature. One very simple means of doing this is to create a "biodiesel blend thermometer" as shown in Figure 16-54. A series of 11 glass vials is filled with mixtures of petro- and biodiesel. The vial at the left contains No. 2-D (straight diesel fuel), the next contains B10 (10% biodiesel/90% petrodiesel), ending with B100 (100% biodiesel) on the right. Each of the

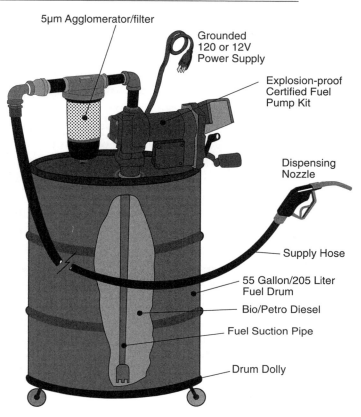

Figure 16-53. This image shows a detail view of a typical fuel dispensing unit. A 55-gallon fuel drum must be purchased or "rented" by paying a drum deposit charge. It is strongly recommended that only a new or recently used diesel fuel drum be adopted for your fuel dispensing system.

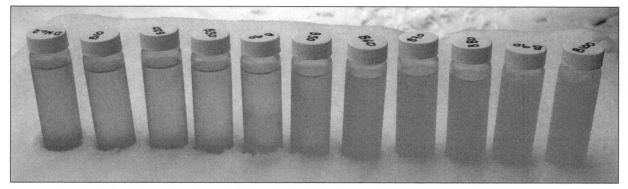

Figure 16-54. Biodiesel is subject to a gelling or freezing condition at its cold-temperature limits as discussed earlier. If you live in an area where cold weather is the norm, it will be necessary to adjust the concentration of biodiesel stored in the fuel dispensing system as a function of temperature. One very simple means of doing this is to create a "biodiesel blend thermometer" as shown here.

lids is marked with the appropriate concentration of fuel mixture.

The vials can be placed in a tin can and left in the same general location, out of direct sunlight, where you would normally store your diesel-powered vehicle. In the example shown in Figure 16-55, the No. 2-D sample (bottom vial) is free flowing and gel free, while the B100 (top vial) is frozen solid at 5°F (-15°C). At this temperature and using refined, cold-pressed, canola-based biodiesel, a concentration of B30 to B50 could easily be used, with greater concentrations possible if the vehicle, generator, or furnace unit was in a garage or heated space.

It is important to make a biodiesel blend thermometer since the ambient local temperature and feedstock oil composition of the biodiesel will greatly affect the concentration of biodiesel that can be mixed into the fuel tank. It is very difficult to accurately calculate the blend levels, and for this reason I keep a spare drum of No. 2-D fuel for splash blending on site.

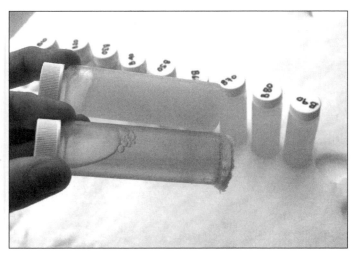

Figure 16-55. In this example, the No. 2-D sample (bottom vial) is free flowing and gel free, while the B100 (top vial) is frozen solid at 5°F (-15°C). At this temperature and using cold-pressed, canola-based biodiesel, a concentration of at least B30 could be used as evidenced by the lack of crystal formation or any sign of fuel thickening.

An additional word of caution: note where your vehicles will be stored during the day. For example, the biodiesel blend thermometer will give an accurate reading assuming your vehicle is stored in a garage at night. However, it may be exposed to colder daytime temperatures if you are parked in a shaded outdoor parking lot all day long. It is always best to err on the side of caution and be a bit conservative when selecting the appropriate biodiesel blend level. There is nothing worse than having to call for a tow truck to pick up a stalled, fuel-starved vehicle because you pushed the blending limits.

Blending Biodiesel with Petrodiesel

Although blending biodiesel with petroleum diesel will increase fossil fuel consumption and greenhouse gas emissions, it should be noted that only areas with very cold weather need to worry about blending.

The National Biodiesel Board (NBB) commissioned a report on cold flow blending issues after the State of Minnesota established a requirement that all on-highway diesel fuels contain at least 2% biodiesel. In response to the need for proper blending and other cold-temperature-related issues, the NBB established a Biodiesel Cold Flow Consortium to study the blending properties of biodiesel and report the results. The study evaluated both "splash" and "proportional" blending techniques, of which only splash blending is used by the micro- and small-scale producer.

The results of the Consortium testing showed that biodiesel must be kept at least 10°F (≈6°C) above its cloud point temperature to ensure successful, homogenous blending with petrodiesel. For those who would like to learn more about the cold flow blending study, a copy of the report can be downloaded from the National Biodiesel Board website.[28]

Summary

The use of waste or recycled oil, including other sustainable feedstocks is a move in the right direction towards creating zero-carbon biodiesel. Unfortunately, feedstock oil makes up only 70.6% of the biodiesel production process, and the other inputs of chemical feedstocks and energy shown in Figure 16-56 must also be converted to carbon-neutral properties.

Electrical energy produced by renewable sources takes care of electricity, and carbon credits can be used to offset natural gas inputs at the commercial plant level.

The issue of alcohol is quite interesting, as approximately 20% of biodiesel is produced using methanol; it could also be called liquid methane, as it is principally derived from natural gas.

The transesterification process does not care which alcohol is used, although methanol has been chosen for its simplicity to work with and low cost. As with anything derived from fossil fuels, carbon valuation will change this picture and alternative alcohols derived from renewable sources will become more interesting. An obvious candidate is cellulosic ethanol, described earlier in this chapter. Although I am still in the early stages of researching this input feedstock, I have discovered that there are no major impediments to the production of biodiesel using cellulosic ethanol.

Looking into the future, it is not possible to say for certain what fuel, biological or fossil, will be used to power advanced high-efficiency vehicles and home-heating systems. The likelihood is that there will be a mix of different choices, but sustainable biofuels will definitely be part of the supply mix.

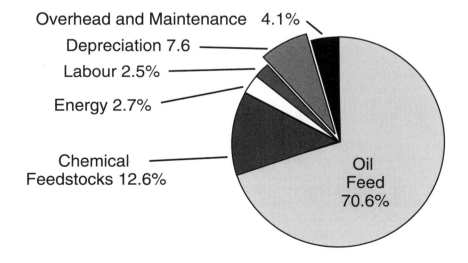

Distribution of Biodiesel Products Costs (%)

Overhead and Maintenance 4.1%
Depreciation 7.6
Labour 2.5%
Energy 2.7%
Chemical Feedstocks 12.6%
Oil Feed 70.6%

Figure 16-56. Although biodiesel can be made into a completely zero-carbon fuel, input energy and chemical feedstocks must be considered. (Source: Jon Van Gerpen et al, Building a Successful Biodiesel Business)[29]

Endnotes

1   Husky Energy, "Better for your car. Better for the environment" (2006), www.huskyenergy.ca/products/downloads/Ethanol.pdf.
2   Canadian Renewable Fuels Association, "Ethanol: Reducing Greenhouse Emissions by the Equivalent of 200,000 Cars," http://www.greenfuels.org/
3   Husky Energy, "Better for your car. Better for the environment," (2006), www.huskyenergy.ca/products/downloads/Ethanol.pdf.
4   Canadian Renewable Fuels Association, "Ethanol."
5   Murtagh & Associates, *The Online Distillery Network for Distilleries and Fuel-Ethanol Plants Worldwide,* (2007), http://www.distill.com.
6   Kurt K. Klein, The Biofuels Frenzy: What's in it for Canadian Agriculture? (March 28, 2007), http://www.aic.ca/whatsnew_docs/Klein%20Final%20%234.pdf.
7   National Biodiesel Board, "Specification for Biodiesel (B100)—ASTM D6751-07b," (March 2007), http://www.biodiesel.org/pdf_files/fuelfactsheets/BDSpec.PDF.
8   National Biodiesel Board, "National Biodiesel Board, DOE, USDA Officials Dispute Biofuels Study," (July 21, 2005), http://www.biodiesel.org/resources/pressreleases/gen/20050721_pimentel_response.pdf
9   Mercedes-Benz, "Mercedes-Benz Debuts High-Tech Diesel Car For Canadian Market," (April 8, 2004), http://www.mercedes-benz.ca/index.cfm?NewsID=144&id=3246.
10 National Biodiesel Board, "Fuel Fact Sheets" (2007), http://www.biodiesel.org/resources/fuelfactsheets.
11 Martin Mittelbach and Claudia Remschmidt, *Biodiesel: The Comprehensive Handbook* (Vienna: Martin Mittlebach, 2004), 1.
12 Martin Mittelbach and Claudia Remschmidt, *Biodiesel: The Comprehensive Handbook* (Vienna: Martin Mittlebach, 2004).
13 Ibid.
14 Ibid.
15 Martin Mittlebach, in discussion with the author, Ottawa, 2005.
16 Note that the ASTM D6751 standard for biodiesel makes reference to the fact that there is little practical knowledge about using high concentrations of biodiesel. While this may be true in the commercial markets, there has been an "underground" economy using biodiesel at 100% concentrations for many years. For further information on this topic, consult the author's book *Biodiesel: Basics and Beyond*, Tamworth, ON: Aztext Press, 2006.
17 Martin Mittelbach and Claudia Remschmidt, *Biodiesel: The Comprehensive Handbook* (Vienna: Martin Mittlebach, 2004), 6.
18 Green Trust, "Biodiesel," (2000), http://www.green-trust.org/biodiesel1.htm
19 Union for the Promotion of Oil and Protein Plants, "Status Report Biodiesel: Production and Marketing in Germany 2005," http://www.biodiesel.org/resources/reportsdatabase/reports/gen/20050601_gen358.pdf
20 National Biodiesel Board, "National Biodiesel Board, DOE, USDA Officials Dispute Biofuels Study," (July 21, 2005), http://www.biodiesel.org/resources/pressreleases/gen/20050721_pimentel_response.pdf,
21 http://www.biodiesel.org/resources/reportsdatabase/reports/gen/19980501-gen-203.pdf
22 Union for the Promotion of Oil and Protein Plants, "Status Report Biodiesel: Production and Marketing in Germany 2005," http://www.biodiesel.org/resources/reportsdatabase/reports/gen/20050601_gen358.pdf
23 Edward English II, in conversation with the author.
24 Every manufacturer has a fuel warranty policy concerning not only biodiesel but any off-specification fuel. Damage caused to a fuel system through the use of biodiesel will not be covered under warranty. Likewise, damage caused by sand, dirt, or other contaminants in petroleum diesel is not covered under warranty either.
25 Automotive Industry Data, (http://www.eagleaid.com/index.htm).
26 Diesel Technology Forum, "Demand for Diesels: The European Experience," (July 2001), http://www.dieselforum.org/fileadmin/templates/whitepapers/EuropeanExperience.pdf.
27 Ibid.
28 http://www.biodiesel.org/resources/reportsdatabase/reports/gen/20050728_Gen-354.pdf.
29 Jon Van Gerpen, et al., Building a Successful Biodiesel Business, January 2005, p. 172, Figure 35.

Chapter 17
LIVING WITH RENEWABLE ENERGY

Solar Thermal Systems

Once a solar thermal system has been installed and commissioned, very little maintenance is required. For the majority of homeowners, sweeping snow from the collectors (assuming they are safely accessible) is the limit of day-to-day maintenance. It is also advisable to check the system pressure either early in the morning or in the evening, before or after the heating day, to ensure consistent day-to-day readings. System leakage is signified by successive pressure readings that indicate a drop in pressure.

The majority of systems installed throughout North America use nontoxic antifreeze known as propylene glycol to transfer heat from the collector panels to the storage tank in the house. Propylene glycol is rated for a maximum operating temperature of 325°F (163°C), which is well above the normal operating temperature of the system. During the summer months, particularly when people are on vacation, stagnation of the propylene glycol may occur as a result of the low heating load of the home. Should stagnation continue for extended periods of time, the glycol solution will become "sticky" and gum up the works.

It is recommended that a trained service technician analyze the glycol solution every second year. During this test, all system components should also be checked to ensure peak operating efficiency. You may also examine the glycol solution for pH and freeze protection using a home test kit as outlined in Chapter 5.3.

Solar Electric Systems

Well, let's get ready to break out the champagne and "throw the switch." All the wiring is done, the circuits have been checked and rechecked, and the electrical inspector has given the OK. It's time to get things running. So where do you start? It can't really be that simple. After all, there's a lot of equipment involved.

Getting Started

The best place to start is with safety and also with some background on operating a system. Each source of renewable energy, whether it's a wood stove or PV panels, has an operating guide supplied with it. All manufacturers provide a list of the safety tools and supplies you will need when working with their products. This equipment has been covered in the previous chapters, so go back and review the material as necessary. The most obvious requirements are common sense and the ability to learn as you work with the various components.

Figure 17-1. The coal-fired grid may become a thing of the past if sufficient renewable energy and efficiency measures are adopted by society at large. (Courtesy Ontario Clean Air Alliance)

Once all of the systems are checked out in accordance with the manufacturers' requirements, we can start things moving. Obviously, the first thing to do is to energize the system as discussed in the following paragraphs.

Step 1.

Record the specific gravity for each battery cell prior to initial operation unless you are using maintenance-free or NiCad batteries, in which case this step is not required. Recording specific gravity will provide a basis for determining the state of the batteries upon receipt and may be required if you have any warranty issues. Use a felt pen or marking plate to give each individual cell a number. If your system is 12V, you will have 6 cells. Likewise 24V and 48V systems will have 12 and 24 cells respectively per battery bank. If you have multiple sets of batteries wired in parallel, such as those shown in Figure 9-19, you will have double the number of cells in your bank.

Step 2.

Energize the main circuit breaker connecting the batteries to the inverter. At this point, the inverter will be ready for startup. Follow the manufacturer's suggested startup procedure to get the unit online. You will need to set up the following basic parameters:

- battery charger rate;
- battery bank capacity;
- battery type;
- over-discharge protection;
- search mode sensitivity.

More advanced inverters require you to set many more options, and this can be done initially or on an as-needed basis. For first-time operators, this would be a good time to have your system dealer available to do a bit of "hand holding."

Step 3.

If your inverter is equipped with a "search mode" option, set the search sensitivity. This setting adjusts the current threshold required to bring the inverter out of the low-power "sleep" mode and into full-power operation. As we learned earlier, the search mode works by sending a brief pulse of power into the power lines. When all circuits are turned off, the flow of electricity is stopped. At the instant the circuit is closed (when a light switch is turned on, for example), the current pulse travels through the load and back to the inverter, signaling it to activate. Sleep mode saves considerable electrical energy when the

Figure 17-2. Saving energy is always more economical than generating additional energy. This is true whether you are talking about an urban homeowner or the electrical utility.

inverter is not required, for example when everyone is sleeping.

Search mode can be tricky to adjust, as the sensitivity differs depending on which load is activated. An 8W CF lamp uses far less power than a toaster. Following the inverter manual instructions, perform this adjustment using a variety of electrical loads in the house.

Step 4.

Program the energy meter according to the manufacturer's instructions. Typical data required include the battery bank voltage and capacity (in amp-hours). With all renewable source supplies turned off, the inverter will be drawing power of varying amounts depending on the electrical loads activated at the time. Try turning a light on or off to see the amount of current draw. Each load will require more or less current (in amps), causing the meter to register accordingly.

Step 5.

Close the circuit breaker for the first renewable-source supply. The input power will vary depending on whether it is sunny and/or windy outside. The energy meter should now start to record the energy going into the battery. A common problem at startup is the current shunt wiring (Figure 14-22) that is connected to the meter backwards. A quick test is to turn off all electrical loads and monitor the meter when energy is being produced. The meter should indicate a "battery charging" condition, with the capacity moving slowly towards the full state. Repeat this step for each renewable source.

Step 6.

Close the circuit breaker for the generator feed inputs. Start the generator (if one is installed) either manually or by activating the autostart function in accordance with the inverter instruction manual. Once the generator has started and stabilized for about twenty seconds, the inverter will "click over" to battery-charge mode.

When this occurs, the batteries should start to charge at the inverter's maximum rate (or as programmed). The energy meter will record the resulting power input and charging will progress.

While generator charging is in progress, verify that the house electrical power is active. All lights and appliances should be able to run as if they were running on the inverter. Note that the generator power may add a "flicker" condition to lights. This is completely normal operation.

Step 7.

If the battery specific gravity is lower than 75% full upon receipt, as determined in Step 1, allow the generator to operate for one charging cycle as directed by the inverter manual. This will ensure that the battery electrolyte is up to 100% capacity as you begin to work with your new system.

Step 8.

Take time to read the energy meter manual to understand the relationship between the meter and batteries. Remember that the battery meter is only a "guesstimate" of actual battery state of charge.

Up and Running

The real beauty of a renewable energy system is its ease of operation once everything is commissioned and running. Correct operation and proper attention to the system will prolong its life. The most neglected part of a system is the batteries, as those unfamiliar with renewable energy systems often do not realize how much energy they are using or understand the relationship between personal energy use and battery depth of discharge.

If there is any point that needs to be made clear from the beginning it is that you do not have a line of credit with your battery bank. There is only so much "cash" in the savings account. Use it up and that's it; the lights go out. Keep draining the bank continually and the bank will close permanently.

Read the battery manual to ensure that you understand how your energy usage, energy meter, and battery specific gravity work together. **CHECK THE SPECIFIC GRAVITY EVERY DAY UNTIL YOU UNDERSTAND THIS RELATIONSHIP.** Once you fully understand the energy meter/specific gravity interaction, you need to check the battery bank only once a month. But until then, use care and follow these steps to the letter:

Figure 17-3. Saving energy is not limited to your home. Hybrid vehicles such as this model from Toyota will reduce your automotive energy bill by 50% and smog-forming emissions by 90%.

- Battery specific gravity will read abnormally low immediately after a full day of charging. This is because of tiny gas bubbles suspended in the electrolyte, causing the density of the fluid to decrease. (Gas bubbles are lighter than water or sulfuric acid.) Wait at least one or two hours after charging before taking the specific gravity reading.

- Likewise, specific gravity reading should not be taken when the batteries are under a full load. Wait until the fridge stops or the microwave is not being used before taking readings.

- Battery electrolyte specific gravity readings that are outside room temperature range will need to be "corrected" in accordance with the manufacturer's data.

- Use a quality hydrometer to take the readings. Look at the scale from straight on to ensure that the reading is correct.

- Compare the specific gravity readings with the battery depth-of-discharge chart in Chapter 9, Table 9-1. If the battery bank indicates that it is 90% full (i.e. 10% depth of discharge) and the energy meter agrees, great! If not, the meter may need some fine-tuning if the readings are too far out of whack. You may wish to look at a figure called the *charge efficiency factor* setting. This is a fancy term for the fudge factor to help calibrate the meter-to-battery setting. Decrease the efficiency factor if the meter thinks the batteries are charged more fully than they really are. Likewise, increase the efficiency factor if the meter reads lower than the specific gravity reading.

- As time goes by, the readings of even the best energy meter will fall out of step with the batteries. At this point, read the meter manual to determine how to reset the meter to agree with the battery capacity. This step is normally accomplished when the batteries are fully charged immediately after an equalization charge, approximately four times per year.

- Speaking of equalization, watch the specific gravity between cells. The readings should all be approximately the same. If the cells are starting to get out of balance by a reading of more than 0.010 (ten points on the scale), it's time to equalize.

- Equalization may be accomplished using grid power, a generator, or better yet, the PV panel or the wind turbine on sunny/windy days. Whatever method you choose, it is necessary to "program" the charge controller and/or inverter to start the equalization process. Many charge controllers have a single button which increases the battery maximum charge voltage, thus enabling equalization mode. The inverter will often have a similar button or setting which is activated when equalization is started from a generator.
- If you use hydrocaps with your battery bank you **MUST** remove them during equalization.

What Else?

There really isn't much else to think about. Equipment manufacturers will give you a list of yearly maintenance steps to follow. Some are absurdly simple, such as spring-cleaning the PV cells to wash off any dirt on the glass. Wind turbines require an inspection, although even this can be pretty simple. Bergey WindPower suggests checking once every year to see if the unit is turning....Honestly, it's right there in the manual!

Of course there are some tricks you will learn over time that help make living with renewable energy easier:

- Watch the battery specific gravity or energy meter when you have a long-term blackout of one day or longer. This is where many newcomers to renewable energy get into trouble with their batteries. Recognize when the batteries are depleted enough to warrant

Figure 17-4. Heating your pool or hot tub with solar energy eliminates fossil-fuel heating, extending the swimming season without increasing the operating cost.

shutting the system down or running a generator.

- During severe lightning storms, consider furling your wind turbine to limit mechanical stress during the storm. Consult the manufacturer's manual for additional details.
- Make seasonal adjustments to a manual PV array: latitude +15° towards the vertical for winter; latitude -15° for summer.
- Remember never to smash ice and snow off the PV panels. Simply brush off the top layer of snow with a squeegee or brush. The sun will quickly take care of the rest.
* Generators require regular servicing, including oil and filter changes as well as battery condition approximately every 500 hours, based on manufacturer requirements. It is typical for generators to run frequently during the winter and not be heard of again for the balance of year. Make sure the unit is started at least once a month, to ensure that it is ready to go when needed.
- Rest easy knowing that you are not the only one living lightly on the planet. Support abounds from dealers, like-minded neighbors, the Web, and publications such as *Home Power* (www.homepower. com). Renewable energy is here to stay.

Most importantly, enjoy your handiwork and marvel at the elegant simplicity of it all.

Figure 17-5a and b. Never smash ice and snow off the PV panels. Simply brush off the top layer of snow with a squeegee or brush. The sun will quickly take care of the rest.

Chapter 18
CONCLUSION

Renewable energy works. Yes, there is a lot of technology to understand before embarking on this path, but the rewards make every bit of effort worthwhile. Lorraine tells me that she now feels more "connected" in her actions and her interaction with the environment.

Once you start harvesting your own energy and seeing what it takes to heat a house or run a light bulb, endless, mindless waste becomes intolerable. As we approach the end of 2009, hundreds of thousands of people are moving toward a more sustainable lifestyle. This movement started with the hippies in the '60s, and after the psychedelic haze left that era environmental concerns and technology started to enter the mainstream. Renewable energy has now arrived.

Companies such as BP, Kyocera, Siemens, and Xantrex are not in the business to help out aging hippies; they are here because renewable energy is business—big business. Governments, NGOs, industry, and grassroots people like you and me are making the transition to a cleaner, more sustainable future for the sake of our children and the planet. Let's hope that the rest of society gets the message before it's too late.

"To see a world in a grain of sand and a heaven in a wild flower, hold infinity in the palm of your hand and eternity in an hour."

William Blake

Appendix 1
Cross Reference Chart of Various Fuel Energy Ratings

Fossil Fuels and Electricity

Heating Fuel	BTU per Unit
Heating Oil	142,000 BTU/gallon (38,700 kj/L)
Natural Gas	46,660 BTU/cubic-yard (37,700 kj/m3)
Propane	91,500 BTU/gallon (26,900 kj/L)
Electricity (resistance heating)	3413 BTU/kWh (3600 kj/kWh)
Coal (air dry average)	12,000 BTU/LB (27,900 kJ/kg)

Renewable Energy Heating Fuels

Heating Fuel	BTU per Unit
Shelled Corn	7000 BTU/lb (16,200 kJ/kg)
	14,000,000 BTU/ton (12,700 MJ/tonne)
Firewood by weight (all types) *	8000 BTU/lb (18,500 kJ/kg)
Hardwood Firewood by volume: * Ash	25,800,000 BTU/cord (27,200 MJ/cord)
Beech	28,900,000 BTU/cord (30,500 MJ/cord)
Red Maple	22,300,000 BTU/cord (23,500 MJ/cord)
Red Oak	27,200,000 BTU/cord (28,700 MJ/cord)
Hybrid Poplar	18,500,000 BTU/cord (19,500 MJ/cord)
Mixed Hardwood (average)	27,000,000 BTU/cord (30,000 MJ/cord)
Mixed Softwood (average)	17,500,000 BTU/cord (18,700 MJ/cord)
Wood Pellets	20,700,000 BTU/ton (19,800 MJ/tonne)
Biodiesel	128,000 BTU/gallon (35,500 kJ/L)

* All firewood has the same heating or carbon content per pound or kilogram of mass. However, the density of softwoods is much lower owing to increased air and moisture content, resulting in lower BTU content per unit mass.

Appendix 2

Typical Power and Electrical Ratings of Appliances and Tools

Appliance Type (Watts)	Power Rating	Energy Usage per Hour, Day or Cycle
Large Appliances:		
Gas clothes dryer	600	500 Wh per dry cycle
Electric clothes dryer	6,000	5 kWh per dry cycle
High efficiency clothes washer	300	250 Wh per wash
Ten-year-old vertical axis clothes washer	1,200	720 Wh per wash
Ten-year-old refrigerator	720	5 kWh per day
New energy-efficient refrigerator	150	1.2 kWh per day
Ten-year-old chest freezer	400	3 kWh per day
New energy-efficient chest freezer	140	0.9 kWh per day
Dishwasher "normal cycle"	1,500	800 Wh per cycle
Dishwasher "eco-dry cycle"	600	300 Wh per cycle
Portable vacuum cleaner	600	600 Wh per hour
Central vacuum cleaner	1,400	1.4 kWh per hour
Air conditioner 12,000 BTU (window)	1,200	1.2 kWh per hour
AC submersible well pump (1/2 hp)	1,150	200 Wh per cycle
DC submersible well pump	80	160 Wh per day
DC slow pump (includes booster pump)	80	160 Wh per day
Small Appliances:		
Microwave oven (0.5 cubic foot)	900	0.9 kWh per hour
Microwave oven (1.5 cubic foot)	1,500	1.5 kWh per hour
Drip style coffee maker (brew cycle)	1,200	1.2 kWh per hour
Drip style coffee maker (warming cycle)	300	0.3 kWh per hour
Espresso/cappuccino maker	1,200	300 Wh per cycle
Food processor	300	50 Wh per cycle
Coffee grinder	100	10 Wh per cycle
Toaster	1,200	150 Wh per cycle
Blender	300	50 Wh per cycle
Hand mixer	100	10 Wh per cycle
Hair dryer	1,500	200 Wh per cycle
Curling iron	600	100 Wh per cycle
Electric toothbrush	2	50 Wh per day
Electric iron	1,000	1 kWh per hour

Appendix 2 Continued

Typical Power and Electrical Ratings of Appliances and Tools

Appliance Type	Power Rating (Watts)	Energy Usage per Hour, Day or Cycle
Electronics:		
Television –12 inch B&W	20	20 Wh per hour
Television –32 inch color	140	140 Wh per hour
Television –50 inch high definition	160	160 Wh per hour
Satellite dish and receiver	25	25 Wh per hour
Stereo system	50	50 Wh per hour
Home theater system (movie)	400	1 kWh per movie
Cordless phone	3	72 Wh per day
Cell phone in charger base	3	72 Wh per day
VCR/DVD/CD component	25	25 Wh per hour
Clock radio (not including inverter waste)	5	120 Wh per day
Computer (Desktop)	60	60 Wh per hour
Computer (Laptop)	20	20 Wh per hour
Monitor - 15 inch	100	100 Wh per hour
Monitor - 15 inch flat screen	30	30 Wh per hour
Laser printer (standby mode average)	50	50 Wh per hour standby
Laser printer (print mode)	600	600 Wh per hour printing
Inkjet printer (all modes)	30	30 Wh per hour
Fax machine	5	120 Wh per day
PDA charging	3	72 Wh per day
Fluorescent desk lamp	10	10 Wh per hour

Appendix 3

Resource Guide

Energy Efficiency Councils and Societies:

American Council for an Energy Efficient Economy

Website: www.aceee.org
Phone: 202-429-8873
Publishes guides' comparing the energy efficiency of appliances

American Solar Energy Society

Website: www.ases.org
Phone: 303-443-3130
ASES is the United States chapter of the world Solar Energy Society. They promote the advancement of solar energy technologies.

The American Wind Energy Association

Website: www.awea.org
Phone: 202-383-2500
The AWEA is the trade association for developers and manufacturers of wind turbine and associated equipment and infrastructure.

California Energy Commission

Website: www.energy.ca.gov
Phone: 916-654-4058
The CEC is the strongest supporter of grid inter-connected renewable energy systems in North America. Their website explores what is happening in California in this regard.

Canadian Standards International

Website: www.csa-international.org
Phone: 416-747-4000
Develops standards for the Canadian marketplace. Tests and administers safety certification work in North America.

Canadian Wind Energy Association

Website: www.canwea.ca
Phone: 800-992-6932
The CAWEA vision is to have 10,000 MW of wind power systems installed in Canada by 2010. They promote all aspects of wind energy and related systems to the industry.

David Suzuki Foundation

Website: www.davidsuzuki.org
Phone: 614-732-4228
Dr. Suzuki is a lecturer and TV broadcaster promoting energy efficiency, global climate change and ocean sustainability. The website provides links and publications on all manner of environmental sciences.

Electro Federation of Canada

Website: http://www.electrofed.com
Phone: 905-602-8877
The Electro Federation is a consortium of electrical manufacturers working in many disciplines of electrical engineering and sales. The micropower-connect division is dealing with small (<50kW) distributed energy producers interconnecting to the grid in Canada.

Energy Star

Website: www.energystar.gov
Phone: 888-782-7937
Their website reviews energy efficient computers and electronics.

The Green Power Network

Website: http://apps3.eere.energy.gov/greenpower/
This website describes the status of utility interconnection guidelines on a state-by-state basis. Also provides information on where to purchase green electricity when connected to the grid.

Appendix 3
Resource Guide (continued)

National Renewable Energies Laboratory (NREL)
Website: www.nrel.gov
Phone: 303-275-3000
The NREL is the national renewable energy research laboratory in the United States.

Natural Resources Canada
Website: www.nrcan.gc.ca
Phone: N/A
The Government of Canada hosts this website which includes the office of energy efficiency. Many resources are presented in this fact filled site.

Rocky Mountain Institute
Website: www.rmi.org
Phone: 970-927-3851
The Rocky Mountain Institute is a think tank regarding all energy efficiency issues. Their website contains a great deal of source information for books and applied research.

Underwriters Laboratories Inc.
Website: www.ul.com
Phone: 847-272-8800
Develops standards for the United States marketplace. Tests and administers safety certification work in North America.

Trade Publications and Magazines:

altenerg.com
Website: www.altenerg.com
Articles about alternative energy sources, wind, solar, geothermal, biomass, biofuels, enerG magazine.

altenergymag.com
Website: www.altenergymag.com/
Alternative energy resources, news, emagazine and library.

Home Energy magazine
Website: www.homeenergy.org
In print or online, the latest news on energy-efficient, durable, comfortable, and green homes..

Home Power Magazine
Website: www.homepower.com
Phone: 800-707-6585
This magazine bills itself as "The hands-on journal of home-made power". Based in Oregon, the magazine deals primarily with a south and west coast flavor. Extensive details related to producing alternate energy.

Renewable Energy World.com
Website: www.renewableenergyworld.com
Renewable energy news, jobs, events, companies and more.

Solar Today Magazine
Website: http://www.ases.org/
Published by the American Solar Energy Society. An award-winning magazine that connects you to the leading solar energy experts Available in print or online.

Appendix 3
Resource Guide (continued)

Manufacturers:

Photovoltaic Panels and Equipment

Photovoltaic panel manufacturers do not supply directly to end consumers, as they rely on their large distribution networks throughout the world. For information purposes, here are some of the major suppliers in this field. Contact the manufacturer or visit their website for distributors in your local area.

BP Solar
Website: www.bpsolar.com
Phone: 410-981-0240

Evergreen Solar Inc.
Website: www.evergreensolar.com
Phone: 508-357-2221

Kyocera Solar Inc.
Website: www.kyocerasolar.com
Phone: 800-544-6466

Schott Applied Power Corporation
Website: www.schottappliedpower.com
Phone: 888-457-6527

Sharp USA
Website: www.sharpusa.com
Phone: 800-BE-SHARP

Siemens Solar Inc.
Website: http://w1.siemens.com
Phone: 877-360-1789

Solardyne Corporation
Website: www.solardyne.com
Phone: 503-244-5815
Manufacturer of small solar modules for charging laptop computers, cell phones, etc.

PV Module Mounts

Array Technologies Inc.
Website: www.wattsun.com
Phone: 505-881-7567
Manufacturer of the Wattsun active tracking system

IronRidge
Website: http://www.ironridge.com/
Office: 707-459-952
Formerly Two Seas Metalworks, provider of solar racking and mounting systems

Sun-Link Solar Tracker
Website: www.northernlightsenergy.com
Phone: 705-246-2073
Manufacturer of Sun-Link active tracking system.

UniRac Inc.
Website: www.unirac.com
Phone: 505-242-6411

Zomeworks Corporation
Website: www.zomeworks.com
Phone: 800-279-6342
Manufacturer of fixed and passive tracking mount systems

Wind Turbines:

Bergey Windpower Inc.
Website: www.bergey.com
Phone: 405-364-4214

Appendix 3
Resource Guide (continued)

Southwest Windpower
Website: www.windenergy.com
Phone: 520-779-9463

Bornay Windturbines
Website: www.bornay.com
Phone: +34-965-560-025
Manufacturer of wind turbines from Spain

Lake Michigan Wind and Sun
Website: www.windandsun.com
Phone: 920-743-0456
New and re-built turbines

Wind Turbine Industries Corporation
Website: www.windturbine.net
Phone: 952-447-6064
Sole manufacturer of Jacob Wind Systems since 1986.

True North Power Systems
Website: www.truenorthpower.com
Phone: 519-793-3290

Windturbine.ca
Website: www.windturbine.ca
Phone: 886-778-5069

Micro Hydro Turbine Systems

Canyon Industries Inc.
Website: http://www.canyonhydro.com/
Phone: 360-592-5552

Energy Systems and Design
Website: www.microhydropower.com
Phone: 506-433-3151

Harris Hydroelectric
Website: n/a
Phone: 831-425-7652

HydroScreen Co. LLC
Website: www.hydroscreen.com
Phone: 303-333-6071
Manufacturer of intake screen for micro-hydro systems

Home Power Systems
Website: www.homepower.ca
Phone: 604-465-0927

Battery Manufacturers

Dyno Battery Inc.
Website: www.dynobattery.com
Phone: 206-283-7450

HuP Solar-One Battery
Website: www.hupsolarone.com
Phone: 208-267-6409

IBE Battery
Website: www.ibe-inc.com
Phone: 818-767-7067

Rolls Battery Engineering (USA)
Surrette Battery Company (Canada)
Website: www.surrette.com
Phone: 800-681-9914

Trojan Battery Company
Website: www.trojanbattery.com
Phone: 800-423-6569

Appendix 3
Resource Guide (continued)

U.S. Battery Manufacturing Company
Website: www.usbattery.com
Phone: 800-695-0945

Hydrogen Recombining Caps

Hydrocap Catalyst Battery Caps
Website: N/A
Phone: 305-696-2504

D.C. Voltage Regulators

Morningstar Corporation
Website: www.morningstarcorp.com
Phone: 215-321-4457

Steca Gmbh
Website: www.stecasolar.com
Phone: N/A

Xantrex Technology Inc.
Website: www.xantrex.com
Phone: 360-435-2220

Inverters

ExelTech Inc.
Website: www.exeltech.com
Phone: 800-886-4683

Out Back Power Systems
Website: www.outbackpower.com
Phone: 360-435-6030

SMA America Inc.
Website: www.sma-america.com
Phone: 530-273-4895

Xantrex Technology Inc.
Website: www.xantrex.com
Phone: 360-435-2220

Backup Power Gensets

Epower
Website:
 www.epowerchargerboosters.com
Phone: 423-253-6984

Generac Power Systems Inc.
Website: www.generac.com
Phone: 1-888-GENERAC

Hardy Diesel & Equipment Inc.
Website: www.hardydiesel.com
Phone: 800-341-7027

Kohler Power Systems
Website: www.kohlerpowersystems.com
Phone: 800-544-2444

Energy Meters

Bogart Engineering
Website: www.bogartengineering.com
Phone: 831-338-0616

Brand Electronics
Website: www.brandelectronics.com
Phone: 207-549-3401

Xantrex Technology Inc.
Website: www.xantrex.com
Phone: 360-435-2220

Appendix 3
Resource Guide (continued)

<u>Miscellaneous</u>

Bussmann
Website: www.bussmann.com
Phone: 314-527-3877
Fuses and electrical safety components

Delta Lightning Arrestors Inc.
Website: www.deltala.com
Phone: 915-267-1000

Digi-Key (Canada and USA)
Website: www.digikey.com
Phone: 800-DIGI-KEY
Supplier of many electrical wiring components

Electro Sonic Inc. (Canada)
Website: www.e-sonic.com
Phone: 800-56-SONIC
Supplier of many electrical wiring components

Real Goods
Website: www.realgoods.com
Phone: 800-919-2400
Suppliers specializing in renewable energy systems

Siemens Energy and Automation Inc.
Website: www.siemens.com
Phone: 404-751-2000
Fused disconnect switches and circuit breakers

Xantrex Technology Inc.
Website: www.xantrex.com
Phone: 360-435-2220
D.C. circuit breakers, battery cables, power centers,
metering shunts, fuses

Appendix 4

Magnetic Declination Map for North America

**MAGNETIC DECLINATION MAP
OF NORTH AMERICA**

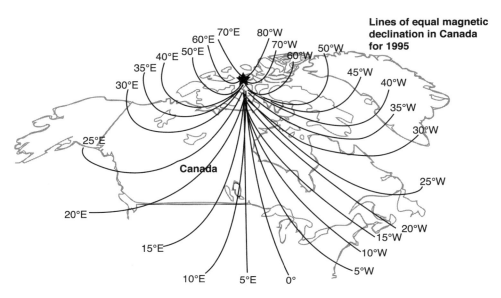

Lines of equal magnetic declination in Canada for 1995

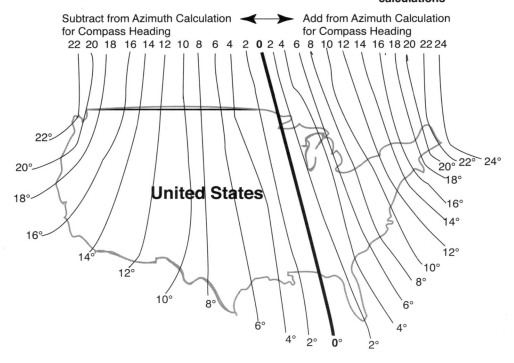

Map of US showing magnetic declatination for Azimuth compass calculations

Appendix 5
Map of Winter Average Sun Hours per Day
for North America

North American Sun Hours per Day (Worst Month)

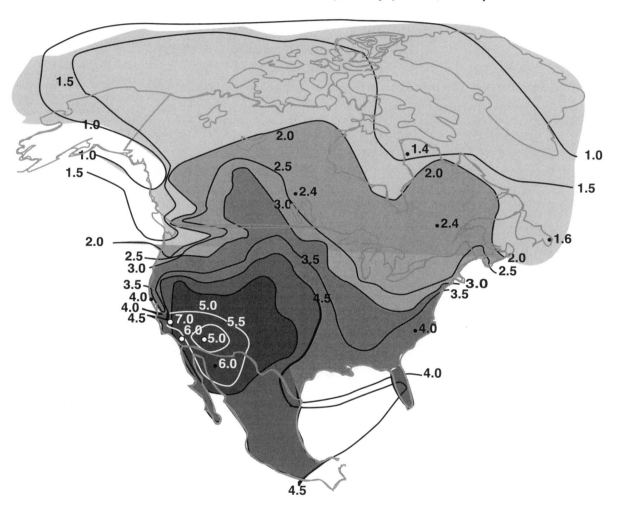

Map indicates the average winter sun hours per day. Use this map when calculating the average energy output of a PV system used all year round.

Appendix 6

Map of Yearly Average Sun Hours per Day for North America

"North American Sun Hours per Day (Yearly Average)"

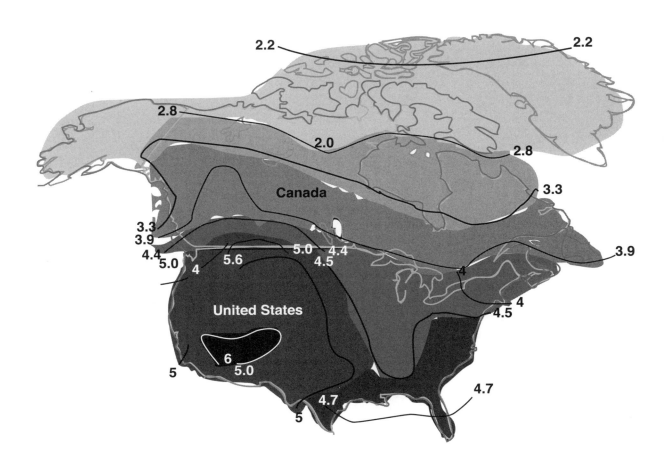

Map indicates the yearly average of sun hours per day. Use this map when calculating the average energy output of a PV system used seasonally during the spring/summer months.

Appendix 7

Electrical Energy Consumption Worksheet

Appliance Type	Appliance Wattage (Volts & Amps)	X	Hours of Daily Use	=	Average Watt-hours Per Day

Total Watt-hours Per Day for all Appliances =

Start by checking your major electrical loads in the house. Anything that plugs into a regular wall socket will have a label that tells you the voltage (usually 120 or 240 for larger appliances) and the current or wattage for that item. Although these labels can over state energy usage, they are a good guide for calculating energy consumption. Enter the wattage data from the labels on the *Energy Consumption Worksheet*. (If the appliance label does not show wattage, multiply amps x volts to calculate watts.)

The next step is to see if you can estimate how many hours you use the device per day. If you use a device only occasionally, try to estimate how long it is used per week and divide this time by 7. Your calculator will quickly give you your "daily average energy usage".

Appendix 8

Average Annual Wind Speed Map for North America

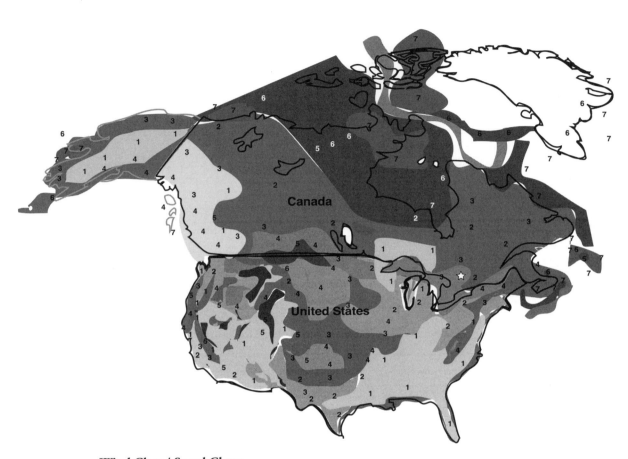

Wind Class / Speed Chart
Class 1: 3.8 m/s (8.5 mph)
Class 2: 4.8 m/s (10.8 mph)
Class 3: 5.4 m/s (12.1 mph)
Class 4: 5.8 m/s (13.0 mph)
Class 5: 6.2 m/s (13.9 mph)
Class 6: 6.7 m/s (15.0 mph)
Class 7: 7.5 m/s (16.8 mph)

Map indicates the average annual wind speed in meters per second averaged over a ten-year period. Elevation of anemometer is 33 feet (10 m). Use this map with care as local wind speed levels will vary. Refer to resource guide for further details on wind mapping for your specific area.

Appendix 9a

Polyethylene SDR Pipe Friction Losses

Pressure Loss from Friction in Feet of Head per 100 Feet (30 m) of Pipe

(Table Courtesy of Energy Systems and Design)

Flow in US GPM	Pipe Diameter in Inches							
	0.5	0.75	1	1.25	1.5	2	2.5	3
1	1.13	0.28	0.09	0.02				
2	4.05	1.04	0.32	0.09	0.04			
2	8.60	2.19	0.67	0.19	0.09	0.02		
4	14.6	3.73	1.15	0.30	0.14	0.05		
5	22.1	5.61	1.75	0.46	0.21	0.07		
6	31.0	7.89	2.44	0.65	0.30	0.09	0.05	
7	41.2	10.5	3.24	0.85	0.42	0.12	0.06	
8	53.1	13.4	4.14	1.08	0.51	0.16	0.07	
9		16.7	5.15	1.36	0.65	0.18	0.08	
10		20.3	6.28	1.66	0.78	0.23	0.09	0.02
12		28.5	8.79	2.32	1.11	0.32	0.14	0.05
14		37.9	11.7	3.10	1.45	0.44	0.18	0.07
16			15.0	3.93	1.87	0.55	0.23	0.08
18			18.6	4.90	2.32	0.69	0.30	0.09
20			22.6	5.96	2.81	0.83	0.35	0.12
22			27.0	7.11	3.36	1.00	0.42	0.14
24			31.7	8.35	3.96	1.17	0.49	0.16
26			36.8	9.68	4.58	1.36	0.58	0.21
28				11.1	5.25	1.56	0.67	0.23
30				12.6	5.96	1.77	0.74	0.25
35				16.8	7.94	2.35	1.00	0.35
40				21.5	10.2	3.02	1.27	0.44
45				26.8	12.7	3.75	1.59	0.55
50				32.5	15.4	4.55	1.91	0.67
55					18.3	5.43	1.96	0.81
60					21.5	6.40	2.70	0.94
65					23.8	7.41	3.13	1.08
70					28.7	8.49	3.59	1.24
75					32.6	9.67	4.07	1.4
80						10.9	4.58	1.59
85						12.2	5.13	1.77
90						13.5	5.71	1.98
95						15.0	6.31	2.19
100						16.5	6.92	2.42
150						34.5	14.7	5.11
200							25.0	8.70
300								18.4

Appendix 9b

PVC Pressure Class 160 PSI Pipe Friction Losses

Pressure Loss from Friction in Feet of Head per 100 Feet (30 m) of Pipe

(Table Courtesy of Energy Systems and Design)

Flow in US GPM	Pipe Diameter in Inches							
	1	1.25	1.5	2	2.5	3	4	5
1	0.05	0.02						
2	0.14	0.05	0.02					
2	0.32	0.09	0.04					
4	0.53	0.16	0.09	0.02				
5	0.80	0.25	0.12	0.04				
6	1.13	0.35	0.18	0.07	0.02			
7	1.52	0.46	0.23	0.08	0.02			
8	1.93	0.58	0.30	0.10	0.04			
9	2.42	0.71	0.37	0.12	0.05			
10	2.92	0.87	0.46	0.16	0.07	0.02		
11	3.50	1.04	0.53	0.18	0.07	0.02		
12	4.09	1.22	0.64	0.20	0.09	0.02		
14	5.45	1.63	0.85	0.28	0.12	0.04		
16	7.00	2.09	1.08	0.37	0.14	0.04		
18	8.69	2.60	1.33	0.46	0.18	0.07		
20	10.6	3.15	1.63	0.55	0.21	0.09	0.02	
22	12.6	3.77	1.96	0.67	0.25	0.10	0.02	
24	14.8	4.42	2.32	0.78	0.30	0.12	0.04	
26	17.2	5.13	2.65	0.90	0.35	0.14	0.05	
28	19.7	5.89	3.04	1.04	0.41	0.16	0.05	
30	22.4	6.70	3.45	1.17	0.43	0.18	0.05	
35		8.90	4.64	1.56	0.62	0.23	0.07	
40		11.4	5.89	1.98	0.78	0.30	0.09	0.02
45		14.2	7.34	2.48	0.97	0.37	0.12	0.04
50		17.2	8.92	3.01	1.20	0.46	0.14	0.04
55		20.5	10.6	3.59	1.43	0.55	0.16	0.05
60		24.1	12.5	4.21	1.66	0.64	0.18	0.07
70			16.6	5.61	2.21	0.85	0.25	0.09
80			21.3	7.18	2.83	1.08	0.32	0.12
90				8.92	3.52	1.36	0.39	0.14
100				10.9	4.28	1.66	0.48	0.18
150				23.2	9.06	3.5	1.04	0.37
200					15.5	5.96	1.75	0.62
250					23.4	9.05	2.65	0.94
300						12.6	3.73	1.34
350						16.8	4.95	1.78
400						21.5	6.33	2.25
450							7.87	

Appendix 10
Voltage, Current and Distance (in feet) Charts
for Wiring (AWG gauge)

(1% Voltage drop shown – Increase distance proportionally if greater losses allowed)

12 Volt Circuit

Amps	WIRE GAUGE									
	10	8	6	4	2	1	1/0	2/0	3/0	4/0
1	48	74	118	187	299	375	472	594	753	948
5	10	15	24	37	60	75	94	119	151	190
10	5	7	12	19	30	38	47	59	75	94
20	2	4	6	9	15	19	24	30	38	48
40	1	2	3	5	7	9	12	15	19	24

24 Volt Circuit

Amps	WIRE GAUGE									
	10	8	6	4	2	1	1/0	2/0	3/0	4/0
10	10	15	24	37	60	75	94	119	151	190
20	5	7	12	19	30	38	47	59	75	94
30	3	5	8	12	20	25	31	40	50	63
40	2	4	6	9	15	19	24	30	38	48
50	2	2	4	6	10	13	16	20	25	31
100	1	1	2	4	6	8	9	12	15	19
125	1	1	2	3	5	6	8	10	13	16
150	1	1	2	2	4	5	6	8	10	13

48 Volt Circuit

Amps	WIRE GAUGE									
	10	8	6	4	2	1	1/0	2/0	3/0	4/0
10	20	30	48	74	120	150	188	238	302	380
20	10	15	24	38	60	76	94	118	150	188
30	6	10	16	24	40	50	62	80	100	126
40	4	8	12	18	30	38	48	60	76	96
50	4	4	8	12	20	26	32	40	50	62
100	2	2	4	8	12	16	18	24	30	48
125	2	2	4	6	10	12	16	20	26	32
150	2	2	4	4	8	10	12	16	20	26

Appendix 11
Wire Sizes Versus Current Carrying Capacity
(Size shown is for copper conductor only)

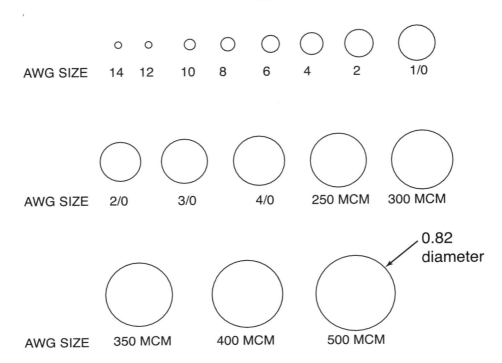

AWG	Maximum Current
14	15
12	20
10	30
8	55
6	75
4	95
2	130
0	170
2/0	195
3/0	225
4/0	260

Appendix 12

A Few Words on Toilets, Greywater, and Waste Management

Like everything else in the city, potable water and waste management are invisible services provided to citizens at an impossibly low price. The average North American takes for granted that clean water magically flows from a tap and that human wastes simply disappear (which of course, is a good thing). Yet we pay a mere dollar or so a day for the privilege, some would say the right, of having these essential services. I placed a phone call to the municipal department of a major city and asked if they could provide me with the cost breakdown for city water. This list included items such as alum, chlorine, filtration, and pumping. I then asked, "So how much do you charge for the actual water?" His reply: "Oh, the water doesn't cost anything. We just pump it. It's the chemicals and purification that cost money." I remember his words every time I go to the developing world and watch people wash food and clothes in fetid puddles, if any are to be found. Southern California and parts of the US Midwest are not too far behind, as wasteful agricultural irrigation methods and household water practices drain the mighty river and aquifer systems millions of people depend on for their survival.

Because humans are so disconnected from the ecosystems that support their very lives, few people consider what they would do if water and sewer were not available and therefore don't make an effort to reduce their current water consumption and waste effluents. Clean water is competing with fossil fuel as the natural commodity headed for extinction.

When you decide to take the plunge and embark on an off-grid life style, these questions become front and center issues, not just academic questions.

We discussed the harvesting and efficient use of water in Chapter 2.5. This appendix will provide a brief overview of wastewater treatment options for an energy-efficient, off-grid, environmentally sustainable home.

For most people, country living, wastewater, and "septic system" are synonymous. True, septic systems work well and are well understood by rural plumbers and tradespeople. However, before we delve into the mechanics of such a system, let's take a moment to review some of the theory behind what we are trying to accomplish.

Defining Wastewater

Wastewater can be divided into two broad categories:
* Greywater: wastewater that has a low pathogen density but contains a combination of organic and inorganic waste materials from everyday activities such as laundry, bathing, and dishwashing;
* Blackwater: water that contains a high-density pathogen load from human organic wastes from toilets.

Wastewater Treatment

The disposal of these waste streams can be treated as two completely different processes, although a typical septic system combines their disposal as one integrated system. The average North American consumes approximately 100 gallons of water per day which breaks down as:

Greywater:	57 gallons/216 liters
Blackwater:	38 gallons/144 liters
Potable water:	5 gallons/19 liters

Provided the greywater is not laden with large amounts of inorganic chemicals, it may be disposed

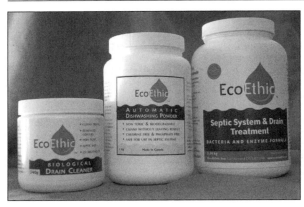

Figure A12-1. The first step in developing any wastewater treatment system is to use water as efficiently as possible and to use biodegradable cleaning products. EcoEthic Inc. (www.ecoethic.ca) provides a full line of home cleaning products that are free of phosphates, petroleum solvents, chlorine, dyes, perfumes, and other environmentally harmful ingredients. EcoEthic products, and others like them, work effectively and can be used even in sensitive wetland areas.

of rather easily by suitable leaching through the earth. The use of biodegradable laundry detergent, bath soap and shampoo and home cleaning products will reduce inorganic compounds dramatically, placing very little stress on the ecosystem as it absorbs and breaks down the waste stream materials.

Blackwater, on the other hand, is loaded with pathogens that are dangerous to human life and must be eliminated. Treatment of human wastes present in blackwater is based on the nutrient cycle of all living organisms. Every organism consumes nutrients (food) and produces a waste by-product. Bacteria decompose these waste materials, turning them back into nutrients and leaving the water in the waste stream to be returned to the ground water table.

Blackwater (toilet waste) is over 99% water, which can be evaporated or clarified and allowed to mix with the greywater stream. The residential septic system is a full sewage treatment plant that processes and purifies both the blackwater and the greywater waste streams simultaneously. The septic system shown in Figure A12-2 allows blackwater and greywater to mix in the main sewer "stack" pipe of the home and feeds this waste stream to the septic tank, where baffles cause the solids to drop to the bottom while turbid water is allowed to flow across the tank baffles and out to the leaching field.

Figure A12-3 shows the connection of household appliances to the main waste stack. Note that the waste stream flows down to the septic tank under the influence of gravity, while "sewer gases" flow upwards to the roof vent. To prevent sewer gases from flowing into the home, plumbing appliances are equipped with either a wet or a dry "gas trap." The wet trap is used in sinks, showers, and washing machines and works by allowing a small amount of water to remain in a U-shaped section of pipe, preventing sewer gas from flowing into the drain and out into the room. A toilet uses a dry trap arrangement, which provides a wet U-trap within the toilet structure itself.

Figure A12-4 shows a detailed view of a septic tank. They may be either concrete or plastic and are buried just below the surface of the earth. Provided the home is used continuously, the tank will not freeze in winter as a result of the heat generated by the waste digestion process and warm waste water flowing into the unit. Cleanout access doors are provided for periodic pumping of residual solids that are not broken down during the digestion stage. Septic tanks must be pumped out every three to five years or solids will flow out to the leaching bed, plugging the system and requiring extensive and costly repairs.

As wastewater flows into the septic tank, internal baffles slow the flow of water and direct solids to the bottom of the tank, forming a sludge layer, while greases and lighter wastes form a floating scum layer. Turbid water remains in the central area of the tank and will periodically flow out to the leaching bed whenever additional contents are added to the tank.

Innumerable bacteria work in all three layers of the septic tank, digesting and breaking down solid

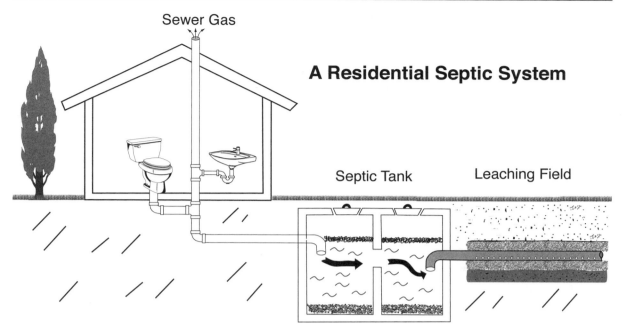

Figure A12-2 The residential septic system is a full sewage treatment plant that processes and purifies both the blackwater and greywater waste streams simultaneously.

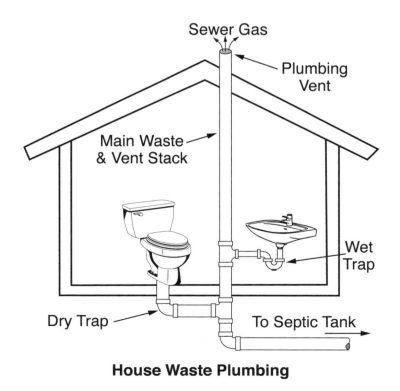

House Waste Plumbing

Figure A12-3. This detailed view shows the connection of household appliances to the main waste stack.

Septic Tank - Side View

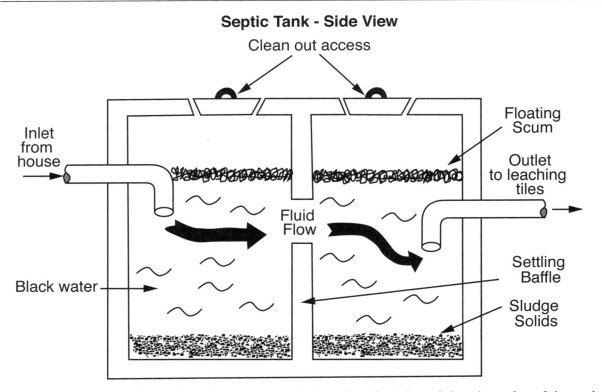

Figure A12-4. Septic tanks may be solid concrete or plastic and are buried just below the surface of the earth. Cleanout access doors are provided for periodic pumping of residual solids that are not broken down during the digestion stage. Septic tanks must be pumped out every three to five years or solids will flow out to the leaching bed, plugging the system and requiring extensive and costly repairs.

effluent matter. About 95% of the solids making up the sludge and scum layers can be broken down by bacteria; the remaining 5% form a solid layer that must be pumped out as discussed earlier.

The leaching bed shown in Figure A12-5 is part of the sewage treatment and is not simply a means of ridding the system of excess water. Partially treated effluent flows from the septic tank into the leaching field where it flows into perforated pipe buried in the soil. Effluent is further broken down by ground-based bacteria and the earth's natural filtration capability as effluent and waste water percolate through the soil. The soil acts as a biological filter, removing harmful materials before they reach the groundwater table which feeds the house water supply.

A properly functioning septic system is a natural process in which ground water is recycled over and over again.

Disadvantages of Septic Systems

You would think that the septic system is the perfect answer to household waste treatment. Why consider anything else? There are a number of reasons:

- a building lot that is very close to fragile wetland ecosystems;
- the cost of a septic system;
- occasional or infrequent use of recreational property;
- a lot that is too small for a full septic system.

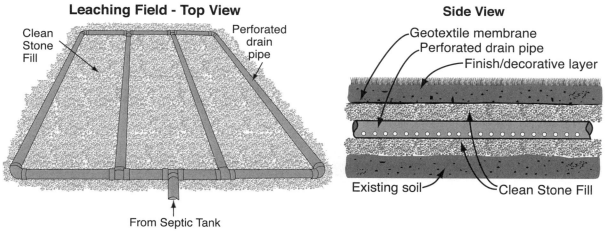

Leaching Field - Top View

Clean Stone Fill

Perforated drain pipe

From Septic Tank

Side View

Geotextile membrane

Perforated drain pipe

Finish/decorative layer

Existing soil

Clean Stone Fill

Figure A12-5. The leaching bed shown in this schematic drawing and photograph is part of the sewage treatment system and is not simply a means of ridding the septic tank of excess water.

In these cases, alternative systems to consider include:
- a full-size effluent holding tank that stores both blackwater and greywater;
- a partial holding tank and leaching bed;
- a composting toilet and leaching bed.

Holding tanks are exactly that: tanks with sufficient capacity to hold several months' worth of total effluent. Holding tanks may be also be downsized if they can be equipped with leaching beds to dispose of excess waste fluids.

Composting toilets have the advantage of decomposing waste solids directly and eliminating the need for periodic pumping. They are also less costly than full septic tanks, although a leaching bed is still required to eliminate greywater. Composting toilets can also be used in boats and RVs where waste management can be very complicated.

A Sun-Mar Centrex composting toilet is shown in Figure A12-6. In this system an ultra-low flush toilet dispenses approximately 1 pint (0.5 liter) of water per flush into a central chamber located in the basement or crawlspace area. A natural "bulking" material similar to peat moss is added to the chamber every two or three days and a handle is rotated, activating the composting material and exposing it to the correct amount of oxygen and moisture. Excess liquid is drained into an evaporation chamber, allowing moisture to escape through the vent system. (Some composting systems allow for the connection of a moisture drain pit, similar to a miniature leaching bed, for the elimination of excess

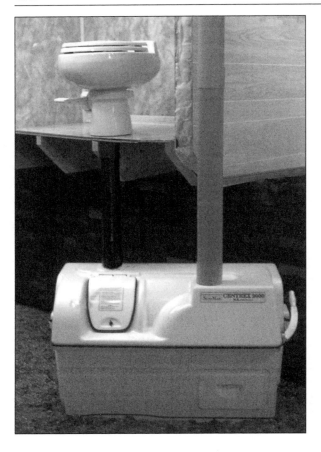

Figure A12-6. Composting toilets have the advantage of decomposing waste solids directly and eliminating the need for periodic pumping. They are also less costly than full septic tanks, although a leaching bed is still required to eliminate greywater. (Courtesy Sun-Mar Corporation)

fluids.) Most models are available as electric or non-electric, allowing them to operate in the off-grid home without any electricity demand.

Periodically, the drum of dry, clean compost is removed from the system and may be added to potting soils or spread on flower gardens just like regular compost.

Composting toilets are also available as self-contained units, having the toilet and composting chamber fitted as one component. These units require very little space and can even be installed in a closet.

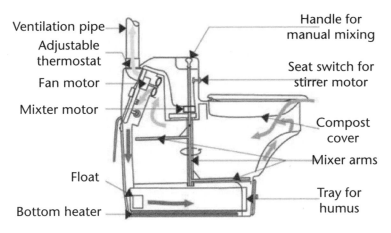

Figure A12-7. A natural "bulking" material similar to peat moss is added to the chamber every two or three days and a handle is activated, rotating the composting material to expose it to the correct amount of oxygen and moisture. Excess liquid is drained into an evaporation chamber, allowing moisture to escape through the vent system. Most models are available as electric or non-electric, allowing them to operate in the off-grid home without any electricity demand. (Courtesy EcoEthic Inc.)

Wastewater Gardens

Nutrients in wastewater are fertilizers. The wastewater garden is an innovative technology that takes advantage of the fact that plants need water and nutrients. The system "grows away" wastewater from residences, eliminating effluent discharge into the ground. The technology is particularly ideal for seasonal locations in proximity to watercourses, areas sensitive to groundwater contamination, or locations where the bedrock is close to the surface.

The system works by accepting the effluent from septic tanks, composting toilet leachate (excess moisture), and greywater and feeding it into a subsurface fluid distribution system or bed, where naturally occurring microorganisms convert the chemical constituents of the wastewater into nutrients for plant growth. Plants use the nutrients as they grow and also take up water and release it into the atmosphere. In areas where freezing occurs, the wastewater garden may be enclosed in a greenhouse or solarium.

The benefits of wastewater gardens include:

- zero discharge and no environmental contamination;
- elimination of holding tank pumpouts;
- suitability for sites with bedrock, shallow soil, poor drainage, or high water table;
- horticultural appeal.

As with all "non-standard" plumbing systems, contact your local public health unit for detailed assistance and regulations regarding waste management in your area.

Figure A12-9. In areas where freezing occurs, the wastewater garden may be enclosed in a greenhouse or solarium, offering free fertilizer to sustain beautiful plants. (Courtesy David Del Porto, Ecological Engineering Group)

Appendix 13

Wind Turbine Comparison Guide

Comparing PV panels is child's play compared with evaluating one wind turbine model against another. With a PV module, all that is required is to review the warranty terms and divide the cost per panel by the rated power output. The model with the lowest "cost/watt" is the model you should choose.

Wind turbines are a different matter, as every manufacturer seems to use different criteria for rating their products. Accordingly, the data listed below provides the expected monthly energy output at a standardized 10 mph (16 kph) wind speed.

When considering the purchase of any turbine, HAWT or VAWT, check to see if the United States National Renewable Energy Laboratory (NREL) has certified the desired model under the small wind turbine certification program. Although this program was just started in 2008 the NREL expects to continue testing wind turbines to standardized conditions, giving potential buyers of these units confidence in the design and quality. To follow the certification work of the NREL visit the website at http://www.nrel.gov/wind/small_wind.html.

Also remember to check for state/provincial and federal incentives as discussed in Chapter 3.3. By way of example, the State of New York provides cash incentives for people installing a new wind turbine. Details of the New York program are available at www.powernaturally.org/programs/wind/incentives.asp. As there are dozens of incentive programs in North America, be sure to discuss this matter with your equipment supplier.

Abundant Renewable Energy

www.abundantre.com

Model:	ARE 110
Rotor Diameter:	11.8 ft (3.6m)
Swept Area:	110 ft^2 (10.2m^2)
Energy Output[1]:	262 kWh
List Price:	$11,800 (off-grid battery charging)
	$12,650 (grid direct)
Warranty:	5 years

Notes:
a) includes turbine controller and diversion load for off-grid unit
b) includes inverter and associated materials for grid-direct model

Abundant Renewable Energy

Model:	ARE 442 (grid direct)
Rotor Diameter:	23.6 ft (7.2 m)
Swept Area:	442 ft^2 (41 m^2)
Energy Output:	1,171 kWh
List Price:	$39,600
Warranty:	5 years

Notes:
a) includes inverter and associated materials for grid-direct model
b) contact factory for battery charging availability

---

1. The energy output per month is based on the manufacturer's data assuming the turbine was installed in an area having an average wind speed of 10 mph (16.1 kph). Using a common average wind speed allows easier comparison between turbine models but may not be indicative of the energy that will be produced at your site.

Bergey Windpower

www.bergey.com

Model:	XL.1
Rotor Diameter:	8.2 ft (2.5 m)
Swept Area:	53 ft² (4.9 m²)
Energy Output:	115 kWh
List Price:	$2,790
Warranty:	5 years

Notes:
a) includes voltage regulator

Bergey Windpower

Model:	Excel (XL-R battery charging)
Rotor Diameter:	23 ft (7 m)
Swept Area:	415 ft² (38.6 m²)
Energy Output:	520 kWh
List Price:	$23,500
Warranty:	5 years

Notes:
a) includes voltage regulator

Bergey Windpower

Model:	Excel (XL-S grid direct)
Rotor Diameter:	23 ft (7 m)
Swept Area:	415 ft² (38.6 m²)
Energy Output:	680 kWh
List Price:	$29,500 (includes inverter)
Warranty:	5 years

Notes:
a) includes synchronous inverter

Eoltec Wind Turbines

www.pineridgeproducts.com

Model:	Eoltec 6 kW (grid direct)
Rotor Diameter:	18.4 ft (5.6 m)
Swept Area:	265 ft² (24.6 m²)
Energy Output:	558 kWh
List Price:	$25,000 (includes inverter)
Warranty:	5 years

Notes:
a) includes synchronous inverter

Endurance Wind Power

www.endurancewindpower.com

Model:	S-250 5 kW (grid direct)
Rotor Diameter:	18 ft (5.5 m)
Swept Area:	256 ft² (23.5 m²)
Energy Output:	383 kWh
List Price:	$ not available
Warranty:	5 years

Notes:
a) induction generator, no inverter required

Endurance Wind Power

Model:	G-3120 35 kW (grid direct)
Rotor Diameter:	63 ft (19.2 m)
Swept Area:	3,120 ft² (290 m²)
Energy Output:	70,000 kWh
List Price:	$ not available
Warranty:	5 years

Notes:
a) induction generator, no inverter required

Natural Power Products Inc.

www.npp.com

Model:	Gryphon 400 (off grid)
Rotor Diameter:	8.2 ft (2.5 m)
Swept Area:	52.8 ft² (4.9 m²)
Energy Output:	not available
List Price:	$800
Warranty:	3 years

Notes:
a) controller and heat dump included

Natural Power Products Inc.

Model:	Gryphon 700 (off grid)
Rotor Diameter:	8.9 ft (2.7 m)
Swept Area:	63.5 ft² (5.9 m²)
Energy Output:	not available
List Price:	$1,400
Warranty:	3 years

Notes:
a) controller and heat dump included

Natural Power Products Inc.

Model:	Gryphon 1300 (off grid)
Rotor Diameter:	9.8 ft (3.0 m)
Swept Area:	75.4 ft² (7.0 m²)
Energy Output:	not available
List Price:	$2,600
Warranty:	3 years
Notes:	

a) controller and heat dump included

Natural Power Products Inc.

Model:	Gryphon 2600 (grid direct)
Rotor Diameter:	19.0 ft (26.0 m)
Swept Area:	283.5 ft² (7.0 m²)
Energy Output:	not available
List Price:	$5,000
Warranty:	3 years

Kestrel Wind Turbines

www.dcpower-systems.com

Model:	Kestrel 800 (grid direct/off grid)
Rotor Diameter:	7 ft (2.1 m)
Swept Area:	38.5 ft² (3.6 m²)
Energy Output:	80 kWh
List Price:	$1,995
Warranty:	5 years
Notes:	

a) no controller or inverter provided

Kestrel Wind Turbines

Model:	Kestrel 1000 (grid direct/off grid)
Rotor Diameter:	10 ft (3 m)
Swept Area:	79 ft² (7.3 m²)
Energy Output:	130 kWh
List Price:	$2,950
Warranty:	5 years
Notes:	

a) no controller or inverter provided

Kestrel Wind Turbines

Model:	Kestrel 3000 (grid direct/off grid)
Rotor Diameter:	10 ft (3 m)
Swept Area:	79 ft² (7.3 m²)
Energy Output:	130 kWh
List Price:	$2,950
Warranty:	5 years
Notes:	

a) controller/dump load provided for off-rid models

b) no inverter provided for grid-direct models

Proven Energy

www.windandsun.com
www.solarwindworks.com

Model:	Proven WT 0.6 (grid direct/off grid)
Rotor Diameter:	8.4 ft (2.6 m)
Swept Area:	55 ft² (5.1 m²)
Energy Output:	83 kWh
List Price:	$4,870
Warranty:	2 years
Notes:	

a) no controller or inverter provided

Proven Energy

Model:	Proven WT 2.5 (grid direct/off grid)
Rotor Diameter:	11.1 ft (3.4 m)
Swept Area:	97 ft² (9 m²)
Energy Output:	293 kWh
List Price:	$9,700
Warranty:	2 years
Notes:	

a) no controller or inverter provided

Proven Energy

Model:	Proven WT 6
	(grid direct/off grid)
Rotor Diameter:	18 ft (5.5m)
Swept Area:	254 ft² (23.6 m²)
Energy Output:	667 kWh
List Price:	$20,500
Warranty:	2 years

Notes:

a)   no controller or inverter provided

Proven Energy

Model:	Proven WT 15
	(grid direct/off grid)
Rotor Diameter:	29.5 ft (9m)
Swept Area:	683 ft² (63.5 m²)
Energy Output:	1,451 kWh
List Price:	$39,000
Warranty:	2 years

Notes:

a)   no controller or inverter provided

Southwest Windpower

www.windenergy.com

Model:	Whisper 100
	(battery charging)
Rotor Diameter:	7 ft (2.1m)
Swept Area:	38.5 ft² (3.6 m²)
Energy Output:	65 kWh
List Price:	$2,475
Warranty:	5 years

Notes:

a)   includes controller

Southwest Windpower

Model:	Whisper 200
	(battery charging)
Rotor Diameter:	9 ft (2.7m)
Swept Area:	63.5 ft² (5.9 m²)
Energy Output:	125 kWh
List Price:	$2,995
Warranty:	5 years

Notes:

a) includes controller

Southwest Windpower

Model:	Skystream 3.7 (grid direct)
Rotor Diameter:	12 ft (3.6 m)
Swept Area:	113 ft² (10.5 m²)
Energy Output:	240 kWh
List Price:	$5,400
Warranty:	5 years

Notes:

a)   includes integrated grid direct inverter

b)   battery charging with optional battery-based inverter

Southwest Windpower

Model:	Whisper 500 (off grid)
Rotor Diameter:	15 ft (4.6 m)
Swept Area:	176 ft² (16.4 m²)
Energy Output:	330 kWh
List Price:	$7,700
Warranty:	5 years

Notes:

a)   includes turbine controller

Southwest Windpower

Model: Whisper 500 (grid direct)
Rotor Diameter: 15 ft (4.6 m)
Swept Area: 176 ft² (16.4 m²)
Energy Output: 330 kWh
List Price: $12,000
Warranty: 5 years
Notes:
a)   includes grid tie inverter

True North Power Systems

www.truenorthpower.com

Model: Wind Arrow 1kW
                    (battery charging)
Rotor Diameter: 6.5 ft (2.0 m)
Swept Area: 33.4 ft² (3.1 m²)
Energy Output: not available
List Price: $3,139
Warranty: 5 years
Notes:
a)   2kW model is due late 2009

Wind Turbine Industries

www.windturbine.net

Model: WTIC 31-20 (grid direct)
Rotor Diameter: 31 ft (9.5 m)
Swept Area: 754 ft² (70 m²)
Energy Output: 1,644 kWh
List Price: $33,900
Warranty: 1 year
Notes:
a)   includes grid tie inverter

INDEX

INDEX

INDEX

INDEX

INDEX

INDEX

About the Author

William Kemp

William (Bill) Kemp, is V.P. Engineering of an energy sector corporation where he leads the development of low environmental impact hydroelectric and agricultural biogas systems. Bill is a leading expert in small and mid-scale renewable energy technologies and is the author of four books including the best selling *The Renewable Energy Handbook.* He is also a co-author of the David Suzuki Foundation report *Smart Generation; Powering Ontario with Renewable Energy.* In addition he has published numerous articles on small-scale private power and is the chairman of electrical safety standards committees with the Canadian Standards Association. He and his wife Lorraine have lived off the electrical grid on their hobby/horse farm in eastern Ontario, for the past 15 years.

Bill Kemp discussing his book with environmentalist Dr. David Suzuki

How to make your own high quality biodiesel using low-cost methods

Biodiesel Basics and Beyond

A comprehensive guide to the production and use of biodiesel for the home and farm

William H. Kemp

588 pages 6"x9"

ISBN 978-0-9733233-3-7

$29.95 Cdn/US

Using vegetable oils as a fuel for home heating and transportation is a hundred years old. Rudolf Diesel's original engine was operated on plant oils due to the lack of fossil fuels. Later, plant and animal oils were converted into a petrodiesel-compatible fuel known as biodiesel: a clean, low-carbon fuel.

In the early 1980s, home brewers discovered they could transform waste restaurant fryer oils into crude biodiesel and use it in automobiles at 100% concentrations at one quarter the cost of petrodiesel. Yet automotive and engine manufacturers insist that late-model vehicles may be damaged when run on high concentrations of biodiesel and will not honor engine warranties where biodiesel fuel has been used.

Biodiesel Basics and Beyond aims to separate fact from fiction and to educate potential home, farm, and co-operative manufacturers on the economic production of quality biodiesel from both waste and virgin oil feedstock. The book includes:

- detailed processes and equipment required to produce biodiesel fuel that meets North American standards
- how farmers can use excess oilseed as a feedstock for biodiesel production
- the use of the co-byproduct glycerin in the making of soap
- a guide to numerous reference materials and a list of supplier data.

This is North America's definitive guide to responsibly producing biodiesel from waste vegetable oil, while minimizing your environmental footprint in the process.

The International Energy Agency says the world has entered its 3rd energy crisis. Could the Zero-Carbon Car offer the solution?

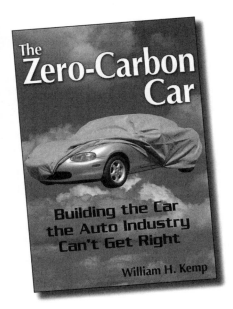

The Zero-Carbon Car

Building the Car the Auto Industry Can't Get Right

William H. Kemp

544 pages 6"x 9"

ISBN 978-0-9733233-4-4

$24.95 Cdn/US

The transportation sector, dominated by the personal automobile, is responsible for over a quarter of North America's total energy consumption. Motor vehicles account for the overwhelming majority of harmful atmosphere emissions. In addition, rising fuel, road and infrastructure costs are beginning to converge, driving us to the end of Autopia.

The Zero-Carbon Car reviews the issues of climate change and carbon rationing, Peak Oil, urban sprawl, and geopolitical and socioeconomic disruption related to fossil fuel use. The book argues that, while there is no way to avoid the eventual demise of the automobile, there is an opportunity for the automotive industry to develop – and governments to support – an ultra-efficient, zero-carbon-emissions automobile.

The second half of the book documents the successful design and construction of a zero-carbon vehicle, proving that the technology is not only possible, it is viable today. For those who wish to fabricate their own vehicle, plans and software are provided in the book and at the accompanying website.

The Zero-Carbon Car is a must-read for automotive enthusiasts, environmentalists and anyone who cares about how their transportation choices affect the planet. The book may not solve the world's fuel, transportation and energy supply problems but it could help pave the way to a cleaner, more sustainable future.

For more information visit www.aztext.com

Grow your own organic vegetables and enjoy a "one hundred-foot diet"!

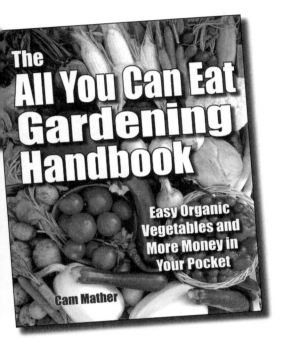

The All You Can Eat Gardening Handbook

Easy Organic Vegetables and More Money in Your Pocket

Cam Mather

250 pages 8"x 10"

ISBN 978-0-9810132-2-0

$24.95 Cdn/US

There's never been a better time to grow some of your own food as we face the converging challenges of the financial crisis, climate change, water shortages, peak oil and 6.5 billion people worldwide competing for a limited amount of food.

Many gardening books are so technical that they are intimidating to the beginning gardener. This book shows how easy gardening can be while providing tips and techniques to ensure success. Whether you live in the country, the city or the suburbs you'll learn:

- how to create lots of rich compost as the basis for vigorous growth
- simple techniques to safely deal with pests
- the most effective techniques for irrigation including harvesting your rainwater
- how to store your harvest using canning, freezing and drying techniques
- how to build and stock a root cellar
- which vegetables pack the most nutritional punch.

This book is a step-by-step guide to turning your lawn into a great source of personal satisfaction and economic independence. Having an organic produce department in your backyard creates your own "one-hundred-foot diet" to reduce your carbon footprint and keep more money in your pocket.

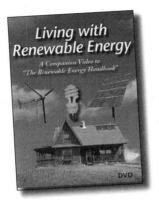
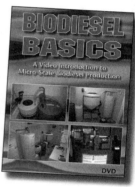